Game theory and political theory: an introduction

Game theory and political theory

An introduction

PETER C. ORDESHOOK
University of Texas at Austin

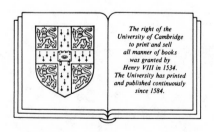

The right of the
University of Cambridge
to print and sell
all manner of books
was granted by
Henry VIII in 1534.
The University has printed
and published continuously
since 1584.

CAMBRIDGE UNIVERSITY PRESS

Cambridge
London   New York   New Rochelle
Melbourne   Sydney

Published by the Press Syndicate of the University of Cambridge
The Pitt Building, Trumpington Street, Cambridge CB2 1RP
32 East 57th Street, New York, NY 10022, USA
10 Stamford Road, Oakleigh, Melbourne 3166, Australia

First published 1986

Printed in the United States of America

Library of Congress Cataloging-in-Publication Data
Ordeshook, Peter C. 1942–
Game theory and political theory.
Bibliography: p.
1. Political games   2. Political science –
Mathematical models.   3. Game theory.   I. Title.
JA74.073   1986   320.5'01'5193   85–30876

British Library Cataloguing in Publication Data
Ordeshook, Peter C.
Game theory and political theory: an introduction.
1. Political science – Mathematical models
2. Game theory
I. Title
320'.01'5193   JA73

ISBN 0 521 30612 4 (hard covers)
ISBN 0 521 31593 X (paperback)

Contents

v

Kramer *Jo Math Soc,*

Preface

Four books mark the beginning of modern political theory: Anthony Downs's *An Economic Theory of Democracy* (1957), Duncan Black's *Theory of Committees and Elections* (1958), William H. Riker's *A Theory of Political Coalitions* (1962), and James Buchanan and Gordon Tullock's *The Calculus of Consent* (1962). These volumes, along with Kenneth Arrow's *Social Choice and Individual Values* (1951), began such a wealth of research that political scientists today have difficulty digesting and synthesizing all but small parts of it. Consequently, the full value of this research often goes unrealized, and teaching it seems increasingly difficult. These problems remain especially true of formal political theory as against, say, approaches that emphasize sociological or psychological perspectives, pure statistical empiricism, or more traditional historical research. Curiously, these problems grow out of the strengths and successes of political theory.

First, because this research seeks to satisfy a rigid definition of "theory," and not some ambiguous criteria of good journalism and insightful comment, it forms a collective whole. Consequently, unfamiliarity with one of its subparts, such as social choice, spatial models of elections, public economics, or game theory, precludes a full understanding of the theory's implications and generality. Second, because political theory itself is closely connected to the discipline of economics and rejects the notion that economic and political activity remain separable, much of the research appears in economics as well as in political science journals. Besides the *American Political Science Review*, the *American Journal of Political Science*, and the *Journal of Politics*, many seminal contributions appear in *Econometrica*, the *American Economic Review*, the *Journal of Economic Theory*, and the *Journal of Political Economy*, as well as interdisciplinary journals such as *Public Choice* and the *Journal of Conflict Resolution*. Indeed, the relevant journals publishing this work would make up a far longer list. Few persons, though, can keep abreast of the journals from several disciplines as well as a constant flow of working papers and the insight of colleagues revealed only at professional meetings. Concurrently, the theory has excited so many scholars from disparate backgrounds that research often proceeds at a pace that overwhelms even the most diligent reader. It is not unusual to find working papers based on working papers based on papers that are only "forthcoming" in journals. This vast literature must seem intimidating to the beginning student.

Third, and most critically, much of this research uses the language of mathematics, which for some is unassailable and for others is an end rather than a means. Political scientists trained in the language of Hobbes, Locke, and Rousseau as well as of Bentley, Key, and Dahl struggle with digesting the first few pages of these professional manuscripts. Others, both mathematicians and pseudomathematicians, present their results in an ever greater spiral of abstraction. Thus, although mathematics serves as an integrating language for some, it serves as a barrier to entry for others. Some scholars, then, turn entirely to other endeavors; some make reference to the theorems and results that form the central part of the theory without fully comprehending their power or limitations; and still others, unable to understand why so many fail to appreciate the value of their research, return to the shelter of mathematical notation.

In spite of these problems of communication and dissemination, there has emerged a theory of politics no less sophisticated and comprehensive than the theory of markets found in economics. This book's challenge is to convey the power and scope of that theory, to make its central results – including the internal logic of proofs – understandable, and to help the reader appreciate the degree to which "formal models of political science" form an integrated whole.

Of necessity, we use mathematics, some of which will be unfamiliar to the beginning student. But the reader will find here few Greek or arcane mathematical symbols. Although understanding requires some minimal perseverance, little more is required than the mathematics found in an introductory statistics text. Nevertheless, this book will not be easy reading for most students of politics, although it should be far easier than reading the essays that provide the sources of many of the reported results and ideas. Whatever the reader's mathematical sophistication, we must all realize that the days are gone in which political scientists could study their craft while relaxing in a comfortable easy chair. Like our colleagues in science, engineering, and economics, we must now study political theory and the cutting edge of our discipline while sitting at a desk.

Many students who encounter this new political theory for the first time believe mistakenly that its theorems and their proofs emerge full-blown from the heads of the scholars who publish them. This characterization certainly does not hold true for the author or for his colleagues. We usually first formulate these theorems geometrically by drawing and redrawing lines and circles on a page until the key insight appears. We then develop the initial result and proof in a most rudimentary form. Up to this point, almost any intelligent person, with assistance at the blackboard, could understand the internal logic of the result. But then comes the "polishing." To ensure publication in prestigious journals, the author restates the results using contemporary and concise notation, and the proofs expand to achieve the greatest possible generality. Greek symbols replace simple words or phrases, and "fixed-point theorems," "binary relations," "upper semicontinuous functions," or "sigma algebras" replace curves and lines. Now comprehension of the result escapes all but a few insiders. Others, less sophisticated in this craft, must wait for interpretation, if it comes at all, at professional meetings and seminars. The proofs that this volume offers take the reader back to those initial scribblings, to ideas once formulated on yellow tablets but now comprehensible only with unpleasant and time-consuming effort. The proofs here are not proofs of which a mathematician would approve, but they are, we hope, more understandable to the rest of us.

The examples and the central theoretical applications concern elections and legislatures. Although these are important political processes and institutions, they do not exhaust the domain of our discipline. But because of Downs's seminal volume, elections became the first significant domain of formal models. Hence, models of electoral institutions, voting, and candidate strategies until recently have dominated the literature of applications on which this volume must rely. Similarly, the structure of legislatures – their prescribed rules and procedures as well as their well-kept records of voting – proved especially amenable to formal study. Thus, with Black's treatise on committees leading the way, legislatures and parliamentary institutions became the second major substantive area of inquiry and application. Provoked largely by Buchanan and Tullock's arguments about the nature of constitutions and the economic imperatives of public policy, a significant "political economy" literature has also emerged, which directly addresses the intersection of economics and politics, restates the classical justifications for the state, and, more generally, rejects the separation of economics and politics into disjoint disciplines. Riker's volume on political coalitions and his formulation of the "size principle" demonstrate the more general application of formal theory to political issues ranging from international relations to elections. In that tradition many scholars have used the theory to gain major insight into such matters as economic and political development (see Mancur Olson's *The Rise and Decline of*

Nations), bureaucracies (see William Niskanen's *Bureaucracy and Representative Government*), interest groups (see Olson's *The Logic of Collective Action*), democratic theory (see Riker's *Liberalism against Populism*), and the origins of cooperation in politics (see Axelrod's *The Evolution of Cooperation*).

Although these diverse applications demonstrate the power of the new political theory, they are not generally the "home" of that theory's central abstract results. Devoid of excessive notation, they focus instead on specific substantive interpretations and implications and are accessible to most students of politics. This volume concentrates on the less accessible parts of this new political theory, that is to say, its theoretical substructure. Hence, although we incorporate as much of this more substantive literature into the presentation as possible, to the extent that crucial and new results reside in models of committees and elections, we are forced at times to limit ourselves to those institutions. Our goal is to present that substructure as a theoretical whole, and thereby to communicate the essential relations among a diverse and growing literature, both formal and interpretative.

Many of the results that we present also reside in game theory. Indeed, the thesis of this volume is that game theory provides the integrating force of the new political theory. The genesis of much that we cover here is the perspective that economics offers all of social science, namely, that any adequate understanding of group choice or action ultimately must be reducible to an understanding of the choices that individual human beings make in the context of institutions for the purpose of attaining individual objectives. This perspective, although an implicit (and sometimes explicit) assumption of such political scientists as Truman, Bentley, and Key, has its roots in decision theory, and game theory is simply the branch of the decision sciences that seeks to explore how people make decisions if their actions and fates depend on the actions of others. This interdependency is what differentiates politics from that most powerful theory of social science, the classical microeconomic theory of perfect competition. Political scientists rarely have the luxury of simplifying their discourse by assuming, as does much of microeconomic theory, that interdependencies are unimportant. We cannot assume, for example, that the number of political producers is large enough so that no single politician can affect the output or price of political goods and services. There are only 435 members of the U.S. House, 100 members of the Senate, 50 states, and rarely more than a few significant political candidates for any specific public office. The underlying assumptions of perfect competition, then, seem out of place in our discipline. As a result, interdependent choice becomes the central fact of our theories. Hence, game theory, the theory of interdependent choice, must provide our central theoretical apparatus.

Game theory is also central because it focuses on concepts of equilibrium. In game theory an equilibrium is a prediction, for a prespecified circumstance, about the choices of people and the corresponding outcomes. This prediction generally takes the form "if the institutional context of choice is . . ., and if people's preferences are . . ., then the only choices and outcomes that can endure are. . . ." Thus, equilibria replace both journalistic interpretations of events and statistical correlations between environmental factors and political outcomes as explanations. In the deepest meaning of the word, the study of equilibria, in game theory, combined with substantive applications, is an attempt to provide *causal* explanations.

This volume, however, will not try to summarize or even survey in part all the developments in game theory. Any such exercise, adequately done, would encompass more than one volume. Game theory, moreover, is very much a living discipline, with new developments and insights constantly appearing on the horizon. For example, we ignore subjects such as stochastic games, and the experimental results that we review are simply those most familiar to the author and probably comprise but a small fraction of the relevant literature. Similarly, the scholarly literature on games of incomplete information and rational expectations, which are becoming important modes of analysis, is simply introduced with some specialized examples. This volume focuses instead on those parts of game theory that have the most immediate relevance to developments in formal political theory. Thus, although we introduce the reader to game-theoretic reasoning, our purpose is to show that the central ideas of "formal political theory," "rational choice models of politics," or of "positive political theory" are an integral whole and that game theory provides the connecting thread.

The political theory that has emerged in the last 25 years or so, then, rejects the notion that we must use one set of concepts to study voting, another to explain conflict in international relations, another to analyze interest-group politics, and yet another to explore legislative politics. It conceptualizes society in such a way that the same set of concepts explains markets; the agencies and laws that are established by legislatures to regulate markets; the lobbying of legislators by interest groups, the bureaucracies, and chief executives that constrain legislators; and, to complete the cycle, the choices of people who are simultaneously voters, taxpayers, and consumers in the market.

Finally, this volume does not provide an exhaustive bibliography, which often simply leads the beginning student to believe that formal theory is too advanced to penetrate, or that years of reading are required before research can begin. This belief is wrong. Research, especially by graduate students seeking to write a dissertation, should begin before one has read "too much." Certainly, it is impossible to "read it all," and those who try often produce little. There is always the danger of reinventing the wheel, but that is why we value colleagues – to guide us to a relevant literature and to steer our thoughts in productive directions. Here, we reference only the key or most readable sources, which, in turn, we can use as guides to further reading.

Acknowledgments

In endeavors such as this volume it is impossible to acknowledge everyone who made an important contribution. Perhaps no one read my manuscript with more care than Peter H. Aranson, and the extent and value of his substantive and editorial suggestions are, in my experience, without precedent. I would also like to thank two other close friends and colleagues, Richard D. McKelvey and Melvin J. Hinich. Aside from contributing to this volume, for 15 years each has helped shape my thinking in fundamental ways by explaining ideas that otherwise would have remained unintelligible, at least to me. Thomas Schwartz not only read and commented on various parts of my manuscript, but he contributed importantly by suggesting alternative proofs and innovative interpretations. The initial outline of this volume followed from extensive discussions with James D. Laing and, later, with Howard Rosenthal, who provided innumerable suggestions and corrections. Similarly, I owe Kenneth Shepsle a special debt for his reading and use of an early draft of my manuscript in a course designed (perhaps not intentionally) to discover as many errors as possible. Three students in that course, Scott Ainsworth, Brian Humes, and Paul Johnson, provided me with an abundance of helpful suggestions. I would like to express my thanks also to Thomas Seung, James Enelow, Ho-Seob Kim, Ken Williams, Michael Greenberg, and, for his careful assistance in helping me with the final preparation of the manuscript, Emerson M.S. Niou.

It is usual to acknowledge those who helped in the preparation of one's manuscript, but this volume, with the exception of the artwork by Wayne Sheeler, was prepared entirely on my home word processing system. This meant that my wife Betty had to be tolerant of her husband's compulsion in more than the usual ways. It is not always easy to maintain a home in which the dining or family room has been seized, converted into a computer lab and (messy) storage area, and made the source of a ceaseless clatter of computer keys and printers. She, as much as I, is happy to see this volume brought to completion.

Finally, I would like to acknowledge the National Science Foundation, without whose support little if any of the research that this volume summarizes would exist.

CHAPTER 1

Individual preference and individual choice

This chapter focuses on the notation and definitions required to represent the theoretical primitives of formal theory – alternatives, outcomes, and preferences. Although this focus may make reading difficult for those who are encountering these ideas for the first time, this material must be mastered if formal political theory is to be understood in more than a superficial way.

We begin with two assumptions: methodological individualism and purposeful action. Methodological individualism holds that we can understand social processes and outcomes in terms of people's preferences and choices. This assumption may seem strange to the student of politics, who is concerned primarily with collectivities such as interest groups, political parties, or legislatures, and for whom the most useful explanatory concepts assume a group consciousness. Such concepts, though, are generally little more than journalistic conveniences that afford us the luxury of not having to delve into the complexities of organizations and institutions. But interest groups cannot lobby: Only their members can opportune legislators. Similarly, a legislature cannot "hold" a norm: Only its members can share a consensus about acceptable standards of action and penalties for violations of those standards.

Obviously, it is convenient to speak of "society preferring clean air," "firms maximizing profits," or "the Office of Management and Budget setting national economic policy." Furthermore, no one should dispute that social interaction conditions people's preferences and choices. The assumption of methodological individualism is but a reminder that only people choose, prefer, share goals, learn, and so on, and that all explanations and descriptions of group action, if they are theoretically sound, ultimately must be understandable in terms of individual choice.

Our second assumption is that people's actions are purposeful, by which we mean "the act or behavior may be interpreted as directed to the attainment of a goal. . . . Purposeless behavior then is that which is not interpreted as directed to a goal." Hence, the concept of purposeful action excludes the types of explanations found in much of social science, those that rely solely on observed correlations between indi-

1

vidual action and, for example, measures of childhood socialization, education, sex, and income. Any comprehensive theory must remain consistent with such correlations. Simply attributing causation to a correlation between environmental variables and choice, though, neglects the fact that people make choices for some purposes and that we must understand those purposes.

The word "rational" is commonly used to summarize our assumptions about these choices. The meaning of this word has been the subject of a lively debate during the past 20 years and has given rise to considerable misunderstanding among the antagonists. Instead of entering this debate, we note that purposeful does not necessarily mean that people carefully and consciously list their alternative actions, map all the relevant or possible consequences of each act, estimate the probability of each consequence, and define precisely their preferences across all consequences. Thus, we cannot ignore habit, instinct, and the use of simple cues and heuristics to uncomplicate complex decisions. Indeed, one of the formal theorist's most important activities is to understand why various heuristics are reasonable responses to complex environments and to the costs of alternative modes of making decisions. The assumption of purposeful choice implies simply that, after taking account of people's perceptions, values, and beliefs, we can model their decisions by asserting that they act *as if* they make such calculations.

1.1   Some fundamental notation and definitions

Politics concerns situations in which the actions that people choose depend on the actions that they think others have taken or will take. The study of politics, then, is the study of interdependent choice. But before we can examine such interdependencies with the detail that they warrant, we must develop an appropriate conceptualization of individual preference and choice. We do so by treating the actions of all other people as exogenous – as a static component of a person's environment. Having thus simplified our problem, we remain with three key concepts: actions, outcomes, and states of nature. By *action* we mean the choice a person makes in a specific context, which is what we are trying to predict. The assumption of purposeful action, though, dictates that we specify the *outcomes* to which a person directs a particular action, the outcomes that people seek to bring about or to avoid by their actions. Finally, linking actions and outcomes, we require the concept of a *state of nature,* which is a sufficient condition for a particular action to yield a particular outcome. That is, a state of nature in conjunction with a particular action yields a specific outcome.

These three concepts are straightforward. To represent them in a way that admits rigorous and deductive reasoning, however, we must introduce some of the notation and mathematical constraints common to formal models. This section discusses that notation. Subsequent sections consider how we represent individual preferences and how we can complicate our descriptions of concepts to admit more substantive content into our discussion. Later we discuss how we treat the uncertainty and risk that ordinarily confront decision makers.

Actions

Beginning with what we are attempting to explain, let a_j denote a specific alternative action – the jth alternative – and, for the person whose decision we are modeling, let $A = \{a_1, a_2, \ldots, a_j, \ldots\}$ be the set of all admissible actions. Now suppose that a person must select exactly one and only one act from A. Hence, A is *exhaustive* (at least one element of A must be chosen), and the elements of A are mutually *exclusive* (the decision maker can choose not more than one element of A). When modeling the choices in an election, for example, if we seek only to explain why a person does or does not vote, we might let A be the simple two-element set $\{a_1$: vote, a_2: abstain$\}$. Even at this crude level of description, A is both exhaustive and exclusive: In a particular election, a person must either vote or abstain; he cannot do both.

The two constraints on A permit considerable discretion as to how we can interpret this set. Identifying two alternative actions may be sufficient to analyze turnout, but for other purposes we may require more details. To analyze the relation between policy predispositions and party preferences in two-party contests, for example, we could let $A = \{a_1$; vote for party 1, a_2: vote for party 2, a_0: abstain$\}$. Or, to study coattail effects, we could elaborate A further to admit split-ticket voting. If we are studying the relation between voting for president and governor, we might identify as an element of A the alternative "vote for the presidential candidate of party 1 but the gubernatorial candidate of 2."

Certainly, other descriptions of alternative acts during an election – such as contributing campaign funds, ringing doorbells, and so on – are useful for some purposes. But specifications of A should not become too complex lest the analysis become intractable. Nevertheless, our description of A is legitimate as long as the acts that we identify are mutually exclusive and together exhaust all possibilities.

Outcomes

Since we shall explain action as purposeful, we must specify purpose. Hence, we require the concept of an outcome – the object to which the person directs his actions. Denoting the set of possible outcomes by $O = \{o_1, o_2, \ldots, o_k, \ldots\}$, where o_k represents a particular outcome, we impose the same restrictions on O as on A. First, O is exhaustive – it must contain any outcome that can occur, and some outcome in O must occur. Second, the elements of O are exclusive – one and only one element of O must correspond to any outcome that can occur.

Once more, research intent dictates the substantive content of O. To analyze voting decisions in two-candidate elections, for example, if we assume that each citizen is concerned solely with who wins the election, we could let $O = \{o_1$: party 1 wins, o_2: party 2 wins, o_3: parties 1 and 2 tie$\}$. If more than one office is at issue, a particular outcome in O could read: "Party 1 wins the presidency, party 2 wins the contested Senate seat, but party 1 wins the governorship." But if citizens are concerned with not only who wins but also by how much, then a generic outcome in O might read, "party 1 receives v_1 votes and party 2 receives v_2 votes." The description of outcomes can also include the historical events that precede them. Some argue, for example, that many people vote because they are socialized to do so. Hence, to model turnout we might distinguish these outcomes: "I voted for my party and it won" and "I did not vote and my party won." Formally, we can identify a new set of outcomes $O' = \{(o_i, a_j)\}$ so that O' corresponds to every feasible combination of acts in A and outcomes in O. Of course, the set O' inherits the properties of O and A – the outcomes (o_i, a_j) are mutually exclusive and exhaustive.

At this point we must introduce some additional notation and concepts. Notice that both A and O are *countable* as represented; that is, we can label all elements in them by using integers. In each of our examples A and O are also *finite*; thus we need not use all integers to label all elements of A and O. We cannot represent many types of actions and outcomes in this fashion, however, without encountering clumsy notation. For example, if we are concerned with how much money people contribute to a candidate's campaign, we might let actions be denoted "a_o: \$0 contributed to candidate 1," "a_1: \$1 contributed to candidate 1," and so on. But such labels can be exceedingly awkward, especially if we consider contributions down to the penny. It is simpler to let A be *infinite* and *uncountable*. For example, if we believe that no one will contribute more than one million dollars, then we could let A be the set of *all* numbers between 0 and 1 million – including such numbers as 5.6678.... Instead of subscripting elements of A by integers, we let $a \in A$ (read "a in A" or "a is a member of A") denote a generic element of A. Thus, $a \in A$ is a specific number that we interpret to mean "$\$a$ contributed to candidate 1." Similarly, we could let $o \in O$ denote a generic element of O, where again O is an infinite and uncountable *subset* of all positive numbers.

One of the virtues of infinite and uncountable sets – and especially of intervals of real numbers – is that we can often represent them with pictures and lean on our geometric intuition to interpret results. Suppose that a party's finance director is allocating some fixed amount of money, say $\$X$, among three candidates competing for different offices, where x_i is the amount of money allocated to candidate i's campaign. We can then denote element of A by $\mathbf{a} = (x_1, x_2, x_3)$, where all admissible combinations of donations must satisfy the party's budget constraint $x_1 + x_2 + x_3 = X$. Thus, each element of A is a *vector* – more specifically, a *3-tuple* – and since each element of the vector $\mathbf{a}$ is in an infinite and uncountable set, A is infinite and uncountable also. (If alternatives or outcomes are explicitly vectors and if we want to distinguish between the vector and its components, then we use boldface type to denote the vector and italicized type to denote a component; but if alternatives or outcomes can be either vectors or simple numbers, and if the possibility for confusion is slight, then we use italicized type.) If we suppose, finally, that each x_i is nonnegative – that a candidate gives no funds to the party – then the set A has a useful geometric representation, called a *budget simplex*. Figure 1.1 portrays a three-dimensional coordinate system in which the first (horizontal) dimension represents the amount of money allocated to the first candidate, the second (vertical) dimension represents the amount allocated to the second candidate, and the third dimension (the dimension pointing out from the page) represents the amount allocated to the third candidate. Hence, any allocation of funds corresponds to a specific point in this coordinate system. If we assume that all of X is allocated to these three candidates, then admissible allocations – those in which the sum to the candidates equals $\$X$ – all lie on the triangle that this figure portrays. Each vertex of the triangle represents an alternative in which $\$X$ is awarded to a single candidate. The sides of the triangle correspond to points in which $\$X$ is divided fully between two candidates, and the third receives nothing. Interior points on the triangle denote alternatives in which each candidate receives a positive amount and the sum is exactly $\$X$.

Our interest in budget simplexes extends beyond a desire to illustrate alternatives geometrically. In later chapters we discuss various formal

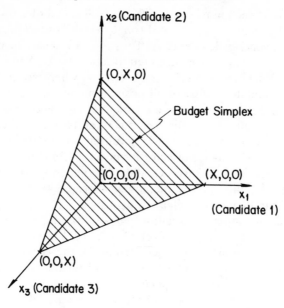

1.1 Budget simplex.

results that are applicable only if A and O satisfy certain conditions. Fortunately, sets of alternatives and outcomes that we can represent by a budget simplex often satisfy those conditions, and thus we know that we can use the corresponding results to model various political resource-allocation problems. To introduce some mathematical definitions that formal theory uses extensively, and that any student should understand if he or she wishes to read professional essays in journals, we review here the specific properties that budget simplexes satisfy.

First, budget simplexes are *bounded* sets: The numbers in them do not "run off" to plus or minus infinity. Examples of unbounded sets include "the set of all integers" or "all numbers greater than zero." Second, since the set of alternatives in Figure 1.1 includes all of its boundary points – a triangle, in this instance – a budget simplex is a *closed* set. Other examples of closed sets include "all numbers between zero and one, including zero and one," and "all numbers greater than or equal to 5." An example of a set that is not closed is "all numbers that are strictly greater than zero"; in this instance, the lower boundary of the set, 0, is not part of the set. For a more fanciful example, suppose that you are walking toward a wall, stopping when you have walked half the distance. If you keep repeating this process, then, mathematically,

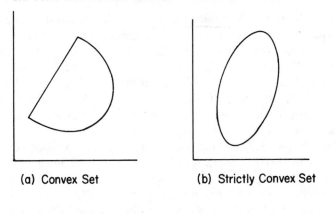

(a) Convex Set   (b) Strictly Convex Set

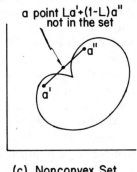

(c) Nonconvex Set

1.2 Convexity of sets.

you can never reach the wall. Some infinitesimal distance always will remain between you and the wall, but on the next iteration you can only cover half this distance. Hence, the set of points on the floor on which you can step is bounded but not closed – it is *open* because it does not include its boundary, the wall.

A third important property that a budget simplex satisfies is convexity. The convexity of sets means that if a' and a'' are any two points in A, then all points on the line connecting a' and a'' are in A as well. For budget simplexes it is evident that if we take any two points in a triangle and connect them with a straight line, then that line must be in the triangle. Formally, if L is any number between 0 and 1 and if a' and a'' are both elements of the set A, then A is a *convex set* if $La' + (1 - L)a''$ is an element of A. We can think of L as the relative weight given to a' as compared to the weight given to a'', so that $La' + (1 - L)a''$ – referred to as *the convex combination* of a' and a'' – is an intermediate value between a' and a''. A set is convex if all such intermediate values are in the set as well. Thus, finite sets cannot be convex: If, for example, $A = \{1, 2, 3\}$, the intermediate values of, say, 1 and 2, which include numbers such as 1.234 or 1.7743, are not in A. Convexity, then, is a convenient way of summarizing the idea that if a' and a'' are two alternatives in A, all alternatives that lie in the geometric representation of A between a' and a'' are also available.

Budget simplexes, however, do not illustrate a stronger version of convexity called *strict convexity*. If $La' + (1 - L)a''$ lies necessarily in the interior of A for all pairs of points in A, then A is strictly convex, so that a strictly convex set has no flat spots on its boundary. Since the boundary of a budget simplex is formed by straight lines, it is convex, but not strictly convex. Figure 1.2 illustrates a convex set (a), a strictly convex set (b), and a set that is neither strictly convex nor convex (c).

Although we have limited our discussion to alternatives, it also applies to any other set, including sets of outcomes. The concepts of bounded, closed, and convex are simply definitions that permit a concise description of a set's properties. But at this point the student of politics might reasonably ask: Of what value are such concepts to the study of, say, the cultural determinants of revolutionary social change, to the instability of democratic institutions, and to the evolution of tastes in an electorate? Measurement is imperfect in political science, especially if it concerns concepts whose definitions are subject to some debate, and can precise mathematics overcome the apparent inherent imprecision of the subject matter? Answers to such questions must concede that our notation's precision often exceeds the things it represents. Nevertheless, in addition to assuring that our assertions follow logically from assumptions, that precision has a simple goal, namely, the greatest possible generality. Thus, although we must introduce concepts that are unfamiliar to most readers, we do so because we want to accommodate as many political phenomena as possible, thereby ensuring that our results do not depend on subsequent refinements of the substantive variables we define and measure. For example, knowing that budget simplexes are closed, bounded, and convex sets tells us that theorems requiring these properties for alternatives and outcomes have some application if key decision variables are constrained. It probably matters little whether the variables concern a precisely measurable item such as money or an arguably imprecise notion such as power. In both instances, if we believe that money or power are in fixed supply, then a theorem that applies to budget simplexes may answer some theoretical question. And learning what happens if these sets do not satisfy such properties informs us about events that might otherwise be unanticipated.

States of nature

Abstract representations of A and O, along with their substantive interpretations, provide an essential but not yet complete description of a decision-making situation. Specifically, we must now link A and O so that we know which outcome prevails when the decision maker chooses a specific alternative. The *state of nature* provides that link. Sometimes, for example, one person's unilateral action can invariably bring about a particular outcome, as when the last legislator in a roll call vote casts a decisive ballot, one that makes or breaks a tie. At other times a person's effect on the outcome is imperceptible and changes only marginally the a priori probability of events, as occurs with voting in large electorates. In both examples, however, other people's decisions or the chance events of nature – as when illness precludes certain legislators from voting or when inclement weather dampens turnout – determine the consequences of action. We call such contingencies states of nature. They identify the sufficient conditions for the selection of a particular action in A to yield a particular outcome in O. We denote individual states of nature by s_k and the set of all feasible states by $S = \{s_1, s_2, \ldots, s_k, \ldots\}$. As with A and O, the set S is exhaustive, and its elements are mutually exclusive.

Returning to elections, suppose that we are interested in knowing which alternative a person chooses in a simple two-candidate, plurality-rule election in which each citizen's alternatives are to vote for candidate 1, candidate 2, or to abstain; and where the relevant outcomes are candidate 1 wins, candidate 2 wins, and the two candidates tie. How do we link alternatives to outcomes? It is evident that the effect of one person's choice on the eventual outcome depends on what all other citizens do – on how many others vote for candidate 1 and candidate 2. If a large majority choose one candidate or the other, then the citizen whose decision we are modeling cannot affect the outcome. But if the election is a tie or within one vote of a tie, then his vote is decisive. Thus, in specifying S, we distinguish states of nature by the size of a candidate's plurality, excluding the vote of a citizen whose decision we are modeling. That is, not counting the citizen under consideration, let v_1 and v_2 be the number of votes that candidates 1 and 2 receive, respectively. We can then identify the states of nature in this example by the following exhaustive and exclusive set:

$$S = \left\{ \begin{array}{l} s_1: v_1 - v_2 < -1 \\ s_2: v_1 - v_2 = -1 \\ s_3: v_1 - v_2 = 0 \\ s_4: v_1 - v_2 = +1 \\ s_5: v_1 - v_2 > +1 \end{array} \right\}$$

Figure 1.3 illustrates one convenient way of summarizing our notation and the relations among actions, outcomes, and states of nature. The rows in Figure 1.3 correspond to a citizen's alternative actions and the columns to states of nature. Each cell entry identifies the election outcome that corresponds to the specific action, a_i, and the specific state of nature, s_k. For example, if s_2 prevails – if candidate 1 is one vote short of a tie among all other voters – then the citizen in question can create a tie if he votes for 1; but if he votes for candidate 2 or abstains, then candidate 2 wins.

1.2   Preferences 2/29

Rarely can we represent political decisions as simply as in Figure 1.3. Even in a legislature whose members are concerned with a single bill, each legislator probably must make several decisions, including how to vote on amendments and whether to commit publicly to a particular position. But before we can study more realistic and interesting models, we must analyze simple situations such as the one depicted in Figure 1.3 and add to the tools we need to understand that situation.

Recall the assumption of purposeful choice, which presumes that people choose particular actions rather than others because those actions yield particular outcomes or avoid the occurrence of others. Purposeful choice, then, presumes some notion of a preference for one outcome over another. Preferences over outcomes, in turn, define preferences over alternative actions, and therefore we use them to predict the action that a person chooses. We cannot define preferences over alternatives, however, until we obtain some additional information about what a decision maker knows about which states of nature will prevail when he acts. With regard to Figure 1.3, if s_1 prevails, then voting is fruitless because candidate 2 wins no matter which act the decision maker chooses; but if s_3 prevails, then the citizen's vote is decisive, and the eventual outcome is determined by the citizen's choice. If the citizen prefers candidate 1, then in the event of s_3, he chooses a_1; but if he prefers candidate 2, then he chooses a_2. Hence, the act a person should choose may depend decisively on the state of nature that prevails.

	s_1:	s_2:	s_3:	s_4:	s_5:
a_1: vote for 1	o_2: 2 wins	o_3: tie	o_1: 1 wins	o_1: 1 wins	o_1: 1 wins
a_2: vote for 2	o_2: 2 wins	o_2: 2 wins	o_2: 2 wins	o_3: tie	o_1: 1 wins
a_o: abstain	o_2: 2 wins	o_2: 2 wins	o_3: tie	o_1: 1 wins	o_1: 1 wins

1.3 Voting in a two-candidate election.

The quality of information that a decision maker enjoys about which state of nature will prevail, however, can range from absolute certainty to a total lack of knowledge. Thus, to predict which act he chooses, we must model his beliefs about the likelihood that a particular state prevails. The simplest possibility is that people know with certainty which state of nature will prevail, in which case they know the consequences of their acts with certainty. We call this possibility *decision making under certainty.* Alternatively, if a person is uncertain about which state will prevail, we might assume that he acts as if he estimates the likelihood that each state of nature prevails, and hence that he estimates the likelihood that specific outcomes follow from specific acts. This assumption generates sentences of the form: If act $a \in A$ is chosen, then $o' \in O$ occurs with probability $p(o')$, $o'' \in O$ occurs with probability $p(o'')$, and so forth. We call this possibility *decision making under risk.*

The distinction between certainty and risk is important because what we must know about people's preferences under certainty to predict their choices requires fewer assumptions than when decisions are risky. Since we want our theory to be as general as possible, we prefer to add assumptions only when we require them. Hence, looking first at decision making under certainty, it is evident that if the decision maker knows the outcome resulting from each act, then a well-defined preference over O yields the simple decision rule: Choose the alternative that yields the most preferred element of O. Simple as this idea might seem, there are important issues to consider in defining "well-defined preference" and "most preferred outcome." Actually, much of what follows will seem like unnecessary notation to some readers since preference is an intuitively well-understood concept. Scholars have found this concept to be more complex, however, and thus have called for a standardized representation to ensure that they are all talking about the same thing. For readers who are concerned principally with learning how they might apply formal theory to their specific substantive problems, we suggest that they adopt tentatively the perspective of learning a language that others use to communicate this sort of information.

The preference relation

We begin with the primitive concept of preference between two outcomes, say o and o', and we use the notation $o \ R \ o'$ to mean, intuitively, that a person likes the outcome o at least as much as o'. If we must distinguish among persons, the notation R_i indicates the preference of person i. Although this description gives some idea of the *weak preference relation*, R, this concept derives its formal meaning from two assumptions that describe its primary characteristics.

  Assumption 1 (Completeness): For every pair of outcomes o and o' in O, either $o \ R \ o'$ or $o' \ R \ o$.

This assumption states that a decision maker acts as if he can order any two feasible outcomes by the weak preference relation.

  Assumption 2 (Transitivity): For any three outcomes o, o', and o'' in O, if $o \ R \ o'$ and $o' \ R \ o''$, then $o \ R \ o''$.

This assumption states that if the first outcome is no worse than the second and the second is no worse than the third, then the first is no worse than the third.

 Before discussing these assumptions in detail, we first define two related concepts.

  Indifference: A person is indifferent between o and o', written $o \ I \ o'$, if and only if $o \ R \ o'$ and $o' \ R \ o$.

A person who is indifferent between two outcomes regards them as equally preferable, but if a person is not indifferent between them, then he must strictly prefer one outcome to the other.

  Strict preference: A person strictly prefers o to o', written $o \ P \ o'$, if and only if $o \ R \ o'$ but not $o' \ R \ o$.

 Although the assumptions of complete and transitive preference seem innocuous, they are neither trivial nor incontestable. Of the two, Assumption 1 seems the least objectionable. The examples raised to question it typically are of the sort: Which do you prefer, angel food cake or a Bach prelude? Such a comparison seems silly, but we suspect that the problem is not with Assumption 1 but rather that the example fails to specify a decision context. For example, if we specify the context of deciding what to do for the next few minutes, then a concert pianist might easily reveal the preference "cake" P "prelude" if he is hungry. We should not interpret Assumption 1, then, as a restriction on preferences, but as a guide to modeling. For example, if after modeling a particular situation, we conclude that the alternative actions are not exclusive or exhaustive, then we would not suppose that these two requirements are restrictive and do not fit reality, but only that we have failed to model the situation correctly. We should take a similar attitude toward the assumption of complete preferences. If, after specifying O, it seems unreasonable to suppose that we can define even a weak preference relation between two distinct outcomes, then we should presume that the fault lies in the description of O.

Assumption 2 is more problematical. Suppose that a person prefers unsweetened tea but cannot detect the difference made by one additional grain of sugar. Consequently, he is indifferent between a cup containing 0 grains and a cup containing 1 grain, and similarly he is indifferent between 1 and 2 grains, ..., and n-1 grains and n grains. Transitivity requires that if "0 grains" I "1 grain," "1 grain" I "2 grains," ..., and "n-1 grains" I "n grains," then "0 grains" I "n grains." Yet, as the number of grains increases, the number n becomes large enough that a person can detect the difference between 0 and n grains, preferring 0 grains and thus contradicting Assumption 2.

Owing to such possibilities, researchers have considered an alternative assumption, which requires only that the strict preference relation be transitive. Admittedly, most of our work assumes transitivity, but we note the existence of alternative assumptions to indicate that we can weaken parts of the analysis in this and subsequent chapters or at least study politics under alternative assumptions. This discussion also shows that although the concept of preference seems intuitively simple, its precise formulation reveals potential ambiguities.

Ordinal utility

The relations R, I, and P are adequate for representing preferences between pairs of alternatives or outcomes if A and O are finite. But it is inconvenient for summarizing preferences over sets that contain an infinite number of elements, such as budget simplexes. Notice, however, that the relations R, I, and P are much like the mathematical notations $\geq$, $=$, and $>$ used to express relations among numbers, and it would be convenient if we could use these notations to represent preferences. To do so, we need the concept of an ordinal utility function.

> *Ordinal utility function*: An ordinal utility function defined over the set of outcomes O, assigns a real number $u(o)$ to each element of O such that, for all o and o' in O, $u(o) \geqslant u(o')$ if and only if $o \, R \, o'$.

Hence, an ordinal utility function summarizes a person's preference order, since nothing is changed if we write $u(o) \geqslant u(o')$ rather than $o \, R \, o'$. Although both representations of preference are equivalent, a utility function can be an especially convenient and compact summary. For example, suppose that O consists of various amounts of money, and $x \in O$ is a specific amount. If a person prefers more money to less, then an acceptable representation of preferences in terms of utility is $u(x) = x$, since the ordinal utility function, $u(x)$, increases as we move up the person's preference order. Similarly, suppose that v_i represents the votes that candidate i receives in a two-candidate election. If each candidate prefers a larger plurality to a smaller one, then we could let $u_1(v_1, v_2) = v_1 - v_2$ and $u_2(v_1, v_2) = -(v_1 - v_2)$, in which the subscript on u denotes the utility function of either candidate 1 or candidate 2. Alternatively, suppose that $v_i = v_i' + \varepsilon$ is a random variable so that candidate 1's probability of winning the election is $\Pr[v_1 > v_2]$, where Pr denotes probability of. Then, given the flexibility in descriptions of O, we could let O be the set of all probabilities (that is, all numbers in the closed interval $[0, 1]$), and assuming that candidates prefer a higher probability of winning to a lower one, we could summarize candidate 1's preferences as $u_1(v_1, v_2) = \Pr[v_1 > v_2]$.

These examples of utility functions illustrate two important aspects of ordinal utility. First, note that in the case of representing preferences for money, we could also have used the functions $u(x) = 3x + 5$ or $u(x) = x^2$. That is, any function that increases with x is an acceptable representation of the presumed preference for more money. And there are infinitely many functions that could serve the purpose of representing a preference order. Stated formally,

> *If u is an ordinal utility function on O and if w is any function such that for all o and o' in O, $w(o) > w(o')$ if and only if $u(o) > u(o')$, then w is an equivalent ordinal utility function.*

The preceding examples are important because they also illustrate how we learn people's preferences. We could, of course, ask them about their preferences directly. Unfortunately, this approach is unreliable. People might lie if they are embarrassed about their preferences, if they do not trust the interviewer, or, in the case of legislators, if they fear that their answers might have undesirable consequences. This approach is also inappropriate if our questions have no meaning for people whose view of the world differs from that implied by our formal conceptualizations.

It is not unusual, then, for researchers to use the technique of *postulated preference*. Here, we insert different assumptions about preference into models until we find the assumption that best predicts outcomes. Examples of postulated preference include the assumption that election candidates prefer winning to losing, that bureaucrats prefer increased budgets for their bureaus to smaller ones, that firms prefer higher to lower profits, and that the desire for reelection provides legislators with their principal motivation. Although we do not, and perhaps cannot, directly measure these preferences, we can suppose that they are a reasonable first approximation for what motivates people in a variety of roles. And although we might also believe that deeper motivations exist, such as a desire to improve public policy or to maximize personal income, objectives such as winning elections are instrumental for these other ends. Later, we may find that a particular assumption about preference too frequently fails to account for actions, in which case our failures may provide clues about alternative hypotheses.

Scholars often postulate a preference for a given class of decision makers not only because they believe it is a good approximation to reality, but also because they are curious about its implications. We might debate the validity of the hypothesis that a legislator's primary objective is getting reelected and that one can rank his or her actions according to the degree to which they contribute to reelection chances. This hypothesis is sufficiently interesting that Mayhew and Fiorina's research arouses our curiosity about the extent to which legislative choice is consistent with it. Similarly, although we might agree that the goals of public bureaucrats vary, Niskanen's assumption that federal bureau heads act to increase the budgets of their bureaus receives considerable attention because this postulated goal is an interesting hypothesis to pursue. Even though Riker's assumption, that federal national governments formed to increase the security of the states involved, greatly simplifies the motivations of decision makers in national politics, it provides a useful basis for understanding the forces behind an entire class of historical events. We can learn about people's preferences, then, in many ways, but often the most fruitful approach is to assume the simplest possibility. As our models become more sophisticated, we can make our assumptions more complex.

1.3   Multiattribute outcomes

The definitions and notation in the previous sections provide the foundation for models in political science that employ the assumptions of methodological individualism and purposeful action. Although this foundation has the virtue of generality, we also need to know how to use those definitions to model specific political processes.

At this point the reader should keep in mind two complementary approaches to research. One approach is to proceed directly to the phenomena of interest, to discern patterns in events and preconditions, and to use contemporary theory as a guide to the construction of explanations. The second approach begins with first principles, building theory in the direction of substantive issues. The first approach is more likely to compromise theoretical rigor in favor of substantive content, whereas the second commonly favors rigor over substance. If done well, each approach complements the other, and thus the researcher's approach depends more on personal taste and expertise than on any philosophical justification. Since this text uses a particular theory, however, we adhere to the second approach, which requires that we expand our horizons slowly and that we resist the temptation to confront the real world of politics prematurely or with naive theoretical constructs.

To extend our theoretical foundations in the direction of substance, then, we can now look more closely at the concept of an outcome and at the preferences that we assume are defined over outcomes. The preceding section treats each outcome as a single entity and suggests that we can ascertain preferences over O by postulating and empirically testing alternatives. Learning people's preferences is difficult, however, if outcomes stand for the complex events that describe most interesting political situations. We can describe few bills that Congress passes, for example, by a single easily identifiable characteristic. Appropriations bills generally allocate monies simultaneously to a plethora of expenditure categories, and amending unrelated measures to a bill is a valuable strategy for a Congress seeking to thwart a president's veto power. Similarly, although it is sometimes convenient to conceptualize the outcome of an election in terms of a simple "left–right" characterization of the winning candidate's platform, those who are concerned with specific issues, such as abortion, social security reform, and arms control, may prefer to characterize the outcome by that candidate's position on each of these issues. If we confront such complexity directly and also maintain generality in our representations of outcomes and preference, we can begin to breathe life into the preceding theoretical foundation.

To begin, suppose that we can describe an outcome by some number, m, of dimensions or attributes, where each dimension consists of a set of mutually exclusive values or possibilities. For example, earlier we raised the possibility that people might value not only the outcome of an election in terms of who wins and who loses but also in terms of whether or not they voted, in which case we expand the set of outcomes to $O = \{(o, a)\}$. Thus, this expanded set is two-dimensional, and the first dimension specifies the winning candidate, while the second dimension specifies how the person voted. Similarly, if O corresponds to alternative appropriations bills concerning m distinct activities, we can represent each element of O by an m-element vector in which the jth element specifies how much money Congress appropriates for the jth activity. That is, we can let each possible bill, $o \in O$, be represented by $(o_1, o_2, \ldots, o_m)$.

It is one thing, of course, to add some additional notation to represent complex events and another to see how preferences are defined over such events. Suppose, then, that the Senate is considering various appropriations for defense and social-welfare expenditures. If o_1 denotes defense expenditures and o_2 denotes social-welfare expenditures, then the two-element vector, or 2-tuple (o_1, o_2), characterizes outcomes. We can then plot these outcomes in a two-dimensional coordinate system, in which the first (horizontal) dimension corresponds to money for defense projects and the second (vertical) dimension corresponds to money for social-welfare programs. To see how we might define preferences in this coordinate space, suppose that there are four outcomes under consideration, o, o', o'', and o''', as Figure 1.4 shows. One possibility is that people attach supreme importance to one dimension. If, for example, a senator considers defense expenditures of paramount importance, then the monies allocated to nondefense programs might become relevant to any decision he might make only if two outcomes allocate the same amount to defense. He is indifferent between two outcomes only if they are identical on both dimensions. Thus, if, ceteris paribus, a senator prefers to spend more on social welfare rather than less but regards defense expenditures as crucial, then the outcomes in Figure 1.4 are ordered o'' P o''' P o' P o. This example illustrates _lexicographic_ preferences.

Lexicographic preferences, however, ignore _tradeoffs_ between dimensions. A senator may strictly prefer o'' to o''' because o'' corresponds to greater nondefense spending and is identical to o''' on the first dimension. But suppose that the senator considers another possibility, o^*, which provides only slightly less spending on defense than either o'' and o''' but considerably more spending on nondefense matters. Thus, because o^* allocates slightly less money to defense, the lexicographic

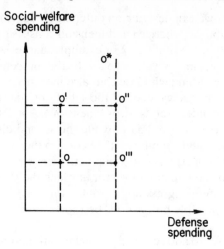

1.4 Multiattribute outcomes.

rule orders **o′″ Po***; however, a senator may be willing to *trade* slight decreases in defense expenditures for large increases in other categories.

Understanding and finding a convenient representation of the tradeoffs people make between components of an outcome move us closer to a general representation of preferences over complex outcomes. For a simple example, then, consider a person who owns x left-footed shoes and y right-footed shoes. If this person prefers matched shoes and as many pairs as possible, then, barring trades with anyone else, he is indifferent between the outcomes ($x = 5, y = 8$), ($x = 5, y = 5$), and ($x = 8, y = 5$), since all correspond to five useful pairs. But he prefers ($x = 6, y = 6$) to any of these three possibilities. Figure 1.5, which represents x on the horizontal axis and y on the vertical axis, portrays these preference and indifference patterns. Specifically, a person is indifferent between any two outcomes on "curve" a, since all outcomes on this curve correspond to one matched pair. (For convenience, curves a through g are drawn as continuous even though it is difficult to imagine fractional shoes.) Similarly, indifference holds among all outcomes on curve b, among all outcomes on c, and so on. But this person prefers an outcome on f to one on e since f corresponds to six distinct pairs, whereas e corresponds to five pairs. Thus, the curves in this figure, referred to as *indifference curves* or indifference contours, model a willingness to make trades and tradeoffs. First, a person should be willing to exchange the outcome (5, 8) for the outcome (8, 5) and vice versa, since these outcomes fall on the same curve and thus utility

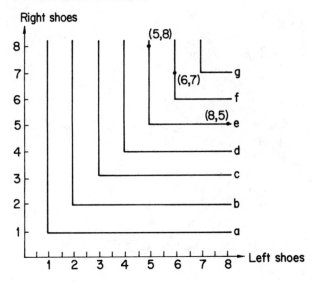

1.5 Indifference contours for shoes.

neither increases or decreases when moving between them. Second, at (5, 8), a person should be willing to trade a right-footed shoe for a left-footed shoe, and attain the more preferred outcome (6, 7) or six pairs; or at (8, 5), to trade a left-footed shoe (or even two such shoes) for a right-footed shoe.

The preceding representation of preference and indifference, of course, is not unrelated to the concept of ordinal utility. In the case of left- and right-footed shoes, for example, we can represent a person's preference order over the set of possible outcomes, $O = \{(x, y)\}$, by the ordinal utility function $u(x, y) = \min[x, y]$, where min means the minimum of. For another example, consider again the situation in which outcomes correspond to the amount of money appropriated for defense and the amount appropriated for social-welfare programs. Suppose that a senator is concerned solely with the total amount of money that government spends, preferring more money to less. If x denotes defense appropriations and y nondefense appropriations, the following ordinal utility function summarizes this preference: $u(x, y) = x + y$. Imposing the easily swallowed assumption that neither x nor y can be negative, we find that Figure 1.6a portrays this function. Specifically, graphing $u(x, y)$ against x and y yields a plane that starts at the origin, $(0, 0)$, and slopes upward to infinity. Now suppose, as Figure 1.6a shows, that we pass a plane horizontal to the x-y axes through the vertical $u(x, y)$ axis at, say,

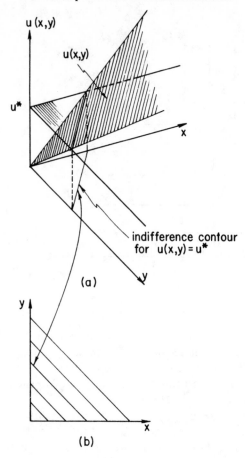

1.6 Derivation of indifference contours.

u^*. This plane intersects the surface representing $u(x, y)$ and cuts a straight line on that surface as shown (since the intersection of two planes is a straight line). If we project this line onto the x-y surface, we get the indifference contour describing all outcomes to which a person associates the ordinal utility u^*. Doing this for several values of u produces an indifference curve "map," as Figure 1.6b shows.

Another senator, however, might believe that as expenditures in either category increase, the money spent is beneficial but the impact of each additional dollar declines. An ordinal utility function consistent with this assumption is $u(x, y) = x^{1/2} + y^{1/2}$, which Figure 1.7a depicts, along with its indifference curve representation in Figure 1.7b.

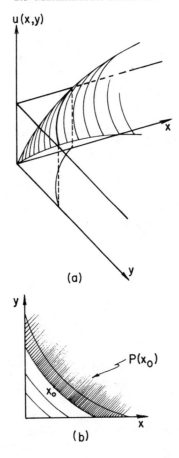

1.7 Derivation of indifference contours.

The preceding examples illustrate preferences that run off to infinity in the sense that people prefer more of everything to less – more pairs of shoes to less and a larger government budget to a smaller one. Economic analysis commonly assumes such preferences because it seems reasonable to suppose that people prefer more money to less, bigger homes, more cars, more food, better health, and so on. But consider a person living in Manhattan, with its approximately square city blocks. Suppose that this person prefers walking as short a distance, the fewest city blocks, as possible, so that the utility associated with a particular outcome (location) is the negative of the minimum distance required to travel to that spot. Figure 1.8 shows these preferences by

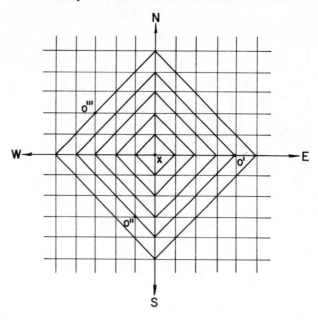

1.8 City-block indifference contours.

letting the horizontal axis denote east–west distance, the vertical axis denote north–south distance, and $\mathbf{x} = (x_1, x_2)$ represent the person's present location. The thin lines in this figure denote streets, and the dark lines represent indifference contours (which, for convenience, we represent as continuous) that entail the same walking distance from $\mathbf{x}$. For example, reaching the points $\mathbf{o}'$ and $\mathbf{o}''$ each require walking four blocks, so these two points lie on the same indifference contour. The point $\mathbf{o}'''$, however, requires walking five blocks and is therefore on an indifference contour further from $\mathbf{x}$ than is $\mathbf{o}'$ or $\mathbf{o}''$. An ordinal utility representation of these preferences is the function

$$u(\mathbf{x}, \mathbf{y}) = -(|x_1 - y_1| + |x_2 - y_2|),$$

in which $|\ |$ denotes absolute value and $\mathbf{y} = (y_1, y_2)$ is an arbitrary location on the grid. Further, since this particular utility function is maximized if $\mathbf{y} = \mathbf{x}$, we refer to $\mathbf{x}$ as the person's *ideal point* – the outcome that stands highest on his preference order.

We call indifference contours like the ones in Figure 1.8 *city block*, and in election models they summarize nicely the intuition that, when comparing candidates, voters look separately at how different each candidate is from their most preferred position on each election issue.

Social welfare

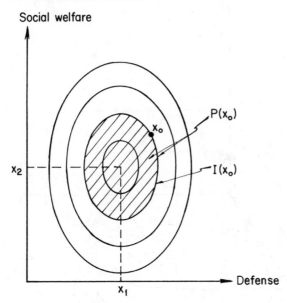

1.9 Euclidean indifference contours.

They then vote for the candidate whose total distance across all issues is the least. Further, since such preferences represent a measure of distance from an ideal point, we call models that use such preferences *spatial models*, and when referring specifically to elections, we call them spatial models of election competition.

For a different kind of spatial preference, imagine a person living on a plain, in which case indifference contours, those of equal walking distance, are concentric circles about his location. For a more relevant example, suppose that a senator prefers neither as much spending as possible nor zero spending. But he believes that if we spend more than x_1 dollars on defense and more than x_2 dollars on social welfare, then money is wasted; his utility declines, though, if less than x_1 and x_2 dollars is spent on defense and social welfare, respectively. Given his preference for defense over nondefense spending, however, he is more sensitive to variations in defense spending. Thus, his indifference contours might look like those in Figure 1.9, in which a one-unit increase or decrease in defense spending away from the ideal point, (x_1, x_2), moves him to a less preferred indifference contour than does a comparable increase or decrease in nondefense spending.

An ordinal utility function consistent with these indifference contours takes the form:

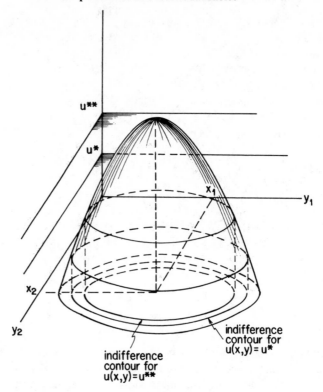

1.10 Derivation of indifference contours from Euclidean utility function.

$$u(\mathbf{x}, \mathbf{y}) = -[a_1(x_1 - y_1)^2 + a_2(x_2 - y_2)^2]^{1/2}, \qquad (1.1)$$

where a_1 and a_2 represent constants that weight the issues and create the elliptical form of the indifference curves. Hence, if dimension 1 counts more heavily than 2, then $a_1 > a_2$.

To understand expression (1.1) better, suppose that the issues are of equal importance ($a_1 = a_2 = 1$), and that this person's ideal point is at the origin ($x_1 = x_2 = 0$). Now let us fix utility at some constant, say $-r$ (that is, $u(\mathbf{x} = \mathbf{0}, \mathbf{y}) = -r$). Expression (1.1) then becomes

$$-r = -[y_1^2 + y_2^2]^{1/2},$$

which, after squaring both sides, yields

$$r^2 = y_1^2 + y_2^2.$$

This is simply the equation for a circle centered at $(0, 0)$ with radius r.

Thus, the level of utility, $-r$, determines the circle's radius. Lower levels of utility, or equivalently, higher values of r, yield circles of greater radius. The ideal point (x_1, x_2) determines the circle's center, and a_1 and a_2 determine whether the "circle" is elliptical. We refer to expression (1.1), then, as the _weighted Euclidean distance_ utility function; and when the weights are all equal we refer to it simply as the _Euclidean distance_ function.

To see better the utility function that expression (1.1) describes, we can consult Figure 1.10, which illustrates the derivation of indifference contours from the ordinal-utility function $u(\mathbf{x}, \mathbf{y})$ after we set $a_1 = a_2 = 1$.

Spatial models of elections and of committees use this functional representation of utility extensively to model the preferences of voters or the members of a legislature. For future reference, then, we note that if we are modeling situations that concern more than two issues, the m-dimensional extension of expression (1.1) is given by

$$u(\mathbf{x}, \mathbf{y}) = -\left[\sum_{i=1}^{m} a_i(x_i - y_i)^2\right]^{1/2} \qquad (1.2)$$

scholars have developed many multidimensional scaling procedures to estimate such preferences from public opinion data. Furthermore, we can prove numerous interesting and important results using expression (1.2). And although we lose some generality when we impose any specific assumption on the form of utility functions, one of the advantages of mathematics is that its precision permits us to explore the possibility of generalization, of expanding the domain of a theory or model. So to secure the initial insight into a problem, we commonly begin by proving a result that assumes the Euclidean representation. Later, we extend the applicability of the result by showing that it holds for less specific assumptions about preferences.

One way to generalize the insight gained from a model that assumes Euclidean preferences is to characterize preferences by the properties of different sets. That is, if the set of feasible outcomes, O, is part of an m-dimensional space, then we can characterize preferences by the properties that we attribute to three sets: $P(\mathbf{o})$, $I(\mathbf{o})$ and $R(\mathbf{o})$, where $\mathbf{o}$ is an arbitrary outcome in O. Since the definitions of $I(\mathbf{o})$ and $R(\mathbf{o})$ follow from $P(\mathbf{o})$, we define $P(\mathbf{o})$ formally,

> _Preference set_: $P(\mathbf{o}) \subset O$ is the preference set of $\mathbf{o} \in O$ if the decision maker strictly prefers all outcomes in $P(\mathbf{o})$ to $\mathbf{o}$. That is, $P(\mathbf{o}) = \{\mathbf{o}' \in O \mid \mathbf{o}' \, P \, \mathbf{o}\}$, or equivalently in terms of utility, $P(\mathbf{o}) = \{\mathbf{o}' \in O \mid u(\mathbf{o}') > u(\mathbf{o})\}$.

In Figure 1.9, the preference set of x_o includes all points inside the indifference contour on which x_o lies. Similarly, in Figure 1.7, $P(x_o)$ corresponds to all points to the northeast of the indifference contour containing x_o. The reader should be able to see how to define the "preferred or indifferent to set" of a point, denoted $R(o)$, using the preference relation R, as well as the "indifference set," denoted $I(o)$, which is defined by the relation I. Of course, if outcomes have a geometric interpretation, a point's indifference set corresponds to outcomes on its indifference contour.

Throughout this book, whenever outcomes or alternatives are a subset of some m-dimensional space such as a budget simplex, and unless otherwise noted, we impose implicitly or explicitly a number of assumptions about $P(o)$, $I(o)$ and $R(o)$. All of the geometric examples of preference considered thus far in this chapter satisfy these assumptions. Therefore, if a theorem refers only to these assumptions, then that theorem holds regardless of whether Figure 1.7, 1.8, or 1.9 describes preferences. Briefly, if O is a closed, convex set, we assume

a. Preference sets – the outcomes in O that the decision maker strictly prefers to o, $P(o)$ – are convex and open (by open we mean that $P(o)$ does not include its boundary). Thus, if o' is strictly preferred to o and, thereby, is an element of $P(o)$, then there exists an arbitrarily small neighborhood around o' – all the outcomes that are no further from o' than some arbitrarily small distance – that is entirely in $P(o)$.

b. The boundaries of preferences sets are indifference sets, $I(o)$, which are closed. Thus the union of $P(o)$ and $I(o)$, denoted $R(o)$, is closed and convex.

c. Indifference sets are "thin"; that is, if the decision maker is indifferent between o' and o so that $o' \in I(o)$, then there exists an arbitrarily small neighborhood around o' such that o is preferred to some of the points in this neighborhood while other points are in $P(o)$. In other words, we can move ever so slightly off the indifference curve in one direction to points that are preferred to o and in another direction to points that o is preferred to.

d. Indifference contours are continuous. This assumption is simply a mathematical convenience that renders figures less complex and permits the use of calculus.

Several of these assumptions are redundant, and assumption (c) warrants substantive comment here. This assumption precludes the possibility that indifference contours have "broad" areas, as when a

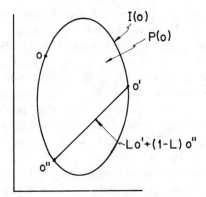

(a) Strictly convex R(o) with
   thin indifference

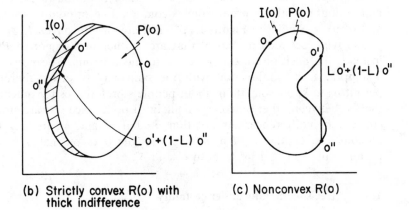

(b) Strictly convex R(o) with
   thick indifference

(c) Nonconvex R(o)

1.11 Assumptions about preference sets.

person is indifferent over ranges of outcomes. Thus, assumption (c) excludes people who are unconcerned with minor changes from some status quo point. Such people have fat rather than thin indifference curves through this point. A senator might believe, for example, that any of a number of possible modifications of the technical details of some bill are inconsequential with respect to the bill's ultimate intent. Hence, he is indifferent among all bills in the region of the original bill reported out of committee. This characterization violates assumption (c).

It is unusual, however, to find an assumption one cannot object to by referring to some example, and we can best discuss the full substantive implications of assumptions in the context of the specific results presented in subsequent chapters. Here we illustrate the meaning of the preceding assumptions by deriving a useful implication. Consider Figure 1.11a, which reproduces part of Figure 1.9, so that the ellipse's boundary corresponds to $I(\mathbf{o})$, $P(\mathbf{o})$ consists of all points inside the boundary, and $R(\mathbf{o})$ equals the union of $I(\mathbf{o})$ and $P(\mathbf{o})$. Now consider any two outcomes, say $\mathbf{o}'$ and $\mathbf{o}''$, that are in $I(\mathbf{o})$. Since $R(\mathbf{o})$ is convex (assumption (b)), the line connecting $\mathbf{o}'$ and $\mathbf{o}''$ (the convex combination $L\mathbf{o}' + (1 - L)\mathbf{o}''$, $0 \leq L \leq 1$) is in $R(\mathbf{o})$. In fact, since $R(\mathbf{o})$ is strictly convex as drawn, this line is in the interior of $R(\mathbf{o})$. But from assumption (c), indifference sets are thin, so the interior of $R(\mathbf{o})$ is $P(\mathbf{o})$. Thus, all points on the line connecting $\mathbf{o}'$ and $\mathbf{o}''$ (excluding its end points) are in $P(\mathbf{o})$; the decision maker strictly prefers all such points to $\mathbf{o}$, $\mathbf{o}'$ and $\mathbf{o}''$. Figure 1.11b, however, illustrates an $R(\mathbf{o})$ that violates assumption (c) (that is, it has a thick boundary), and thus the decision maker does not strictly prefer convex combinations of points such as $\mathbf{o}'$ and $\mathbf{o}''$ to $\mathbf{o}$. Similarly, Figure 1.11c illustrates a nonconvex $R(\mathbf{o})$ and points lying between $\mathbf{o}'$ and $\mathbf{o}''$ that are neither in $I(\mathbf{o})$ nor in $P(\mathbf{o})$; instead, the decision maker prefers $\mathbf{o}$ to those points. The preceding assumptions, in conjunction with the convexity of $R(\mathbf{o})$, therefore permit us to know something about people's preferences for outcomes that lie between other outcomes, namely: If preference sets are strictly convex, if indifference sets are thin, and if $\mathbf{o}'$ and $\mathbf{o}''$ are any two outcomes in O, then $\mathbf{o}' \, I \, \mathbf{o}''$ implies that the decision maker strictly prefers $L\mathbf{o}' + (1 - L)\mathbf{o}''$ to both $\mathbf{o}'$ and $\mathbf{o}''$.

1.4   Decision making under certainty

The preceding section discusses some of the commonly imposed assumptions about the mathematical functions and geometry that we use to represent preference over complex outcomes. Anyone familiar with introductory microeconomics will recognize the indifference contours in Figures 1.6 and 1.7, since they are common representations of preferences for consumers in markets. Preferences like the ones depicted in Figures 1.8 and 1.9, however, find extensive application in models of elections, legislatures, and the like, and they are unfamiliar to most students of economics. Thus, by exploring the relationship between the two types of preference representations, we can begin to appreciate some of the differences between economics and politics. Briefly, what we intend to show is how the preferences commonly assumed in economics give rise to spatial preferences if we interpret the decision makers in our models as voters or legislators, rather than

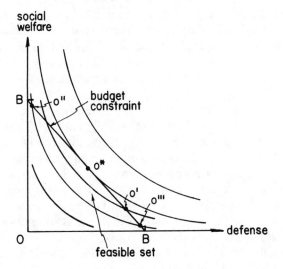

1.12 Inducing an ideal point on a budget-constraint line.

consumers in markets, who are forced to consider the budgetary considerations that commonly constrain government action.

Constrained economic decisions

To begin, recall that under certainty, the decision maker acts as if he knows the relevant state of nature and thereby believes that each act in A yields a unique outcome in O. Thus, it appears that a person should choose the act that yields the most preferred outcome in O that the particular state of nature allows. That is, given the state of nature, a person should choose the alternative that maximizes his ordinal utility function. But we must be careful: A person who prefers more money to less has no best choice in the problem "choose any amount of money." If a legislator prefers spending as much as possible in accordance with the preferences indicated in Figure 1.7, and if any spending program is attainable, then there is no most preferred outcome.

In reality, people's and government's budgets are constrained, which limits attainable outcomes. Maximizing utility given the prevailing state of nature, then, requires that people choose to maximize utility over the *feasible set* of outcomes. Consider Figure 1.12, which shows indifference curves similar to the ones in Figure 1.7 in conjunction with a budget constraint, B, represented by a straight line that corresponds to the condition "defense plus nondefense spending must sum to B dollars."

Hence, if appropriations are nonnegative, then the attainable outcomes must all lie on the line *BB*. Notice that there now exists a unique most preferred attainable outcome, o^*. Since any other feasible outcome, such as o' or o'', lies on a lower (less preferred) indifference curve, in choosing a feasible allocation between defense and domestic spending, the senator chooses o^*.

The properties of the budget constraint and the assumptions that we introduce in the previous section about the sets $P(o)$ and $R(o)$ guarantee the existence of a unique most preferred outcome. Indeed, if the truth of some proposition requires the existence of a unique most preferred feasible outcome, then these assumptions often provide an indirect way of assuring the existence of such an outcome. To see this, note, first, that the feasible set, a line, is convex (if we take any two points on the line and form their convex combination, which consists of all points on the line connecting those two points, that convex combination must also lie on the original line). Second, the indifference curves here define preference sets, $P(o)$, and preferred or indifferent to sets, $R(o)$, that are strictly convex. Because indifference curves are also thin, if the budget-constraint line, *BB*, and some indifference curve are tangent, then they are tangent at a unique point, which is o^*.

The preceding geometric discussion is inconvenient for more than two dimensions or if complex analytic functions specify utility functions. More generally, then, ascertaining whether a unique most preferred alternative outcome exists proceeds in two steps: (1) checking whether $R(o)$ is convex and, if it is convex, (2) computing o^*. For example, suppose that $u(x, y) = xy$, where x and y are nonnegative variables that together describe outcomes (for example, defense and social-welfare spending). Suppose that we wish to maximize this function subject to the budget constraint $x + y = B$. (We could impose the constraint $x + y \leq B$ to indicate that no more than the amount B can be spent. But since our decision maker derives no pleasure from saving and since his utility increases as x and y increase, we know that he will spend his entire budget and for him the relevant constraint is $x + y = B$.) To illustrate the general method of checking whether the sets $R(o)$ defined by $u(x, y)$ are strictly convex, let $o' = (x', y')$ and $o'' = (x'', y'')$ be two arbitrary outcomes that lie on the same indifference curve; that is $u(o') = u(o'') = u^*$. Thus, $R(o')$ is identical to $R(o'')$. From the definition of a strictly convex set and referring back to Figure 1.11a, $R(o')$ is strictly convex if any point on the line connecting o' and o'' is in the interior of $R(o')$, that is, in $P(o')$. Hence, $R(o')$ is strictly convex if the utility of the convex combination of o' and o'', $u(Lo' + (1 - L)o'')$, exceeds u^*. To see that this is true for the utility function of our example, notice first that since $o' = (x', y')$ and $o'' = (x'', y'')$,

$$u(L\mathbf{o}' + (1 - L)\mathbf{o}'') = u(Lx' + (1 - L)x'', Ly' + (1 - L)y''),$$

which, from the fact that $u(x, y) = xy$, becomes

$$= [Lx' + (1 - L)x''] [Ly' + (1 - L)y''].$$

After multiplying the two terms in brackets and rearranging, we get,

$$= L^2x'y' + (1 - L)^2x''y'' + L(1 - L) [x''y' + x'y''].$$

If we substitute into this expression the assumption that $\mathbf{o}'$ and $\mathbf{o}''$ yield the same utility, that is, $x'y' = x''y'' = u^*$, then,

$$= [L^2 + (1 - L)^2]u^* + L(1 - L) [x''y' + x'y''].$$

Adding and subtracting $2L(1 - L)u^*$ to this expression doesn't change its value, so

$$= [L^2 + 2L(1 - L) + (1 - L)^2]u^* - 2L(1 - L)u^*$$
$$+ L(1 - L) [x''y' + x'y''].$$

Notice that the first bracketed term can be rewritten as $[L + (1 - L)]^2$, but since $L + (1 - L) = 1$, this term is 1 and our expression becomes

$$u(L\mathbf{o}' + (1 - L)\mathbf{o}'') = u^* + L(1 - L)[x''y' + x'y'' - 2u^*].$$

So $R(\mathbf{o})$ is strictly convex if this new bracketed term is positive. To show that this is true, let $x'' = x' + \Delta$, where Δ accounts for the difference between x'' and x'. Since $x''y'' = u^*$, then $y'' = u^*/x'' = u^*/(x' + \Delta)$. Substituting these values for x'' and y'' into the bracketed term, and substituting $x'y'$ for u^* in that term as well, yields, after some algebraic manipulation and cancellations, $\Delta^2y'/(x' + \Delta)$. But this term must be positive since all x's and y's are positive; therefore $u(L\mathbf{o}' + (1-L)\mathbf{o}'')$ must equal u^* plus some positive term. Thus, it must be the case that this utility number is strictly greater than u^* alone, just as we wanted to establish. Therefore, any point on the line connecting $\mathbf{o}'$ and $\mathbf{o}''$ lies in $P(\mathbf{o}')$ and, hence, $R(\mathbf{o}')$ is strictly convex.

This argument, which uses the definition of strict convexity, tells us that there exists a unique most preferred feasible outcome. It does not tell us, however, what that outcome is; to find it or otherwise characterize it, we must use calculus, which may be unfamiliar to some readers (and who therefore may prefer to skip the next few paragraphs). Briefly, we proceed by adding the term $-H(x + y - B)$ to the utility function $u(\mathbf{o}) = u(x, y) = xy$, to form the expression

$$w = u(x, y) - H(x + y - B) = xy - H(x + y - B), \quad (1.3)$$

where H is an unknown constant called a *Lagrange multiplier*. Notice that by adding the term $H(x + y - B)$ we have changed nothing since this term, when the budget constraint is satisfied (when $x + y = B$), equals zero. The usual way to maximize a function is to compute its first derivative with respect to both x and y, set the results equal to zero, and solve for x and y. (Generally, we must take more care to ensure that a derivative exists, that the function is being maximized and not minimized, and that it is not maximized at the end points of the budget constraint, that is, that the maximum is not a "corner solution." We have chosen our examples to avoid such complications.) Taking the first derivative,

$$\frac{dw}{dx} = y - H \text{ and } \frac{dw}{dy} = x - H,$$

which, after being set equal to zero, requires that $x = y = H$. Substituting the identity $x = y$ into the expression of the budget constraint, we find that utility is maximized if $x = y = B/2$. That is, utility is at a constrained maximum if one-half the budget goes to defense and one-half to social-welfare spending.

We do not offer the preceding discussion of the utility function $u(x, y) = xy$ because this function serves any special role in the models that this volume reviews. Rather, it simply helps us illustrate the meaning of convexity and the application of Lagrange multipliers. Convexity facilitates the proof of a great many results, results that do not hold unless this assumption is satisfied. And because we assume that people maximize utility, predicting their actions necessarily requires that we deduce the acts or outcomes that maximize the mathematical representation of their preferences. Convexity and Lagrange multipliers, then, are important concepts and anyone unfamiliar with them should devote some effort to understanding them.

Constrained political decisions

Thus far, our analysis does not differ much from what we might find in most introductory economics texts, in their discussions of consumers' choices in markets, given the budget constraints that their incomes impose. After developing a parallel model for firms, those texts then proceed to discuss markets in general. At this point, however, the analysis of economic and political systems must diverge. Consumers, contemplating the things that they can purchase in supermarkets and department stores, given their budget constraints, are dictators over their choices of commodity bundles. If Figure 1.12 models a consumer and the dimensions correspond to the units of bread and milk to be purchased, then that consumer presumably can select o^* directly. In the

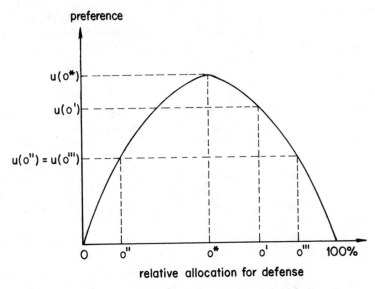

1.13 Single-peaked preference on a budget-constraint line.

context of our legislative example, it is unlikely that a senator unilaterally can impose government policy. Minimally, he or she must give some consideration to the preferences of other legislators. That is, unlike the model of a consumer in a market, a model of a single legislator's preferences cannot predict the outcome that eventually prevails.

If each legislator's preferences over defense and social-welfare spending yield the same value of o^*, then conflict among legislators would not exist and government policy could be chosen by unanimous consent. But because the values of x and y that maximize $u(o) = u(x, y)$ depend on the specific functional form of utility, this almost certainly guarantees some conflict. We know that if $u(x, y) = xy$, then the legislator most prefers an equal mix between defense and social-welfare spending. Other functional forms of $u(x, y)$, though, can yield different most preferred outcomes. Hence, even if all members of a legislature prefer as much spending as possible, they need not agree on the relative share that they should allocate to defense. Variation in o^* among legislators, in turn, gives rise to the concept of a political issue. Briefly, we can think of the budget line in Figure 1.12 as an issue in which alternative outcomes on the line correspond to alternative appropriations of the government's budget between defense and social welfare. If we lay this line out horizontally, as in Figure 1.13, then legislators who prefer to spend more money on defense than on social welfare will have their ideal points on the right side of the line, and those who prefer to spend relatively more on social welfare will have their ideal points on the left side of the line. A legislator with the utility function $u(x, y) = xy$ will have an ideal point in the center of the line, at the point "50% for defense."

Since no legislator is dictator, the eventual outcome depends on the procedures that the legislature uses to make a group decision. Although we are not prepared to discuss these procedures in this chapter, it is evident that, to assess the implications of even simple ones such as majority rule, we must know something more about preferences than merely what each legislator would choose if he or she were dictator. How a legislator responds to bargaining and vote trades, for example, depends on what preferences look like over other feasible outcomes. Hence, anticipating models that seek to predict the outcome that a legislature chooses, we characterize each legislator's preferences over the entire issue, which leads to the preferences depicted in Figure 1.13.

To derive such a characterization, recall that the indifference curves in Figure 1.12 satisfy assumption (c) and define strictly convex preference sets. Thus, as defense appropriations increase or decline from o^*, we necessarily move to lower (less preferred) indifference contours. Equivalently, as we move in either direction from o^* along the horizontal line in Figure 1.13, the legislator's utility declines monotonically. Hence, if o' and o''' are two outcomes to the same side of o^* and if o''' is further from o^* than is o', then o' is preferred to o'''. Figure 1.13 shows an ordinal utility function over the feasible outcomes that is consistent with this property.

The general shape of this derived utility function has widespread application in political theory. Its specific form will depend on the exact form of the utility function in the unconstrained space. The reader can confirm, for example, that if $u(o) = u(x, y) = xy$, then the derived utility along the budget constraint is a function that is symmetric about the person's ideal point. More generally, if $u(o)$ yields strictly convex preference sets in its indifference curve representation, then the preference function that a budget constraint induces is necessarily *single peaked*: A person's ordinal utility function on the constraint has a unique ideal point and declines monotonically to either side of this point.

Preferences on such constraints are interesting because of their substantive interpretations. Our example focuses on governmental defense and social-welfare spending. Many other important political issues also consist of weighing the relative importance of two alternative public programs, such as choosing between the attention that government should give to space exploration compared with food stamps, to teachers' salaries compared with the salaries of police and fire personnel, and to operating expenditures compared with capital projects. These choices often correspond to ideological or left – right issues, and the preceding analysis yields a basis for deriving preferences on such issues from preferences over the specific goods and services that people consume.

The utility function in Figure 1.13 is simply the one-issue analogue of the two-issue utility function that Figure 1.10 illustrates. It is instructive to see how Figure 1.13 relates to the indifference curves that economists ordinarily use, since political debate often concerns more than two programs or policies. Creating a public budget, even in its simplest form, requires choosing different expenditures for different programs *and* choosing an overall budget or tax rate. To complicate the previous example, then, suppose that besides choosing an allocation between defense and nondefense spending, the legislature also chooses the budget constraint by setting a tax rate. Suppose, further, that the tax rate, t, must lie in the interval $[0, 1]$; if $t = 0$, then there are no taxes, and if $t = 1$, then the government expropriates all wealth in society. We can now denote an outcome by the 3-tuple (x, y, t), where x is the amount spent on defense, y is the amount spent on social welfare and t is the tax rate. But now $x + y$ cannot exceed tW, where W is the total wealth or earned income in society. Hence, if we ignore the macroeconomic effects of tax policy and assume that W is a constant, then a budget simplex in three dimensions represents the set of feasible outcomes. The first dimension denotes x, the second dimension corresponds to y, and the third dimension corresponds to spending in the private sector of the economy, which equals W if $t = 0$ and equals 0 if $t = 1$. Figure 1.14 shows this simplex.

Suppose that a legislator, without taking constraints or taxes into account, prefers as much defense, social-welfare and private sector spending as possible (ceteris paribus, his utility increases with an increase in any of these measures), and that his indifference contours among x, y, and private consumption look like the three-dimensional analogue to the contours in Figures 1.7b. This may be asking a bit much of our conceptual abilities, and it is useful to think of these indifference contours as a set of nested mixing bowls whose rounded bottoms all point toward the origin. Suppose that the smallest mixing bowl (highest indifference contour) that we can reach while remaining on the budget simplex is tangent to this simplex at o^*. This point corresponds to the legislator's most preferred and feasible government expenditure and tax policy. In Figure 1.14, this point corresponds to a tax rate of t^* and

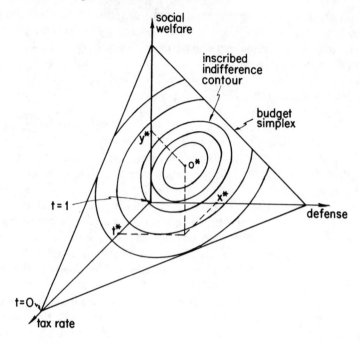

1.14 Indifference contours induced on a budget simplex.

allocations of x^* and $y^* = t^*W - x^*$ between defense and domestic spending. As we move away from the point $\mathbf{o}^*$, however, we necessarily move to larger mixing bowls (lower indifference contours), so the legislator's utility declines. Note, however, that any specific and less preferred bowl intersects the budget simplex and describes a circle, ellipse, or similar figure on the simplex and that larger (less preferred) contours describe larger circles or ellipses that circumscribe the intersections of more preferred contours. Hence, a new set of indifference contours are inscribed on the budget simplex about $\mathbf{o}^*$, as Figure 1.14 illustrates. These contours, then, are the legislator's indifference contours in the feasible set. It is possible to show, moreover, that if the original preference sets are (strictly) convex, then the preference sets induced on a budget simplex are (strictly) convex as well.

If we take the budget simplex portrayed in Figure 1.14, with its inscribed indifference contours and lay it flat on the page, those contours will look much like the contours in Figures 1.9 and 1.10. Thus, we can rationalize the contours in those figures, which are intended to describe utility functions with satiation points, in terms of constrained decision making. Indifference contours of this sort, then, are extremely important for modeling the decisions of citizens or legislators coping with public policy issues in an environment of constrained or limited resources, or analyzing the implications of alternative tax policies for public spending decisions.

1.5   Decision making under risk

The assumptions about preference and the various geometric representations of utility that the previous section reviews are sufficient for modeling many important political processes, including two-candidate elections (Chapter 4) and the implications of different decision procedures that legislative committees might use (Chapter 6). But uncertainty is missing from our discussion. The preceding structure provides a basis for representing preferences if the decision maker knows the state of nature and, thus, the consequences of his actions. But precise mathematical certainty seldom characterizes decision making, especially in situations that involve the interactions of many people in complex institutional environments. Decision making under certainty, then, is an abstraction that simplifies analysis, and before we can appreciate what we gain or lose by using it, we must consider decision making if a person knows only how a choice of an alternative in A affects the probability that each of several outcomes occurs.

Our example in Figure 1.3 of a simple voting decision is an appropriate place to begin. It seems unreasonable to suppose there that people know with certainty which state of nature will prevail. Some people may believe that candidate 1 is ahead, whereas others believe that 2 is leading, but few may be willing to suppose that the election is a forgone conclusion. Thus, suppose instead that people have estimates of the probability, $p(s_k)$, that each state prevails, and that these probabilities satisfy the usual constraints imposed on such numbers. This means, first, that $0 \leq p(s_k) \leq 1$ for all states of nature. Second, since the states of nature are exclusive,

$$\text{Prob}[s_k \text{ or } s_j] = p(s_k) + p(s_j), \quad k \neq j \tag{1.4a}$$
$$\text{Prob}[s_k \text{ and } s_j] = 0, \quad k \neq j. \tag{1.4b}$$

And since the states of nature are exhaustive,

$$\sum_S p(s_k) = 1,$$

where $\sum_S$ means summed over all elements of S, over all possible states of nature.

Referring to Figure 1.3, notice that if a particular act in $A = \{a_1, a_2, a_o\}$ is selected, the probabilities that various states of nature prevail induce probabilities over the elements of O. If the citizen votes for candidate 1 by choosing a_1, for example, outcome o_2 occurs if the state of nature s_1 prevails, o_3 occurs if s_2 prevails, and o_1 occurs if s_3, s_4 or s_5 prevails. Thus, if a_1 is chosen, o_2 occurs with probability $p(s_1)$, outcome o_3 occurs with probability $p(s_2)$ and the probability that outcome o_1 occurs equals the probability that either s_3, s_4, or s_5 prevails, which, from expression (1.4a), must be equal to $p(s_3) + p(s_4) + p(s_5)$.

Keeping in mind that the probability that a particular outcome occurs depends on which act is selected, suppose the decision maker has made his choice and that q_1, q_2, and so on denote the probabilities that outcomes o_1, o_2, and so on, respectively, occur. Since these probabilities are derived directly from the probabilities defined over S, and since O is exclusive and exhaustive, the q's inherit the properties of the p's. First, $0 \le q_j \le 1$ for all j. Second, since some outcome must occur, $\Sigma q_j = 1$. Finally,

$$\text{Prob}[o_j \text{ or } o_k] = q_j + q_k, j \ne k$$
$$\text{Prob}[o_j \text{ and } o_k] = 0, j \ne k.$$

The preceding discussion, then, assumes that each alternative in A gives rise to a *simple lottery*, denoted $\mathbf{q} = (q_1, q_2, \ldots)$, that satisfies these conditions but in which different alternatives may give rise to different lotteries. A person confronting a risky decision, then, must choose the "best" lottery. Our problem is to extend the concepts of preference and utility so that they model preferences over alternative lotteries.

We can view lotteries as complex outcomes, those in which we give some weight to each event that might occur. And just as we assume that we can define an ordinal utility over O if preferences over O are complete and transitive, we begin by supposing that there exists an ordinal utility function over the set of all simple lotteries that satisfies completeness and transitivity.

> *Assumption 3*: The preference order over all simple lotteries is complete and transitive.

To be certain that we do not exclude any possibilities unnecessarily, we must strengthen this assumption by extending it to complex lotteries as well as simple ones. We can imagine, for example, a two-stage lottery in which, in the first stage, a simple lottery, $\mathbf{q} = (q_1, q_2, q_3)$, among three outcomes is selected by nature with probability p and a different simple lottery, $\mathbf{q}' = (q_1', q_2', q_3')$, between the three outcomes is selected with probability $1 - p$. Similarly, many state lotteries select the contestants for some grand prize from the winners of a prior lottery.

These examples illustrate a *compound lottery*. Compound lotteries can have more than two stages: The outcome of a lottery at any stage except the last determines the lottery played in the next stage, and so on.

We can use some simple algebra and probability theory to reduce any compound lottery to a simple one. In the first example, the probability that the lottery **q** is selected in the first stage equals p. If **q** is selected, then o_j is the final outcome with probability q_j. Hence, the probability that **q** is selected and that o_j occurs equals the product of the probabilities of these two events, pq_j. Similarly, the probability that the lottery **q**′ is selected is $1 - p$, so that the probability that **q**′ is selected and o_j occurs equals $(1 - p)q_j'$. Overall, the probability that the compound lottery ultimately results in o_j, whether through **q** or **q**′, equals the sum $pq_j + (1 - p)q_j'$. That is, the two-stage compound lottery is equivalent to a one-stage simple lottery in which o_j occurs with probability $pq_j + (1 - p)q_j'$. The next assumption, which assumes, in effect, that people do not derive any intrinsic pleasure from the complexity of the lottery itself but rather that they are concerned only with the likelihood of outcomes, states that people are indifferent between a compound lottery and its simple lottery equivalent.

> *Assumption 4* (Reduction of compound lotteries): A person is indifferent between any simple lottery **q** on O and any other lottery that yields each outcome in O with probability **q**.

Assumptions 1 through 4 are sufficient to justify using an ordinal utility function, $u(\mathbf{q})$, defined over all simple lotteries to represent preferences over all lotteries, simple or complex. For those of us who are used to manipulating probability numbers algebraically, the two new assumptions, 3 and 4, seem innocuous. But not all action is consistent with them. Given the opportunity, some cannot resist casino gambling, even though the odds favor the house. Such action is not inconsistent with Assumptions 3 or 4, since it merely reveals a temporal preference for various gambles. But Assumption 4 implies that probabilities dictate preferences and not the lottery's complexity itself. In choosing between blackjack, craps, roulette, or a slot machine, each of us should choose the game and the bet that best matches our taste for a gamble over simple lotteries. Many, if not most persons, though, will move from table to table and play several games in an evening, enjoying the change and the variety of ways to place bets. Assumption 4 excludes this kind of action.

Our assumptions about preferences over lotteries, then, will be more or less applicable, depending on the situation under study. But we should not take the view that sometimes these assumptions are right and sometimes they are wrong. Instead, they serve as approximations, and the approximation in Assumption 4 permits considerable simplification in the representation of preferences over lotteries. Hopefully, the benefit from that simplification will offset the errors that it introduces in those situations that most concern us. In this spirit, there is a consensus that much is to be gained by adding two additional assumptions that permit the description of preferences over lotteries on the basis of preferences over the outcomes in O. Ascertaining preferences over outcomes is difficult enough. Measuring and representing them over the set of all possible lotteries, especially if the description of outcomes themselves is complex, may be impossible. Hence, to formulate a more parsimonious theory, we must look for ways to describe $u(\mathbf{q})$ more concisely.

To see what we mean by concise, suppose that you must decide whether to pay five dollars for a gamble that returns nothing with probability .3 and ten dollars with probability .7. In making your decision, you might reason thus: The gamble costs five dollars, but its *expected value* is .3 times $0 plus .7 times $10, or seven dollars. Hence, you accept the gamble, since its net expected value is positive, two dollars. Note that the particular feature of this reasoning is that in evaluating the gamble, in ascertaining your preference, you calculate the *probability weighted average* of the several possible prizes.

Suppose that we can perform the same calculation with utility. In particular, suppose that we can express the utility of any lottery over O as the probability-weighted average of the utilities for its prizes, the outcomes in O:

$$u(\mathbf{q}) = \sum_{oj \in O} q_j u(o_j). \tag{1.5}$$

The advantage of this form is that we need only the utility numbers associated with outcomes to calculate a person's preferences over lotteries. If we can deduce preferences for lotteries directly from the utility numbers assigned to nonprobabilistic outcomes, if we can calculate preferences over lotteries by expression (1.5), then we can introduce considerable parsimony into our models.

Assumptions 1 through 4, though, cannot justify such a simplification. For example, suppose that we assign the utility numbers $u(o_1) = 1$, $u(o_2) = .5$, and $u(o_3) = 0$ to represent the preference order $o_1 \, P \, o_2 \, P \, o_3$. From expression (1.5), it seems a person should be indifferent between $\mathbf{q}' = (0, 1, 0)$ and $\mathbf{q}'' = (.5, 0, .5)$. That is, $u(\mathbf{q}') = u(\mathbf{q}'')$, which implies indifference between realizing o_2 with certainty and a fifty-fifty lottery between the first and third outcomes. But the utility numbers $u'(o_1) =$

2, $u'(o_2) = .5$, and $u'(o_3) = 0$ also summarize preferences over O, and these numbers do not equate $u'(\mathbf{q}')$ and $u'(\mathbf{q}'')$ when substituted into (1.5).

The implications of this example are that only certain utility functions summarize preferences over lotteries, and we need additional assumptions to justify expression (1.5). We require two additional assumptions in particular, and since both can be stated in terms of simple lotteries over only the most and least preferred outcomes in O, denoted o_1 and o_m, respectively, we represent such a lottery more compactly as

$$\mathbf{p} = (p_1, p_m) = (q_1, 0, 0, \ldots, 0, q_m).$$

The first additional assumption states that if two simple lotteries concern only the most and least preferred outcomes, then a person prefers the lottery that yields the greater probability of the more preferred outcome:

> *Assumption 5* (Monotonicity): If the decision maker strictly prefers the most preferred outcome in O, o_1, to the worst outcome in O, o_m, then he strictly prefers the lottery (p_1, p_m) to (p'_1, p'_m) if and only if $p_1 > p'_1$.

Recall that by Assumption 4, a person is indifferent among several lotteries if they yield the same "final" probabilities for each of the m outcomes. That is, a person derives no pleasure or pain from the conduct of the lottery itself, but only cares about the probabilities of the various outcomes in O and the importance that he attaches to these outcomes. Assumption 5 extends this argument by asserting that a person's utility for a lottery between the best and the worst outcomes increases uniformly as the odds associated with the best outcome increase.

We can extend this argument further to yield a simple implication: *Every outcome is equivalent to a unique lottery on the extreme (most and least preferred) outcomes.* That is, for every outcome o_j in O, there exists some lottery between a person's most and least preferred outcomes such that he is indifferent between o_j and that lottery. To understand this possibility, we start with the lottery $(0, 0, \ldots, 0, 1)$, meaning "the worst possible outcome with certainty." Presumably, if the person prefers o_j to the worst possible outcome, then he prefers o_j to this "lottery." Now increase the chances of getting o_1 at the expense of o_m. Assumption 5 states that the resulting lotteries increase in the person's preference order until, in the limit, we arrive at $(1, 0, \ldots, 0, 0)$. Somewhere in between, though, the order of the lottery passes o_j and at that point we say that the person is indifferent between o_j and a simple lottery on the extremes. We denote this lottery by $\mathbf{p}^j = (p_1^j, p_m^j)$.

The final assumption that we require to model decision making under risk concerns the substitutability of one lottery for another. Let $\mathbf{q}$ be a simple lottery $\mathbf{q} = (q_1, \ldots, q_j, \ldots, q_m)$. Now suppose that, rather than awarding o_j as a prize, we award the equivalent lottery between the most and least preferred outcomes, $\mathbf{p}^j = (p_1^j, p_m^j)$. The outcome o_j, then, can never occur in this revised lottery. Aside from this substitution, the revised lottery is identical to $\mathbf{q}$. Assumption 6 is that a person is indifferent between $\mathbf{q}$ and the revised lottery:

> Assumption 6 (Substitutability): A person is indifferent between a lottery that awards o_j and an otherwise equivalent lottery that awards $\mathbf{p}^j$ instead of o_j.

The following result justifies representing a person's utility for any lottery as a probability weighted average of the utilities on outcomes in O.

> 1.1. (Expected utility theorem): *If a person's preferences satisfy assumptions 1–6, then there exists a number $u(o_j)$ for each o_j in O, such that the utility function summarizing this person's preferences over simple lotteries corresponds to expression (1.5).*

The numbers posited by Theorem 1.1 are called *cardinal utility numbers*, and the corresponding choice rule for decision making under risk is: *Choose the act that maximizes expected utility* [as defined by expression (1.5)].

1.6   The meaning of cardinal utility

Theorem 1.1 tells us that Assumptions 1 through 6, in conjunction with expression (1.5), are consistent. But it does not communicate the power and limitations of the cardinal utility concept. First, we must understand that applying expression (1.5) does not require direct measures of cardinal utility. If this requirement held true, then cardinal utility would be of little use to political theorists, because it is unlikely that we could ever measure the cardinal utilities of legislators, political candidates, and national leaders. Often, though, simply knowing that expression (1.5) is legitimate is sufficient.

To illustrate, consider the voting problem that Figure 1.3 summarizes. Suppose that a respondent to a survey indicates a preference for candidate 1 over 2 so that, regardless of whether we assume cardinal or ordinal utility, $u(o_1) > u(o_2)$. We might then ask, should a person ever consider voting for his second choice in a two-candidate election –

candidate 2 in this instance? An answer of NO, of course, seems obvious. To check whether our intuition is correct, refer back to Figure 1.3 and let p_k be the probability that state s_k prevails, in which case Theorem 1.1 justifies the following calculations to evaluate the expected utilities of voting for candidates 1 and 2, respectively:

$$u(a_1) = p_1u(o_2) + p_2u(o_3) + [p_3 + p_4 + p_5]u(o_1)$$
$$u(a_2) = [p_1 + p_2 + p_3]u(o_2) + p_4u(o_3) + p_5u(o_1).$$

The voter chooses candidate 1, then, if $u(a_1) > u(a_2)$. If we use these two expressions, this inequality is equivalent (after we cancel common terms) to

$$p_2u(o_3) + [p_3 + p_4]u(o_1) > [p_2 + p_3]u(o_2) + p_4u(o_3).$$

After we rearrange terms, this expression becomes

$$p_2[u(o_3) - u(o_2)] + p_3[u(o_1) - u(o_2)] + p_4[u(o_1) - u(o_3)] > 0.$$

But if the voter prefers candidate 1 to candidate 2, and if we make the reasonable assumption that the utility of a tie falls somewhere between the utilities associated with 1 winning and with 2 winning, then every term in this expression is positive, and the inequality is necessarily satisfied. Thus, the expected utility of voting for one's preferred candidate always exceeds the expected utility of voting for one's second choice in two-candidate contests. Furthermore, this conclusion follows regardless of the exact numbers that one assigns to $u(o_1)$ and $u(o_2)$, no matter what probabilities one assigns to states of nature. This result confirms our intuition and also shows that we do not require problematical measurements of utility to justify that intuition. (The reader might want to confirm that this conclusion does not extend necessarily to three candidate contests. Voters might prefer to vote for their second choice if their first choice has little chance of winning.)

Justifying expression (1.5) by using Theorem 1.1 is useful, but often we must know something more specific about utility. Suppose that you are confronted with the following gamble (referred to as the *St. Petersburgh paradox*): A fair coin is tossed until a head appears, and the total number of tosses is recorded. The possible outcomes, then, are of the form "tails occurs on the first n-1 tosses and a head appears on the nth toss." Suppose that you are also offered 2^n dollars (read "2 times 2 times 2 $\cdots$ n times") if a head does not appear until the nth toss. Suppose, finally, that you are asked how much you are willing to pay to play this lottery. Our experience is that undergraduates are willing to pay at least $4 but few will pay more than $50. Suprisingly, all are pikers

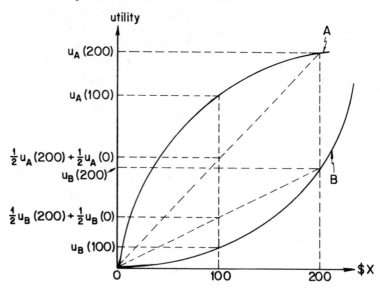

1.15 Risk aversion and risk acceptance.

– if money is the only measure of value, then people should be willing to pay any amount to play.

To see that this claim is true, notice that the probability that a head occurs on the first toss is $\frac{1}{2}$. The probability of a tail on the first toss and a head on the second is $\frac{1}{2}$ times $\frac{1}{2}$, or $\frac{1}{4}$. In general, the probability of $n - 1$ tails followed by a head is $(\frac{1}{2})^n$. Thus, the expected dollar value of this lottery is

$$2(\tfrac{1}{2}) + 4(\tfrac{1}{4}) + \cdots + 2^n(\tfrac{1}{2})^n + \cdots$$
$$= 1 + 1 + \cdots + 1 + \cdots = \text{infinity},$$

in which case, if dollars are your only consideration, the expected dollar value of the lottery is infinite and you should be willing to pay any amount to play.

Since hardly anyone is willing to pay even one hundred dollars to play this game, something must be amiss with the analysis. That something appears to be the measure of value. The preceding calculation equates money and utility, that is, $u(\$x) = x$. But consider Figure 1.15. Both curves there are monotonically increasing and are suitable representations of the *ordinal* preference "more money is better than less." Suppose, however, that a person is confronted with a choice between $100 with certainty and an even chance lottery between $0 and $200.

Applying the expected utility formula to the curve marked A, we see that the person prefers \$100 with certainty to the lottery: that is, $u_A(100) > u_A(0)/2 + u_A(200)/2$. For the curve marked B, on the other hand, the person prefers the lottery to \$100: that is, $u_B(100) < u_B(0)/2 + u_B(200)/2$. Thus, curve A models *risk aversion* whereas B models *risk acceptance*.

More formally, note that any lottery between two outcomes is simply a convex combination between those outcomes. Thus, if a person is risk averse, it must be the case that

$$u(Lo_1 + (1 - L)o_2) \geq Lu(o_1) + (1 - L)u(o_2), \qquad (1.6)$$

which we call a *concave utility function*. In Figure 1.15, the dashed straight line connecting the point $(0, 0)$ and $(200, u_A(200))$ corresponds to the right side of the inequality in expression (1.6). Notice that this line lies below the utility curve u_A, in accordance with (1.6). On the other hand, if a person is risk acceptant, then

$$u(Lo_1 + (1 - L)o_2) \leq Lu(o_1) + (1 - L)u(o_2), \qquad (1.7)$$

which we call a *convex utility function*. Thus, the line connecting the points $(0, 0)$ and $(200, u_B(200))$ – which corresponds again to the right side of expression (1.7) – is always above u_B, in accordance with (1.7). If equality holds, the person is *risk neutral* and the utility function that models these preferences is a straight line.

Clearly, there are people who love to gamble and those who avoid any risk. Cardinal utility permits us to represent both kinds of people. With it we can model the political candidate who adopts platforms that increase the chances of electoral defeat and at the same time increase the chances of a landslide victory as well as the candidate who chooses a campaign strategy that will greatly upset no one but that also leaves the electorate unexcited. And with cardinal utility we can also explain why people are unwilling to pay any sum of money to play the game of tossing a coin until a head appears. Although cardinal utilities may differ greatly in a region feasible for most people – say \$0 to \$1,000,000, or even \$1,000,000,000 for the few super rich – for larger numbers the meaning of added wealth becomes unimaginable. Consequently, the difference between, say, \$1,100,000,000 and \$1,000,000,000 is not the same as the difference between \$100,000,000 and \$0. A risk-averse person might value the first \$100 million more highly than the eleventh \$100 million. In cardinal utility terms, then, increases in utility flatten out as additions to unimaginable wealth become a possibility. In terms of the coin-toss lottery, the difference between \$$2^n$ and \$$2^{n+1}$ for large values of n is slight in terms of utility (but not in terms of dollars), and

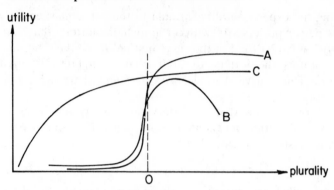

1.16 Possible preferences for plurality.

this difference becomes smaller still as n increases. This reduction in added utility also reduces the expected value of the game, so that its expected utility is not an infinite sequence of ones, but an infinite sequence of smaller and smaller numbers. The sum of such a sequence is finite, and for most of us it may be the equivalent of $5 to $50.

This is not to say that people are all either risk averse or risk acceptant. Figure 1.16 illustrates some possible utility functions that candidates might hold for plurality. Curve A shows the utility of a candidate who is most concerned simply with winning, and hence his utility rises sharply in the region of a tie. This candidate is risk acceptant for pluralities less than zero and risk averse for pluralities greater than zero. Curve B illustrates the utility of a candidate who is concerned that if his plurality becomes too large, expectations of his performance in office will exceed his capabilities. He, too, is risk averse and risk acceptant in the same regions of plurality as is the first candidate. Finally, curve C illustrates the utility of a candidate who is happy simply to run, although he prefers a greater plurality to a smaller one. This candidate is everywhere risk averse.

Cardinal utility models many aspects of choice, but it requires more assumptions than ordinal utility. Consequently, we should assume cardinal instead of ordinal utility only if it becomes necessary to do so, only when risk is present and people must choose between alternative lotteries. Of course, simply stating that ordinal and cardinal utility differ because one entails stronger assumptions does not necessarily communicate the difference between these two concepts. To appreciate this difference, recall that an ordinal utility function is unique up to any positive monotonic transformation. Thus, if a person prefers more money to less, and if the amounts of money considered are positive, then suitable representations of the ordinal utility for x include x, x^2, $x^{1/2}$, $4x + 10$, and so forth. Cardinal utility, though, is unique only up to any positive *linear transformation*.

1.2. *If the utility function, u, satisfies assumptions 1 through 6, then only functions of the form u' = a + bu convey equivalent information about preferences, where a and b are constants and b > 0.*

The theory of choice under risk assumes that people act to maximize their expected utilities. The preceding result implies that a person chooses the same act whether it maximizes u or u'. For utility functions that model decisions under risk, then, we cannot transform u by any positive monotonic function. Such transformations do not ensure that utility functions retain their concave or convex properties and the information that they convey about people's attitudes toward risk. For example, if a person is risk neutral, then we could let $u(\$x) = x$, but the transformation $u'(\$x) = x^{1/2}$ yields a curve like A in Figure 1.15, which models risk aversion, whereas $u'(\$x) = x^2$ yields a curve like B, which models risk acceptance.

One important limitation on the use of cardinal and ordinal utility remains with respect to linear and monotonic transformations. Specifically, we cannot use this theory to justify statements that compare one person's utility for an outcome to that of another person. We can best develop an intuitive understanding of the formal problem through the use of an analogy. Temperature scales (like cardinal utility) are unique only up to a positive linear transformation. For example, we can transform the Fahrenheit scale into the Celsius scale by the transformation $°C = -\frac{160}{9} + \frac{5}{9}°F$. Utility is analogous to each person's private temperature scale. There is a crucial difference, however, between different persons' utility scales and alternative temperature scales. We can ascertain the relation of one temperature scale to another by using both scales to measure the same events, such as the freezing and boiling points of water. Knowing these two facts, we can then deduce the appropriate parameters in the linear function to transform one scale to the other. But what events calibrate people's utility functions for comparison? Certainly, for any pair of events, we can always find two persons who will rank them differently, and this possibility destroys any hope of calibration.

Suppose that we measure two persons' utilities for money in the range [$0,$10] and conclude that $u(\$5) = \frac{1}{2}$ for the first person and $\frac{1}{4}$ for the second. Should we conclude that the first person derives twice as much utility from $5 as does the second? The answer is NO. Because we can add, subtract, or multiply a person's utility numbers by a positive constant, we can arbitrarily multiply the second person's utility numbers by, say, 4, thereby setting the utility of $5 for him equal to 1. We can no more conclude that he now values $5 twice as much as does the first person than we can say the first values $5 twice as much as does the second.

The arbitrariness of utility scales places much of normative philosophy outside the realm of our theory. But we cannot ignore the fact that in matters of public policy, people consider statements such as "this helps you more than it helps me." Much of what governments do entails the redistribution of income or wealth with justifications referring to terms such as fairness, equity, and equality. Here, though, our theory is descriptive, and we require simply that people be free to make any comparisons that they choose. Even so, such comparisons should find reflection only in the utility functions of the persons making them. Thus, if you prefer an equitable distribution of income, as analysts we might differentiate between the outcomes "you earn $15,000/year and so does everyone else" and "you earn $15,000/year while some earn less and others earn more" and assign them to different locations in your preference order. Utility theory is not inconsistent with the empirical fact of interpersonal comparisons. But, while it can model a person's individual comparisons, it cannot find use to justify one comparison, one set of values, over another.

1.7   Cardinal utility and subjective probability

Allais's paradox

Cardinal utility provides a valuable basis for modeling preferences, but not all preferences are consistent with the assumptions that lead to Theorem 1.1. Earlier we discussed people's tastes for different games of chance at a casino. For perhaps a more striking example, suppose that O consists of three outcomes: o_1, $5 million; o_2, $1 million; and o_3, nothing, and that the ordinal preferences of everyone are $o_1 \, P \, o_2 \, P \, o_3$. Reasoning that a bird in the hand is worth two in the bush, suppose that a person strictly prefers o_2 to the simple lottery $\mathbf{q} = (.10, .89, .01)$. In terms of expected utility, this means that

$$u(o_2) > .10u(o_1) + .89u(o_2) + .01u(o_3),$$

or equivalently, that

$$.11u(o_2) > .10u(o_1) + .01u(o_3). \tag{1.8}$$

Suppose now that we must choose between the following two lotteries: $\mathbf{q'} = (0, .11, .89)$ and $\mathbf{q''} = (.10, 0, .90)$. Our experience in the classroom is that many persons will choose $\mathbf{q''}$ over $\mathbf{q'}$, including those who agree with the bird-in-the-hand principle. Using expected utility, $\mathbf{q''} \, P \, \mathbf{q'}$ im-

$$.11u(o_2) + .89u(o_3) < .10u(o_1) + .90u(o_3),$$

which simplifies to

$$.11u(o_2) < .10u(o_1) + .01u(o_3).$$

But this expression contradicts inequality (1.8). Known as *Allais's paradox*, this choice situation and the revealed preference of many people pose a paradox for our theory.

preference reversal

Several hypotheses might resolve this paradox, but note that the crucial difference between receiving o_2 with certainty and the lottery $\mathbf{q}$ is the trade of some chance of \$5 million compared with an increase from zero to .01 in the chance of getting nothing. A similar trade occurs between $\mathbf{q'}$ and $\mathbf{q''}$, except that here the chance of getting nothing, although again added to by only .01 in probability terms, begins at .89. It appears that people treat these two probability increments differently even though both are equal to .01 and thus are mathematically equivalent. In the first instance, something that is impossible becomes possible, whereas between the second pair of lotteries, something that is likely becomes only a bit more likely.

see Lichtenstein + Slovic

The calculus of voting

The Allais paradox suggests that the meaning of probability numbers can vary, for the same person, across decision contexts. This is a serious problem for the hypothesis of cardinal utility and for the theory of decision making under risk in general, and it is a problem with which scholars today grapple. But we should not suppose that differences between *subjective* and *objective probabilities* necessarily impede inquiry. To illustrate, consider again the voting problem that Figure 1.3 represents, but now let us analyze the decision to vote or to abstain. We know already that if a person votes, he should vote for his preferred candidate, who we suppose is candidate 1. Hence, a person votes rather than abstains in two-candidate elections if and only if $u(a_1) > u(a_0)$. Before setting up our inequalities, however, note that although we can reasonably interpret abstaining as an act that costs nothing in terms of time and energy, voting is costly. It is fair to argue, moreover, that voting differs from abstaining because people associate certain psychic rewards with fulfilling their obligations of citizenry no matter who wins or loses and regardless of the effect of one's vote on the outcome. That is, the utility associated with candidate i's victory if you vote, $u(o_i,a_1)$, is not equal to the utility of i's victory if you abstain, $u(o_i,a_0)$.

One way to model this hypothesis is to suppose that we can separate the utility of the act, voting or abstaining, from the utility associated with the victorious candidate. If we suppose that voting, regardless of the outcome, yields the utility $u'(a_1)$, and that you associate no pleasure or cost with abstaining, then we can let

$$u(o_i,a_1) = u(o_i) + u'(a_1)$$
$$u(o_i,a_0) = u(o_i).$$

After we substitute these identities into the appropriate expected utility expressions derived from Figure 1.3 and after we perform some algebraic manipulation, keeping in mind that p_2 and p_3 represent the probabilities that the voter can create or break a tie, and assuming that a coin toss decides ties and that $u(o_3) = (u(o_1) + u(o_2))/2$, the condition for voting, namely that $u(a_1) > u(a_0)$, becomes

$$[u(o_1) - u(o_2)] \frac{(p_2 + p_3)}{2} + u'(a_1) > 0. \tag{1.9}$$

What many scholars find bothersome about this expression is that, in elections with millions of voters, the probabilities that voters can make or break ties, p_2 and p_3, are objectively small, if not zero. It appears, then, that we can explain the decision to vote only if $[u(o_1) - u(o_2)]$ is unreasonably large, if subjective estimates of $p_2 + p_3$ greatly exceed objective calculations, or if $u'(a_1) > 0$. Because both the psychic rewards and the costs component of $u'(a_1)$ are difficult to measure, to explain voting by saying that citizens vote because $u'(a_1) > 0$ seems nearly tautological. Do people vote merely because they are socialized to do so?

Even today the implications of expression (1.9) are the subject of a lively debate. We do not know how to satisfactorily resolve that debate, but we should recognize that this expression serves as an inspiration for a discussion that sharpens the analysis of why people vote and why turnout varies across elections and electoral institutions. The hypothesis that the discrepancy between subjective and objective probabilities cannot be so great as to render the first term of expression (1.9) significant draws our attention to the second term, $u'(a_1)$, and to the sources of its variation. Thus, even though we might believe that objective and subjective probabilities differ greatly, a simple decision-theoretic analysis helps to focus our research in valuable ways.

1.8   Summary

In the course of this chapter, we have encountered a great many new concepts that, at best, relate only dimly to the subject of politics and to the task of modeling political processes. The following is a list of the most important concepts, and readers should review the relevant parts of the text if they are insecure about any of them.

action	outcome	state of nature
exhaustive	exclusive	countable set
finite set	infinite set	uncountable set
subset	budget simplex	bounded set
closed set	open set	convex set
convex combination	strict convexity	certainty
risk	weak preference relation	completeness
transitivity	indifference	strict preference
R, I, and P	ordinal utility	postulated preference
lexicographic	tradeoffs	indifference curve
city block	spatial model	ideal point
Euclidean distance	preference set	$P(o)$, $I(o)$, $R(o)$
open set	feasible set	Lagrange multiplier
single peaked	simple lottery	compound lottery
expected value	probability weighted average	monotonicity
cardinal utility	St. Petersburgh paradox	risk aversion
risk acceptance	concave function	convex function
risk neutral	linear transformation	Allais's paradox
subjective probability	objective probability	

Most of these concepts concern the representation of the intuitively simple idea of individual preference. The abstractions that we encounter, however, are necessary for two reasons. First, because our theory is formal, we must be precise in our meanings. We distinguish between decision making under certainty and under risk, for example, because we must know more about preferences under risky choice if we are to model that choice. To be certain that our models follow from first principles, we must state those principles precisely. Second, since we want our theory to be general, we prefer to use the weakest assumptions possible. We do not want to require cardinal utility if ordinal assumptions will suffice, and the virtue of formalism is that it permits us to see the relationships among assumptions and their relative generality.

By formalizing the notions of "purposeful" and "self-interested action," moreover, we see that our approach is neither so narrow as to preclude the possibility that people are concerned about the welfare of others nor so general as to be tautological. Some models that use this structure exclude the possibility of altruism, but they are simply specialized subparts or applications of the general paradigm. And the treatment of lotteries as well as the assumption that preferences are transitive removes the paradigm from the realm of tautological theory. Nor does this paradigm ignore the sociological or communal determinants of preference and choice. Exactly the opposite is true. A formal approach yields the substantive generality we seek. Utility functions are simply abstract representations, and the a's, o's, x's, and y's that express them await interpretation.

CHAPTER 2

Individual preference and social choice

Before we can develop a theory of politics we must have a framework for representing individual preferences. Individual human beings form the body politic, no matter what the society or political system under consideration looks like. Interest groups, parties, nations, legislatures, and politburos consist of individuals who shape the character of these institutions. Even the procedures, traditions, and constitutions that guide and regulate people's choices within collectivities are the product of individual decisions. The maintenance and longevity of these things, moreover, depend on the willingness of people to tolerate outcomes under them compared with the outcomes wrought by revolution, political activism, and alternative procedures. A framework that focuses solely on individual choice is inadequate, however, because people in politics choose in an environment made up of other persons' choices, and these choices affect and condition each other. By and large, "nature" in Chapter 1 is neither a deity nor some unthinking creature, but other persons who pursue their goals. A theory that does not incorporate this reality cannot say anything about what political scientists study.

Our central concern is the study of public outcomes such as the laws and regulations that legislatures authorize, the policies that victorious election candidates implement, the interpretation of laws that courts render, the actions that public bureaucrats take under the authority that legislators and executives give them, and the decisions to go to war or the conditions of peace that nations establish. But this chapter seeks to dissuade readers from believing that such outcomes follow in some simple, obvious way from individual preferences. Our argument begins with the observation that institutions mediate between all public outcomes and individual preferences. For example, a democratic system requires elections, legislatures, and laws governing referenda and petitions, to solicit statements of individual preference and then, by their internal structure, to transform those statements into policy, into specific resource allocations, into victorious candidates, or even into modifications of the institutions themselves.

A classic example of the importance of institutions illustrates these

53

remarks. Farquharson, in his *Theory of Voting*, recounts a situation described in the letters of Pliny the Younger in which "the consul Africanius Dexter had been found slain, and it was uncertain whether he had died by his own hand or at those of his freedmen. When the matter came before the Roman Senate, Pliny wished to acquit them [alternative x], another senator moved that they should be banished to an island [alternative y]; and a third that they should be put to death [alternative z]" (pp. 6–7). Farquharson infers that different senators held one of three sets of preferences over these three alternatives:

1. Those, like Pliny, who preferred acquittal and believed the freedmen innocent, in which case $x P_i y P_i z$
2. Those who, perhaps because of the uncertainty of the situation, preferred banishment but least of all preferred death, in which case $y P_i x P_i z$
3. Those who believed the freedmen guilty and sought the severest punishment, death, in which case, presumably, $z P_i y P_i x$

The first and, as we shall see, most important decision confronting the Roman Senate was choosing an appropriate procedure for deciding the freedmens' fate. One commonly used procedure, called a *binary procedure* because it divides the alternatives into two sets that the assembly must vote on, requires first that the Senate decide whether the freedmen are guilty or innocent (that is, whether they should be acquitted or whether the Senate should debate between banishment and death). In this instance, since a majority of the senators prefers either banishment or death to acquittal, the freedmen presumably would be judged guilty. In the second vote, over the form of punishment, it is apparent that a majority prefers banishment to death. Thus, with a binary procedure, banishment is the final outcome. Pliny, however, insisted upon *plurality rule*, in which case all alternatives must be put to a vote simultaneously and the alternative with the greatest number of votes wins. Pliny's preference for this procedure can be explained by the fact that those ranking acquittal first slightly outnumbered those who most preferred banishment or death, so plurality rule resulted in acquittal.

This example shows that the institutional context of decisions, in this case the voting rule, as well as the individual preferences that operate within them, can affect profoundly what outcome people choose. But this lesson is deceptive if it tempts us to infer that we can understand institutions in a purely mechanical way, as black boxes into which we plug preferences and out of which emerge outcomes or "social preference." In fact, if our black box reveals a complete preference order for a group based on individual preferences, then we might perhaps be anthropomorphic about things and talk about a group (for example, society), under a particular institution, maximizing its social preference order. In the example of Pliny, if we pair all alternatives against each other in a series of majority votes, y defeats both x and z, and x defeats z. Hence, we could infer that the *social preference order* under this institution is the transitive relation $y \, P \, x \, P \, z$. That is, we might assume that institutions are like people with preferences that conform to the structure that the previous chapter describes, in which case our principal research objective would be to find out how particular institutions transform individual preferences into social preferences. Unfortunately, with this approach we encounter several obstacles that show why the study of interdependent choice in politics requires more than a naive application of the ideas in Chapter 1.

This chapter examines some of those obstacles by reviewing three important theorems. The first theorem, Arrow's general impossibility result, shows that we cannot guarantee that any institution will aggregate individual preferences in a way that permits an anthropomorphic reference to social preferences. Specifically, even if all individual preference orders are transitive, there is no guarantee that the social preference order induced by formal- or informal-group processes will be transitive as well. Thus, groups are unlike people in a fundamental way, and to utter theoretically sound propositions about "their" decisions we must explore their inner structure. Focusing first on groups that use various voting procedures to aggregate individual preferences, we give several examples of how such procedures can yield perverse results or are subject to manipulation.

Looking closer at the mechanism of simple *majority rule*, we then discuss a theorem that shows that if preferences are spatial, as are the ones in Figure 1.10, then one can manipulate majority rule to produce *any* outcome. Thus, although Arrow's theorem tells us that being anthropomorphic about groups is theoretically unsound, and that we should not treat institutions as black boxes, this result demonstrates that procedural details may be decisive for outcomes. Since we should not suppose that legislators, politburo members, committee chairmen, and interest group leaders remain ignorant of the possibility of manipulation, it also tells us that such details are an important strategic variable in politics.

A third general result shows that every institution is subject to manipulation by someone who misrepresents his preferences. Once again, we see why we cannot accept a black-box approach. Such an

		Person		
		1	2	3
Preference order:	1st	x	z	y
	2d	y	x	z
	3d	z	y	x

x d y
z d x

2.1 Intransitivity of social preferences.

approach ignores the strategic complexity of all institutions in that no institution can preclude the possibility that someone will have an incentive to disguise his true preferences. As an example, we examine vote trading in legislatures to see when trading yields a determinate outcome and when misrepresentation can yield preferred outcomes for some subset of participants. The central theme of this chapter, then, is that social outcomes follow from individual preferences in anything but trivial ways.

2.1   Arrow's impossibility result

To appreciate more fully the difficulties of a black-box approach to individual choice in a social or institutional setting, suppose that three persons must choose one alternative from the list $\{x,y,z\}$ as the social outcome, and that individual preferences are like those in Figure 2.1. Now consider a formal voting procedure in which people vote on each pair of alternatives one at a time, and the majority winner in each pairwise competition is ranked higher in the social order. Hence, if we pair x and y, x beats y by two votes to one (persons 1 and 2 vote for x, and 3 votes for y). Similarly y beats z and z beats x. But then the social preference order is intransitive. Indeed, the intransitivity here is more severe than the intransitivity we used in Chapter 1 to illustrate why people might violate the assumption of transitive individual preference (our example of sweetened vs. unsweetened tea). Specifically, majority rule reveals a *cycle*: x is socially preferred to y, y to z, and z to x.

This example, known as the *Condorcet paradox*, uncovers an important possibility about social institutions; although we assume that the individual preferences that operate within them are transitive, there need not exist a transitive social preference.

We cannot understate the importance of this example because it undermines fundamentally any approach that treats institutions and collectivities as though they are people. We are all accustomed to journalistic references to "nations acting," "legislatures choosing,"

See Vickery .

"interest groups lobbying," and so forth. To act, to choose, or to lobby, however, implies the pursuit of some goal. But how can we speak of a legislature, an interest group, or a nation, or any other collectivity "maximizing its utility or welfare" when the preferences of that group, as its institutional structure reveals, are intransitive? Further, terminology such as "the public interest" and "community goals" becomes immediately suspect. The literature of politics is replete with references to the public interest, either as justifications for specific policies or as explanations for specific events. Similarly, people often appeal to community goals in discussions of urban renewal, zoning, and highway construction, while reference to national interests is common to the international relations literature. The Condorcet paradox raises the possibility that none of this journalistic shorthand has a proper place in any adequate theory of political processes.

Even if we maintain an individualistic perspective, the Condorcet example is profound because of what it implies about the difficulty of transposing the concept of individual utility maximization into a social setting. Consider the dilemma that two election opponents face if persons 1, 2, and 3 are the voters, if alternatives x, y, and z are possible election platforms, and if the candidates' utility functions take on the values of one, zero, or minus one, depending on whether they win, tie, or lose. Since each platform can be beaten by another, what platform should a utility-maximizing candidate choose? For example, if a candidate believes that his opponent is most likely to choose x, then he should choose z because z beats x in a majority vote. But if he thinks that his opponent might second-guess him by choosing y instead, to counter z, then he should not choose z in the first place, but should choose x instead. Of course, if the opponent thinks this problem through, then he will choose x, ..., and so on, and so on. Thus, the assumption that people act to maximize some utility function that we ascribe to them is not sufficient to render predictions about choices.

Before we proceed too far with drawing possible implications from the Condorcet paradox, however, we must recognize that it is only an example. Before we let it shape our theoretical perspectives, we must establish that it is not a curious and unique situation that a specific configuration of individual preferences creates, and that it is not a peculiarity of majority rule. If it is only a unique example, then it is of minor concern. But if it portends a general possibility, that intransitive social preferences are a feature of other institutions and configurations of individual preference, then we must study it further.

There are two approaches to studying the generality of this paradox. The first, which maintains the focus on simple majority rule, measures

	Number of voters							
Number of alternatives	3	5	7	9	11 ⋯	25 ⋯	49 ⋯	∞
3	.056	.069	.075	.078	.080	.084	.086	.088
4	.111	.139	.150	.156	.160	.169	.172	.176
5	.160	.200	.215	.224	.229	.242	.246	.251
6	.202	.251	.271	.281	.228	.303	.309	.315
7	.239	.296	.318					.369
8	.271	.334	.359					.415
⋮								⋮
49								.841

2.2 Probability of Condorcet paradox.

the frequency with which the paradox occurs. Specifically, suppose n persons, confronting m alternatives, determine the social preference order by exhaustive pairwise voting among the alternatives. Suppose, further, that we assign preferences over the m alternatives randomly to each person. We can then compute the likelihood that a pairwise vote among all possible pairs of alternatives reveals a cyclic social preference order and thereby gain some insight into the magnitude of the problem that Condorcet's paradox occasions.

Figure 2.2 summarizes the results of such a calculation, which is based on computer simulations of voting bodies. It reveals, first, that with a random assignment of preferences, the likelihood of the paradox is never zero. Thus, the paradox is not an artifact of three-person electorates and three-alternative agendas – many configurations of individual preferences exhibit cyclic social preferences, and these intransitivities can encompass a great many alternatives. Second, the likelihood of the paradox increases as either the number of alternatives or the number of voters increases, with an apparent limit of one as n and m approach infinity. Hence, in large societies confronted with numerous alternatives, the likelihood that the paradox occurs is great.

The numbers in Figure 2.2 rest on the assumption that all preference orders are equally likely. But if everyone shares the same (transitive) preference order, no paradox is possible, and thus in majority-rule institutions, it may be that we can avoid the paradox if individual preferences correlate highly, owing to the forces of socialization. We might speculate, then, that the paradox is less likely to arise if decisions are limited to issues that tap into commonly held values, and it is more likely to occur if issues are new and divide people into several camps.

Although Figure 2.2 suggests interesting hypotheses, it does not answer questions about the generality of the paradox. Is it a curiosity of pairwise majority voting or does it extend to other institutions as well?

Answering this question requires a different approach, since we cannot exhaustively examine all imaginable institutions. The approach that has proved most fruitful in this regard is the *axiomatic method*. Instead of examining institutions and voting procedures one at a time, this approach specifies conditions (axioms) that any reasonable institution ought to satisfy, including those entailing informal as well as formal procedures. For example, if we wish to exclude dictatorships from consideration, we can first define a dictatorship formally and then with an axiom assume that the class of institutions under consideration are not dictatorships. Or, if we want to require transitive social orders, we state axiomatically that intransitive orders are prohibited. After the list of axioms is complete, we check it for consistency. If the axioms are consistent, institutions and procedures that one or more of the axioms do not explicitly rule out are possible; but if the axioms are inconsistent, no institution can satisfy all of them simultaneously. In the event of an inconsistency, if we select the axioms carefully so as not to exclude institutions or procedures that we deem reasonable, then all such reasonable possibilities must be impossible.

Proceeding formally, let O be the set of outcomes that a group is evaluating, let R_i denote person i's preferences over O, and let R be a relation that tells us the group's preference between pairs of outcomes in O, as some institution that operates on individual preferences identifies. R may or may not be like the individual preference relation in that it may or may not be complete and transitive; we have not yet made any assumptions in this regard. But we can think of any institution in two ways. First, we can think of a formal or informal institution as a mathematical function, G, whose arguments include a vector of preference profiles $\mathbf{R} = (R_1, R_2, \ldots, R_n)$ that summarizes each group member's preferences over O, and includes O itself, and whose "output" is a relation, R, that orders pairs of elements of O, that is, $G(\mathbf{R}, O) = R$. Thus,

> *Social preference function*: $G(\mathbf{R}, O) = R$ is a social preference function if it defines the preference relation, R, between every pair of outcomes in O.

Describing social preferences is different from predicting a final outcome, however, and often we are interested only in what the members of a group or society eventually choose. To avoid additional notation, and if there is no confusion, we sometimes refer to G as a social choice function.

> *Social choice function*: $G(\mathbf{R}, O) \to O$ is a social choice function that takes individual preferences and selects outcomes in O as the social outcome. If G selects a unique outcome in O, then G is *resolute*.

There is a close relationship between social choice and social preference functions. If a particular configuration of preferences over O within society, according to the social choice function, G, produces a particular o_j in O, then we can interpret this to mean that o_j stands highest in the social order that the social preference function, G, implies. More precisely, suppose $G(\mathbf{R}, \{o_1, o_2\}) = o_1$. That is, suppose that if people are presented with a choice between o_1 and o_2, the social choice function, which reflects the rules people use to make choices, selects o_1. Then we can write $o_1 \, P \, o_2$ to indicate that o_1 is "socially preferred to" o_2. And if $G(\mathbf{R}, \{o_1, o_2\}) = \{o_1, o_2\}$, then we can write $o_1 \, I \, o_2$ to indicate that, for example, the two outcomes tie in a vote.

There is an important difference, however, between social preference and social choice functions. A social preference function describes preferences and does not constitute a prediction about outcomes; after all, if social preferences are intransitive, then, even if we want to speak in anthropomorphic terms, we have no theoretical basis for predicting that the group chooses one outcome over another. Social choice functions, on the other hand, constitute predictions, since for a particular configuration of individual preferences over a set of outcomes, they tell us what outcome or set of outcomes the corresponding institution implies.

Suppose, now, that individual preferences satisfy both the complete and transitive properties that we assume in Chapter 1 (Assumptions 1 and 2). Here, though, the Condorcet paradox motivates our discussion, and we want to identify, if it is possible to do so, the kinds of institutions (functions G) that yield complete social orders that satisfy these same assumptions. Following the development of Arrow's theorem, we assume as our first axiom:

A1. (*Collective rationality*): The social preference function satisfies Assumptions 1 and 2.

The second axiom addresses the issue of whether to permit any restrictions on individual preferences, aside from Assumptions 1 and 2.

A2. (*Unrestricted domain*): Every individual preference relation that satisfies Assumptions 1 and 2 is admissible.

If individual preferences take a particular form, it is commonplace for some institutions to filter or otherwise modify the outcomes of other institutions. For example, the U.S. Supreme Court reviews the constitutionality of congressional enactments. But we want G to be wholly general and to operate for any constellation of real institutions. Assumption A2 requires simply that some process of social decision exists for any configuration of individual preferences. That is, no deity rules that a particular preference order placed in G is illegitimate. This does not mean, however, that particular customs or institutions cannot violate A2 in practice. Socialization and shared values may make certain preference configurations unlikely. Hence, not institutions themselves but the evolutionary structure of our culture may limit A2's relevance, and in subsequent chapters we explore specific violations of this axiom.

A third axiom, the Pareto principle, provides one connection between individual preference orders and social preference orders.

A3. (The *Pareto principle*): If $x \, P_i \, y$ for all persons i in a group, then $x \, P \, y$ in the social preference order.

As a restriction on the institutions that we consider, the Pareto principle seems simple enough. It imposes the minimal requirement that if everyone unanimously prefers some alternative or outcome, x, to some other alternative, y, then x stands higher than y in the social preference order. That is, if $x \, R_i \, y$ for all i, then $G(\mathbf{R}, \{x, y\}) = x \, R \, y$. Actually, this axiom summarizes two conditions that perhaps better reveal its content. The first, referred to as *citizen sovereignty*, states that for any pair of alternatives, x and y, there must exist some set of individual preference profiles over O such that G yields $x \, P \, y$ as the social preference. Thus, no external entity imposes a social preference order to the exclusion of $x \, P \, y$. The second condition, *nonperversity*, states that, for a given set of preference profiles, if the social preference is $x \, R \, y$ and if one (or more) member(s) raises x in his (their) preference order(s) without other members of society changing the relative positions of x and y in their orders, then either $x \, R \, y$ (as before) or $x \, P \, y$. This condition prohibits systems in which some person's preferences are weighted negatively. Arrow's justification for this condition is that "Since we are trying to describe social welfare and not some sort of illfare, we must assume that the social welfare function is such that the social ordering responds positively to alterations in individual values or at least not negatively" (p. 25).

A4.   (*Independence*): If **R** is a profile of individual preferences over some set of alternatives that includes x and y, if $G(\mathbf{R}, \{x, y\}) = x \, P \, y$ and if **R**' is another preference profile such that each person's preference between x and y is the same in **R**' as in **R**, then $G(\mathbf{R}', \{x, y\}) = x \, P \, y$.

Put simply, A4 requires that if the set of feasible outcomes is restricted to the pair x and y, if x and y are the only two outcomes that the group can consider, then the social preference between x and y depends only on individual preferences over $\{x, y\}$ and not on individual preferences over any larger set, including sets of outcomes that have not even been offered to the public for evaluation. In effect, this condition prohibits expressions of intensity of preferences over the set $\{x, y\}$ by referring to some alternative not under consideration. Although this might not seem distasteful, outcomes can be manipulated, as we argue later, by persons strategically misrepresenting their true intensities.

The last axiom requires the concept of decisiveness. A group, g, (or person i) is *decisive* for x over y if $G(\mathbf{R}, \{x, y\}) = x \, P \, y$ for every admissible profile, **R**, in which the members of g (or i) all prefer x to y and the "non-g's" all prefer y to x. The last axiom, then, is

A5.   (*Nondictatorship*): No person i is decisive for every pair of outcomes in O.

This condition, which is perhaps the simplest one to appreciate, requires that no person can always get his way whenever everyone else opposes him.

Although we might debate whether institutions ought to satisfy these conditions, Arrow's theorem reveals their exceptional quality:

2.1.   (*Arrow's impossibility theorem*): *If O consists of three or more outcomes, the only institutions, G, that satisfy conditions A1 through A4 are institutions that violate condition A5, nondictatorship.*

An equivalent restatement of this result, which relates more directly to our earlier discussion, is that nondictatorial institutions that satisfy A2 through A4 carry no guarantee that they yield transitive social preference orders. We cannot overemphasize this theorem's importance; the rest of this chapter explores its many implications and the research that it has stimulated. First, however, we outline its proof.

The proof proceeds in stages. First, we deduce three consequences from Arrow's first four axioms and then show that these consequences contradict axiom A5. The first consequence is

1. If $x \, R \, y$, given a profile **R** in which i prefers x to y whereas everyone else prefers y to x, then $x \, R \, y$ for *every* profile **R'** in which i prefers x to y whereas everyone else prefers y to x.

This consequence follows directly from axiom A4 since everyone's preferences between x and y are being held constant in both profiles.

2. If i is decisive for x vs. y, then i is decisive for x vs. any $z \neq x$.

By axiom A2, the following preference profiles are admissible:

person i: $x \, P_i \, y \, P_i \, z$
each member, k, of $N - \{i\}$: $y \, P_k \, z \, P_k \, x$

Since i is decisive for x vs. y, $x \, P \, y$, whereas $y \, P \, z$ from the Pareto principle, axiom A3. Hence, from axiom A1, $x \, P \, y \, P \, z$. So, from consequence 1, i is decisive for x over z.

3. If i is decisive for x vs. y, then i is decisive for any $z \neq y$ vs. y.

Again from Axiom A2, the following preference profile is admissible:

person i: $z \, P_i \, x \, P_i \, y$
each member, k, of $N - \{i\}$: $y \, P_k \, z \, P_k \, x$

From the Pareto principle, it must be that $z \, P \, x$, whereas since i is decisive, $x \, P \, y$. Thus, from A1, $z \, P \, y$, in which case, from consequence 1, i is decisive for z vs. y.

4. If i is decisive for x vs. y, i is decisive for *any* z vs. *any* $w \neq z$.

By the assumption of the theorem, there exists an alternative t that is distinct from x and z. By hypothesis, i is decisive for x vs. y, and, therefore i is decisive for x vs. t (from consequence 2); therefore i is decisive for z vs. t (from consequence 3); therefore i is decisive for z vs. w (from consequence 2).

5. In violation of A5, someone is decisive for every pair of distinct alternatives.

Let g be a smallest decisive set for some pair, say, x and y. That is, $g - \{i\}$, for any i in g, is not decisive. From the Pareto principle, A3, g is not empty since N is decisive. From A2, the following preferences are admissible:

a particular member, i, of g: $x \, P_i \, y \, P_i \, z$

each member, j, of $g - \{i\}$: $z \, P_j \, x \, P_j \, y$
each member, k, of $N - g$: $y \, P_k \, z \, P_k \, x$

By the decisiveness of g, $x \, P \, y$. From consequence 1, if $z \, P \, y$, then $g - \{i\}$ is decisive for z vs. y, which contradicts the assumption that $g - \{i\}$ is not decisive. Hence, $y \, R \, z$. But from A1, this means that $x \, P \, y \, R \, z$ and $x \, P \, z$. From consequence 1, i is decisive for x vs. z. From consequence 4, i is decisive for every pair of distinct alternatives, which proves 5 as well as Arrow's theorem.

Three features of Arrow's theorem warrant emphasis. First, although we state the theorem here to show that axioms A1 through A4 imply dictatorship, it is equivalent to the assertion that if *any* four of the axioms are satisfied, the fifth is violated. Thus, Theorem 2.1 can be restated to say that if axioms A2 through A5 are satisfied, we cannot be assured that Axiom A1 is satisfied.

Second, this theorem is often interpreted to mean that transitive social preferences cannot occur, but clearly if everybody unanimously prefers $xPyPz$, any reasonable institution (see Arrow's third axiom) will yield this order as the social preference order. Similarly, since the probabilities in Figure 2.1 are never 1 in any cell, it is also true that many profiles yield transitive orders. Theorem 2.1's actual assertion is that, barring dictatorship, it is always possible to find *some* preference profile that yields an intransitive social preference. Our analysis of institutions, then, will also explore restrictions on individual preference that preclude intransitive social orders.

Third, another common misinterpretation of Arrow's theorem entails a confusion between intransitive and *cyclic* social preference orders. The social preference order that corresponds to the Condorcet paradox is cyclic, and thus intransitive. But not all intransitive orders are cyclic, and Theorem 2.1 does not guarantee the possibility of cycles. The social preference order $x \, I \, y \, I \, z \, P \, x$ violates transitivity and is one kind of order that Arrow's theorem says we cannot avoid short of allowing dictatorship. This order, however, is not a cycle, and many people would not find anything perverse in the selection of z as the social outcome. Nevertheless, Arrow's theorem and its proof have inspired considerable research. Thomas Schwartz, for example, has shown how to modify Arrow's proof to establish an impossibility theorem concerning cycles. If we assume, instead of transitivity, that social preferences are *acyclic* (that is, if $x_1 \, P \, x_2 \, P \, x_3 \, P \, \cdots \, P \, x_m$, then $x_1 \, R \, x_m$), and if we add two conditions to Arrow's axioms (that there are at least as many people as alternatives and that, if all but one person prefers x to y, then x and y are not socially indifferent), then an equivalent impossibility theorem holds.

Despite this extension, when examining specific institutions we should ascertain whether intransitivities entail cycles or "less troublesome" types of perverse social orders.

2.2   Paradoxes of voting

It is not stating the case too strongly to say that Arrow's theorem and the research that it inspired wholly undermine the general applicability or meaning of concepts such as "the public interest" and "community goals." The theorem makes no reference to any specific institution for aggregating individual preferences into a social preference or choice. Hence, any institution – any legislative procedure, voting rule, or social or committee process – that satisfies conditions A2 through A5 can yield social preferences that are paradoxical in the sense of being unlike individual preferences. Our interest, however, extends beyond the normative problem of deciding what is "best" for society or a group. We are also concerned with predicting group and individual choices in social settings, and the relevance of Arrow's theorem is that paradoxical social preferences imply that individual utility maximization in politics is a nontrivial task. People commonly view institutions simply as mechanisms for rendering social choices and allow that, although one procedure may advantage or disadvantage certain persons, once a particular institution or procedure is established, the final outcome follows in an almost mechanical way from individual preferences. Some simple examples of paradoxes that research related to Arrow's result has uncovered readily dispel this belief.

The agenda paradox

One widespread belief is that majority rule is an unambiguous procedure for resolving social disputes and that it is somehow fairer than other procedures. We often find an aura of legitimacy cast over outcomes selected by majority rule. But consider our earlier, three-voter, three-alternative example, and suppose that instead of using an exhaustive round robin of voting, people make decisions in a parliamentary setting, with a chairman who determines a specific order of voting on the alternatives, an *agenda*. For example, the chairman might set x and y against each other with the winner going against z. This agenda yields the outcome z, because x beats y, but z beats x (see Figure 2.3a). Thus, the apparent social preference is $z \, P \, x \, P \, y$, with z the social choice. But the chair could also first pair z and x, with the winner put to a vote against y (see Figure 2.3b), and in this case z beats x but loses to

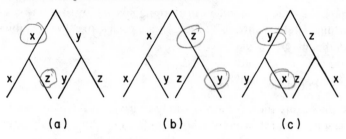

2.3 Three amendment agendas.

y, so that the apparent social preference is $y\ P\ z\ P\ x$, with y the social choice. Finally, the chairman could first place z and y against each other, and then place that vote's winner, y, against x (see Figure 2.3c), so that now the apparent social preference is $x\ P\ y\ P\ z$, with x the social choice.

Thus, playing on the preference cycle that a specific configuration of individual preferences occasions, a committee chairman can, by selecting an appropriate agenda, control the final outcome. Later, we consider how other members of the committee might respond to a chairman's attempt to manipulate the agenda, and we must also consider that the particular agendas in our example, *amendment agendas*, illustrate but one kind of procedure that might be employed to affect the final outcome. Our point here is that with cyclic preferences, the strategic possibilities embedded in a straightforward application of a simple procedure, majority rule, seem profound.

The contrived Condorcet paradox

Our previous examples all involve three alternatives. Although three is the minimum number of alternatives for an intransitive order, we should not suppose that committees confronting only two motions are without strategic interest. Referring to the previous example, suppose that the committee's deliberations are limited initially to x and y, in which case x defeats y. The opponents of x, however, might now contrive to introduce the third alternative, z, to take advantage of any fixed order of voting. For example, suppose that x is some bill that a congressional committee reports to the floor, and that y denotes the status quo, no bill. If z is an amended version of x, then the parliamentary rules of the U.S. Congress require that the chairman must first pair z and x against each other, with the winner voted against the status quo, y. Thus, z beats x but loses to y, and the status quo is preserved – an outcome that may very well have been the intent of those who introduced z in the first place. We might appropriately refer to z, then, as a "killer amendment."

Such events actually occur. As many as six motions may be on the floor of Congress at the same time: the status quo (a_0), a bill as reported out by committee (a_1), an amended bill (a_2), an amendment to the amendment (a_3), a substitute bill (a_4), and an amended substitute (a_5). Congressional rules require that a_2 be paired first against a_3, followed by a vote that pairs a_4 against a_5. The survivor of this vote is then paired against the survivor of the first vote. The survivor of this third vote is then paired against a_1. The eventual survivor of this process is paired against the status quo a_0, with the winner becoming the new status quo. Such a process yields ample opportunities for people to attempt to manipulate outcomes by the timely introduction of alternatives into the agenda. In the early part of this century, for example, there was apparently majority support in Congress for the Seventeenth Amendment to the Constitution – for electing all senators by popular vote. If a_0 is the status quo and a_1 the Seventeenth Amendment, it seems evident that a majority preferred a_1 to a_0. But opponents of the Seventeenth Amendment introduced the DePew amendment, which required federal supervision of elections, an anathema to southern Democrats. Denoting the Depew amendment by a_2, we can thus infer the following preferences:

> Republican supporters of a_1: $a_2 \ P_i \ a_1 \ P_i \ a_0$
> Republican opponents of a_1: $a_0 \ P_i \ a_2 \ P_i \ a_1$
> Democratic supporters of a_1: $a_1 \ P_i \ a_0 \ P_i \ a_2$

With these approximately equal-sized factions, and with the specific predetermined order of voting, the Depew amendment defeated the unamended version of the Seventeenth Amendment ($a_2 P a_1$), but the amended bill lost to the status quo. Thus, by the simple ruse of introducing a killer amendment, opponents of popular elections of senators found a way to have the status quo prevail even though a majority of the senators preferred the original amendment.

The dominated-winner paradox

Arrow's third axiom requires pairwise Pareto-optimality, but it seems reasonable to extend this idea and to suppose that a group should not select a particular outcome if there is some other outcome that everyone likes more, and specifically, that a legislature will not pass a bill if there exists an alternative that its members unanimously prefer. But before

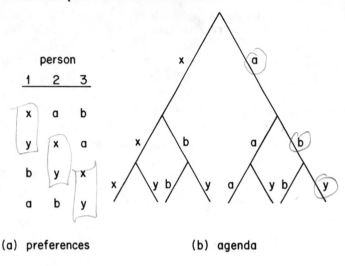

(a) preferences   (b) agenda

2.4 Dominated-winner paradox.

we can make such a prediction, we should consider the configuration of preferences in Figure 2.4. Under majority rule, these preferences yield the cyclic social order $x\,P\,y\,P\,b\,P\,a\,P\,x$. Suppose, however, that the agenda is the one also shown in Figure 2.4, which sets x against a, the winner against b, and the survivor against y. In this instance, a beats x, b beats a, and y beats b, so that y is the social choice. All three voters, however, unanimously prefer x to y, so that the final outcome is *Pareto-dominated* – majority rule fails to choose an alternative that everyone prefers to the final outcome.

The inverted-order paradox

Unless one is willing to vote repeatedly over pairs of alternative outcomes, majority rule may be an impractical procedure for making a social choice when large collectivities of people must select from among more than two alternative outcomes. The *Borda count* is one possible procedure that might be efficiently applied in this circumstance. Briefly, the Borda count requires that each person order his preferences among the m alternatives, and that the committee chair assign m points to an alternative each time it is ranked first, $m-1$ points each time it is ranked second, and so on. The alternative that receives the greatest number of points is taken to be the social choice. Consider, then, the seven-person preference profile in Figure 2.5.

Person

1	2	3	4	5	6	7
x	a	b	x	a	b	x
c	x	a	c	x	a	c
b	c	x	b	c	x	b
a	b	c	a	b	c	a

x 22   *a 13*
a 17   *b. 14*
b 16   *c 15*
c 15

2.5 Preferences for inverted-order paradox.

Person

1	2	3	4	5	6	7
a	b	c	a	b	c	a
b	c	x	b	c	x	b
c	x	a	c	x	a	c
x	a	b	x	a	b	x

c 20   *a 15*
b 19   *b 14*
a 18   *c 13*
x 13

2.6 Preferences for winner-turns-loser paradox.

In this instance, x receives three first-place votes (12 points), two second-place votes (6 points) and two third-place votes (4 points), for a total of 22 points. Similarly, alternative a receives 17 points, b 16 points, and c 15 points. Hence, x wins. But suppose that x is deleted for some reason. It seems logical for the committee to switch to alternative a, which received the second highest point total in the preceding vote. If, on the other hand, a new vote is taken over the remaining three alternatives, a receives 13 points (6 points for two first-place ballots among three alternatives, 4 points for two second-place rankings, and 3 points for three last-place rankings), b receives 14 points, but c receives 15 points. Thus, paradoxically, the alternative that ranked lowest in its score now ranks highest.

The winner-turns-loser paradox

Suppose that we reverse the preferences of the preceding example so that they look like those in Figure 2.6. In this case, x is the clear loser, with only 13 points; a receives 18 points, b 19 points, and c, the winner, 20 points. Now suppose that x, a, b, and c are political candidates but that after announcing these results, x dies. A recalculation awards alternative a 15 points, b 14, and c 13. The original winner is now the loser and a loser, a, is the winner.

Person

1	2	3	4	5	6	7
a	a	a	b	b	c	c
b	b	b	c	c	d	d
c	c	c	a	d	a	a
d	d	d	d	a	b	b

2.7 Preferences for truncated point-total paradox.

The truncated point-total paradox

Instead of awarding m points for a first-place ranking, suppose that we award $k < m$ points, with $k - 1$ points awarded for a second-place ranking, ..., 1 point for a kth place ranking, and zero thereafter. Note that if k equals m, then we have the Borda procedure, but if k equals 1, then this procedure is equivalent to simple plurality rule. Now consider the preference orders in Figure 2.7. Under plurality rule ($k = 1$), a is the clear winner, with three first-place votes. Under the Borda procedure ($k = 4$), c is the winner with 20 points, compared with 19 for a, 19 for b, and 12 for d. But with $k = 2$, the winner becomes b: b receives 7 points, a and c each receive 6 points, and d receives 2 points. Thus, the choice of a winner varies with the choice of k. We might keep this paradox in mind at faculty meetings when we decide which candidate should receive an offer.

The majority-winner paradox

Both the Borda procedure and majority rule can yield paradoxical results, and astute chairmen can manipulate them to do so. But there is also no reason to suppose that these two procedures yield consistent results. Consider the preference profiles in Figure 2.8, in which a is the clear majority winner. Under the Borda procedure, however, a receives 15 points, b receives 16 points, and c receives 11 points: b is the Borda winner.

This example should not surprise us. Farquharson's example is another in which clever people thwart a majority winner; banishment (y) in this instance by plurality rule, is the eventual outcome. It is not unusual, in fact, for three-candidate plurality-rule elections to produce as a winner a candidate that either opponent would have defeated in a pairwise vote. That an outcome defeats all others in a pairwise comparison is no guarantee that, for an arbitrary procedure, that

Person

1	2	3	4	5	6	7
a	a	a	b	b	b	c
b	b	b	c	c	a	a
c	c	c	a	a	c	b

a P b : 4/7

a P c 4/7

2.8 Majority-winner paradox.

outcome prevails. Later, when we examine various institutions in detail, one question that we will address is whether and under what conditions those institutions permit majority winners to prevail.

2.3   The power of agendas

The preceding examples underscore the lesson of Farquharson's example and show that the opportunity to control procedure can be decisive for one outcome over another. The Speaker of the House and the Rules Committees of Congress dictate whether or not amendments to bills are allowed from the floor, and they determine in part whether voting proceeds by secret ballot or by roll call and whether bills require simple or special majorities for passage. The importance of these powers should be obvious to students of the legislative process. But the preceding examples, formulated as paradoxes, suggest that even after the selection of a general method for aggregating individual preferences, seemingly innocuous proposals can dictate outcomes. The amendment agenda paradox is especially relevant, since it concerns a simple and commonly applied committee rule, pairwise voting. Not only can people manipulate outcomes by selecting this procedure over another (for example, plurality rule), but they can also manipulate pairwise voting itself by choosing the order in which motions are voted on, and, as in a contrived paradox, by introducing motions such as killer amendments, which will fail in the final vote against the status quo.

Examples with a *finite* number of outcomes or alternatives, though, can be deceptive. Arrow's theorem is important not because the judicious selection of individual preferences yields a paradoxical social preference, but because it compels us to acknowledge that such paradoxes can prevail with any preference aggregation mechanism. Similarly, although the preceding example of agenda manipulation is one in which *any* motion can win through the selection of an appropriate agenda, we must ask whether a fortuitous selection of preferences determines the agenda's decisiveness.

Preference order

Person	1	2	3	4	5	6	7	8
1	a	d	b	f	c	e	g	h
2	c	e	a	h	b	g	d	f
3	b	g	c	h	a	d	f	e

2.9 Top cycle.

There is a close relationship, of course, between cycles and the dependence of final outcomes on agendas, although we must be careful here about asserting equivalences. If social preferences do not cycle but are intransitive, agenda manipulation is still possible (if $x\ P\ y\ P\ z\ I\ x$, if the chair breaks ties in favor of the alternative that stands lowest in the alphabetical order, then the agenda "x against z, the winner against y" yields y, whereas the agenda "y against z, the winner against x" yields x). But if social preferences are transitive, then unless someone introduces a new alternative to create a cycle, agenda manipulation is more difficult, and as we shall see in the case of amendment agendas, it will fail. But not every cyclic social order yields a situation in which people can achieve any outcome by the selection of an appropriate agenda. Consider the preferences over eight alternatives in Figure 2.9, and note that although the alternatives $\{a, b, c\}$ cycle and those in $\{d, e, f, g, h\}$ cycle, each alternative in the first set defeats each alternative in the second set. Thus, although we can select an agenda to yield any outcome in $\{a, b, c\}$ – called the _top-cycle set_ – we cannot use majority rule to move from, say a, to any outcome in $\{d, e, f, g, h\}$. And if the motions $\{a, b, c\}$ differ only slightly in their substantive significance, then we might not regard this opportunity to manipulate outcomes as important.

Cyclic or intransitive social preferences, then, are necessary but not sufficient for most agendas to be decisive over the set of outcomes. If we add some additional structure to individual preferences, however, then we can answer such questions as: Under what conditions can agendas lead "anywhere" in contrast to final outcomes that are restricted to a subset of possibilities? When debating an appropriations bill, for example, can an agenda, in theory, lead to any allocation of resources, or can people use agendas only to modify slightly a bill reported out of committee? These are important questions, because their answers indicate the complexity that we are likely to encounter when studying committees.

The particular structure that we add to individual preferences is the

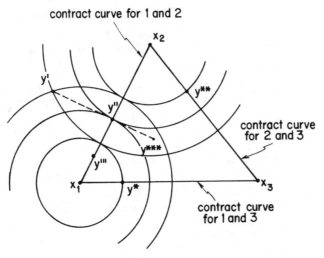

2.10 Contract curves.

characterization of preferences by indifference contours in an m-dimensional issue space. Specifically, suppose we confront a committee with m issues, and that we can summarize preferences on these issues by simple Euclidean distance, as defined in Chapter 1 by expression (1.3). And to simplify discussion, let us summarize each person's preferences over O by circular indifference contours centered about each one's respective ideal point, denoted x_i for person i. Hence, we might once again imagine that we are considering a legislature that must decide the relative weight given to defense versus social-welfare spending, as well as the overall size of the government's budget (the tax rate). By representing outcomes and preferences geometrically, we are now in a better position to ascertain the importance of agendas. By incorporating the corresponding notion of distance, we give greater meaning to questions about the importance of agendas and how far they can lead from a status quo. Suppose a particular point is the status quo, and a first agenda leads the group to adopt the motion close to it, but a second agenda leads to a point off the page. This variance may represent a significant public policy difference, and we certainly want to know how far different majority rule agendas can lead if preferences take this geometric form.

Consider the two-dimensional preferences configuration for three voters in Figure 2.10, and notice the lines connecting the ideal points of each pair of voters. First, since indifference curves are circles, geometry tells us that the straight line connecting two ideal points corresponds to the locus of tangencies of the indifference curves of the two voters involved. Second, this line, called a *contract curve*, corresponds to all alternatives that are *Pareto-optimal* for those two persons.

> *Pareto-optimal*: The outcome o in O is Pareto-optimal for the set of persons N' if there does not exist another outcome o' in O such that everyone in N' likes o' at least as much as o, and someone in N' strictly prefers o' to o.

Figure 2.10 illustrates this definition. For example, y' is not on the contract curve for voters 1 and 2, and both 1 and 2 prefer y'' to y' since moving from y' to y'' takes each of them to indifference curves that are closer to their ideal points. Thus, y' is not Pareto-optimal for voters 1 and 2. But *any* movement from y'' to another point, even one on the contract curve, such as y''', would harm at least one person – person 2 in this case. Hence, y'' is Pareto-optimal for 1 and 2. Points that are Pareto-optimal for one set of persons, though, are not necessarily Pareto-optimal for another set. Although y'' is Pareto-optimal for 1 and 2, it is not Pareto-optimal for 1 and 3 or for 2 and 3. Both 1 and 3 prefer y^* to y'', whereas 2 and 3 both prefer y^{**} to y''. Thus, the notion of Pareto-optimality is specific to the people under consideration.

The geometry of circular contours also permits us to identify, for any point such as y', what other points relative to a contract curve each person prefers to y'. Because y' is not Pareto-optimal for 1 and 2, various points to the other side of 1 and 2's contract curve, such as y^{***}, are closer to their respective ideal points and, hence, they prefer them to y'. The particular property of y^{***}, given circular indifference curves, is that the line connecting y^{***} and y' is perpendicular to 1 and 2's contract curve and y^{***} is closer to that line than is y', thereby ensuring that y^{***} is closer to both x_1 and x_2 than is y'.

This discussion shows the advantage of assuming circular indifference contours. With them, simple geometric intuition tells us a great deal about preferences. We emphasize, though, that such contours do not necessarily characterize preferences in, say, a legislature. Even with a spatial conceptualization of outcomes, the issues may be weighted unequally, and different legislators may give greater weight to different issues. But the geometry of circles is especially suited for formulating examples and counterexamples, and may help us develop some sense of what a particular model can accomplish. Then we can generalize to other indifference contour shapes.

With the lessons of this discussion of contract curves, we can now illustrate the power of agenda control in committees. Consider the two

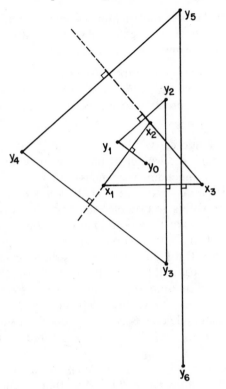

2.11 An agenda with three voters.

outcomes y_o and y_6 in Figure 2.11, which reproduces the ideal points and contract curves in Figure 2.10. Clearly, all three persons prefer y_o to y_6 and thus any reasonable institution "ought" to yield y_o as against y_6. Indeed, y_o is Pareto-optimal for this three-person group taken as a whole: No other point makes anyone better off without hurting someone else. But consider this amendment agenda: First we pair y_o against y_1, where both points fall on a line perpendicular to 1 and 2's contract curve. Since we have selected a y_1 that is closer to 1 and 2's contract curve than is y_o, y_1 wins in a majority vote. But if we pair y_1 against y_2, now choosing a y_2 that is closer to 2 and 3's contract curve than is y_1, then y_2 wins. Similarly, y_3 defeats y_2, y_4 defeats y_3, y_5 defeats y_4, and finally y_6 defeats y_5. This specific agenda leads the group from y_o to an outcome that everyone likes less.

What makes agendas so powerful here is that the preferences in this example share a property with the finite preferences in the example of the agenda paradox from the previous section. In each example some other point defeats every possible outcome. Put differently, neither example enjoys a Condorcet winner, which we define thus:

> _Condorcet winner_: If O is the set of alternative outcomes, then $x \in O$ is a Condorcet winner if, for all other $y \in O$, the number of people who strictly prefer x to y exceeds the number who strictly prefer y to x.

No motion can defeat a Condorcet winner in a majority vote and no agenda that sequentially pairs the winner of the previous vote against the next outcome on the list (amendment agendas) can lead away from it. To confirm that the situation in Figure 2.11 does not have such a point, note that with three voters and simple majority rule, two voters can upset any outcome if it is not Pareto-optimal for them – they can find some other point on their contract curve that they both prefer to any point not on their curve. But Figure 2.11 reveals that regardless of which point we choose, it cannot lie on all three contract curves simultaneously. Hence, from any motion we can move to some other that lies on a new contract curve, and which thereby defeats the original motion. Thus, the group can cycle endlessly throughout the issue space in much the same way as it cycled in our first three-alternative three-person example.

We must exercise some care in drawing a parallel between our infinite and finite outcome examples, however, and it is here that we encounter important questions concerning cycles and agendas with spatial preferences. If we augment the three-alternative example with a fourth motion that ranks last on everyone's preference order, then certainly no agenda can reach it. So we need not concern ourselves with the possibility that some agenda setter can lead us to such an inferior choice. Similarly, Figure 2.9 illustrates two cycles, one of which we cannot reach from the other. Is this possible with spatial preferences? Does the cycle remain unconstrained, or is it limited to part of the issue space? If it is constrained, then it may be of minor significance, but if it is unconstrained, then agendas might lead anywhere. Furthermore, is cycling peculiar to three-person committees or to committees that involve only two issues? Does cycling require circular indifference contours? Finally, what are the necessary and sufficient conditions for the existence of a Condorcet winner?

Richard McKelvey provides a theorem that gives answers to all but the last of these questions. McKelvey's theorem shows the pervasiveness of cyclic social preferences under majority rule in the context of spatial preferences:

2.2 (McKelvey): *For n-dimensional majority rule systems, if prefer-*
ence sets are convex over O and if a Condorcet point does not
exist, the social preference is wholly cyclic over O in the sense
that for any two points x and y, there is a finite amendment
agenda leading from x to y and back again to x.

The proof of this theorem contains two essential steps. First, we
designate some arbitrary point as the origin, so that, if the theorem is
correct, then we can take *any* other point as the status quo, and if a
Condorcet winner does not exist, then there must be some outcome that
defeats it. Suppose – and this is the crucial step – that among the
outcomes that defeat the status quo, we can show that there necessarily
exists a point that is further from the origin than is the status quo. If we
relabel this new point the new status quo, then we can reapply our proof
to show that a third point exists that defeats the second and takes us
further still from the origin. Thus, beginning at any outcome, we can
move as far from the origin as we want. The second step of this proof is
simple. Suppose that we seek an agenda that takes us from y_o to y_n,
where in a direct vote, y_o defeats y_n. By the first step, we know that we
can get as far from y_o as we choose, say to y_{n-1}. Certainly, then, we can
get to an outcome, y_{n-1}, that is so far from each and every person's ideal
point that all unanimously prefer y_n to y_{n-1}, which completes the
agenda.

Since the proof of the crucial first step illustrates some ideas that
subsequent chapters use, we outline it here in the context of the
two-dimensional five-voter example from Figure 2.12. In that figure we
connect the ideal points of three voters – 1, 2, and 3 – and we assert that,
with circular contours, the resultant triangle and its interior constitutes
the set of Pareto-optimal outcomes for the set of voters $\{1, 2, 3\}$. That
is, although the Pareto-optimal outcomes for two people correspond to
the contract curve between their ideal preferences, we can describe the
Pareto-optimals for three persons by the triangle formed by the three
contract curves between the respective pairs of ideal points. Furthermore, as in the three-voter example, if a point lies outside this triangle,
outside of the Pareto-optimals for the particular majority $\{1, 2, 3\}$,
then there exists at least one point (and usually many) on or inside the
triangle that defeat such a point by the margin of at least three votes
(voters 1, 2 and 3) to two.

To see that the configuration of preferences in Figure 2.12 does not
yield a Condorcet winner, and thus the configuration satisfies the
precondition of Theorem 2.2, note that such a point must be Pareto-
optimal for every three-person collection of voters (for every majority

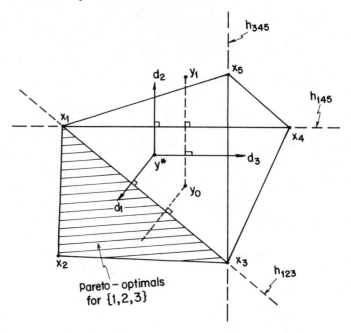

2.12 Three directions for moving away from y_o.

collection). But as Figure 2.12 reveals, there is no common intersection for the Pareto-optimals of the collections $\{1, 2, 3\}$, $\{1, 4, 5\}$, and $\{3, 4, 5\}$, nor for any other majority coalition. We can easily find outcomes that defeat any point. Consider the line h_{123} in Figure 2.12, which extends the contract curve between voters 1 and 3. Now consider any point above and to the right, or "wrong" side, of this line [since the ideal points of 1 and 3 are on this line, the "wrong" side of h_{123} is the side not containing the third voter's (voter 2's) ideal point], such as y_o. Drawing a perpendicular from y_o to h_{123} and beyond, any point on this perpendicular that is closer to h_{123} than is y_o is necessarily closer to the ideal points of voters 1, 2, and 3, so it defeats y_o in a majority vote.

We can now prove the crucial first step of Theorem 2.2. First, we must choose a point as the origin. Without proof we assert that, since a Condorcet winner does not exist, there must be a point that is outside the Pareto-optimals of at least three majority groups of voters (in the illustration, $\{1, 2, 3\}$, $\{1, 4, 5\}$, and $\{3, 4, 5\}$), but that is surrounded by those regions. In Figure 2.12, y^* is such a point. Next, construct the three directions $d1$, $d2$, and $d3$, which represent perpendicular moves from y^* toward the Pareto-optimals of $\{1, 2, 3\}$, $\{1, 4, 5\}$, and $\{3, 4, 5\}$,

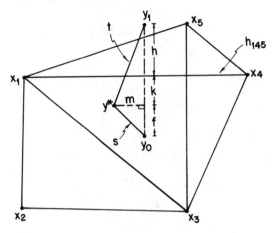

2.13 Proving that $s < t$.

respectively. (By perpendicular moves, we mean that a move in the direction $d1$, for example, takes us to the Pareto-optimals of $\{1, 2, 3\}$ by the shortest distance.) Referring to Figure 2.12, note that if we move from y_o in any of these directions, say $d2$, then points along this move, such as y_1, defeat y_o. The line connecting y_o and y_1 is perpendicular to h_{145}; y_1 is closer to h_{145} than is y_o; and y_o is on the "wrong" side of h_{145}. We now show that there exists a y_1 that defeats y_o that is further from y^* than is y_o.

If y_1 is to defeat y_o, the distance from y_1 to h_{145}, which we denote by h, must be less than the distance between y_o and h_{145}, which we denote as $k + f$, where k is the distance of y^* from h_{145} and f is as indicated in Figure 2.13. From the geometry of right-angle triangles (the Pythagorean theorem), and using the notation in Figure 2.13, we find that

$$s^2 - m^2 = f^2 \tag{2.1}$$
$$(h + k)^2 = t^2 - m^2. \tag{2.2}$$

Suppose that y_1 is only slightly (by the amount e) closer to 1 and 4's contract curve, h_{145}, than is y_o (which, of course, it must be if y_1 is to defeat y_o). That is, let $h = k + f - e$. Solving expression (2.2) for h and setting the result equal to $k + f - e$ yields

$$f = [t^2 - m^2]^{1/2} - (2k - e). \tag{2.3}$$

From expression (2.1), however, we can solve for f and, after combining the result with expression (2.3), obtain the identity,

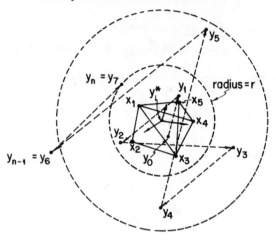

2.14 An agenda from y_o to y_n.

$$[s^2 - m^2]^{1/2} = [t^2 - m^2]^{1/2} - (2k - e).$$

Squaring both sides of this expression, we find that

$$s^2 = t^2 - 2(2k - e)[t^2 - m^2]^{1/2} + (2k - e)^2$$

or, if we rearrange the terms, this becomes

$$t^2 = s^2 + (2k - e)\{2([t^2 - m^2]^{1/2} - k) + e\}.$$

Or equivalently, since $h + k = \{t^2 - m^2]^{1/2}$,

$$t^2 = s^2 + (2k - e)(2h + e).$$

Note that as long as we keep e less than $2k$, $(2k - e)(2h + e)$ is positive. We can, of course, make e as small as we choose, so it follows that if we make e nearly zero, $t^2 > s^2$, or equivalently, $t > s$. Thus, we can find a y_1 that defeats y_o in the direction $d2$, which is further from y^* than is y_o. Applying this algorithm any number of times by appropriately choosing one of the three directions each time leads us further and further from the origin, y^*; that is to say, we can get as far from y^* as we choose by some agenda.

To form an agenda that leads from y_o to y_n, we know from the previous argument that an agenda exists leading to y_{n-1}, and that we can choose y_{n-1} so that it is sufficiently far from all ideal points to ensure that y_n defeats y_{n-1}. Referring to Figure 2.14, if we draw a circle about the origin y^* with radius r that encompasses y_n and all ideal points, then

y_n cannot be further than $2r$ from any voter's ideal point. Since we can find an agenda that leads as far from y^* as we want, let y_{n-1} be further than $4r$ from y^* so that y_{n-1} must be greater than $2r$ from any voter's ideal point. Hence y_n defeats y_{n-1} unanimously, and the agenda is completed.

We emphasize that a general proof of Theorem 2.2 does not require circular indifference contours. We use such contours here only to facilitate exposition. Thus, this theorem, although limited to majority rule and convex preference sets, is an important addition to the research inspired by Arrow's result. Arrow's theorem, 2.1, asserts that every nondictatorial institution can yield intransitive social preferences, but it does not establish that these intransitivities are always of substantive significance. We might regard as insignificant the cycle that the preference profiles in Figure 2.9 occasion, for example, if, substantively, outcomes $\{a, b, c\}$ do not differ by much. Predicting a top-cycle set may be wholly adequate for some purposes. Theorem 2.2 shows, however, that *for the case of spatial preferences, if the majority rule relation "breaks down" in the sense that it cannot produce an unambiguous winner, then it breaks down completely because the entire alternative set becomes involved in the group's preference cycle.* Put differently, the top cycle consists of the entire outcome space.

The subsidiary questions that this theorem raises are important. First, under what conditions will a Condorcet winner exist in majority rule spatial situations? In subsequent chapters we suggest that the conditions for such winners in a spatial context are exceedingly fragile, so that nearly all such committee situations are susceptible to manipulation of the agenda, of the ways in which majority rule is implemented. Second, if the conditions for a Condorcet winner are sufficiently delicate, then what prevents a committee chairman from misrepresenting his preferences to eliminate any chance of a Condorcet winner existing, so that he can manipulate the agenda thereafter? Third, if agendas are so important in determining final outcomes, then how do people choose agenda setters? Fourth, can the members of a committee do anything to protect themselves against someone who controls an agenda? Finally, if agendas can lead anywhere, then what other features of committees can we consult to say something more specific about the outcomes that eventually prevail?

We must formulate carefully the answers to these questions. Some interpret Theorem 2.2 mistakenly to mean that "anything can happen," that an astute chairman or agenda setter can lead a majority rule committee to any outcome. In his original essays, however, McKelvey is careful to limit the implications of his analysis. First, Theorem 2.2

simply describes the properties of the social preference relation under majority rule with spatial preferences. The theorem does not provide a prediction about events, and thus it does not necessarily describe a social choice function. Certainly, even a naive observer of Congress would not believe that a chairman can manipulate agendas for very long, to the detriment of all others, before he loses the authority to set agendas. In fact, several of our earlier questions suggest that we do not yet know the whole story about how committees choose. Theorem 2.2 does not suppose that members of the committee respond to the agenda setter by, say, forming coalitions or by hiding their preferences from him so that he cannot accurately judge the consequence of an agenda. As we proceed through this book and develop tools to analyze how people in committees respond to the power of chairmen, we shall reexamine Theorem 2.2 and its consequences.

2.4   Misrepresentation of preferences

One interpretation that we can give Theorem 2.2 is that it reveals the black box of political institutions to be subject to the strategic manipulation of elites, of those who control the specific ways institutions and procedures are implemented. But then we should ask: What about the rest of us? If institutions are complex because they are subject to strategic manipulation in this way, are those of us who are asked merely to register our preferences by voting also capable of manipulation? If the answer to this question is YES, then we must conclude that the way in which institutions transform individual preferences into actual outcomes is indeed a complex process.

Notice that although the substantive implications of Theorems 2.1 and 2.2 are profound, their common objective of describing social preferences requires that they take individual preferences as given. To interpret Theorem 2.2, then, as offering predictions about events requires, beyond a specific model of institutional procedures, the assumption that people reveal their preferences truthfully. But we should not assume that in actual political processes people do not have incentives to disguise their preferences. Referring to Figures 2.1 and 2.3, recall that with the agenda "x vs. y, and the winner against z," if everyone votes sincerely – that is, if everyone votes for their more preferred alternative at every stage – then z emerges as the eventual winner. But if person 1, who is aware of the consequences of this particular agenda, votes for y rather than x in the first stage, then y emerges as the winner in this first vote and subsequently defeats z, an outcome that 1 prefers to the consequences of *sincere voting*.

As a substantive illustration of this possibility, consider the 1956 federal school construction aid bill (alternative a_1), which a majority of both houses of Congress appeared to prefer to the status quo of no aid (alternative a_o). After this bill was reported to the House floor, certain members offered an amendment with a proviso that federal funds be awarded only to states with schools "open to all children without regard to race in conformity with the requirements of the United States Supreme Court decisions." Although this motion, called the Powell amendment (alternative a_2), may not have been proposed with any strategic purpose in mind, it nevertheless opened the door for those who opposed federal aid to schools. William Riker, after a careful examination of the various votes taken on the bill and of the public record of key legislators, concludes that opponents of aid strategically contrived an intransitive social order, $a_2 P a_1 P a_o P a_2$, to defeat the original bill. Since the proscribed voting order is "a_2 vs. a_1, the winner against a_o," this order implied that no federal aid would pass. The specific conjecture that Riker offers, and that some political observers believed at the time, is that the social order with sincere preferences was the transitive order, $a_1 P a_o P a_2$, but that some 97 Republican representatives who, contrary to their true preferences, voted for the Powell amendment (a_2) when it was paired against the unamended bill (a_1), contrived a paradox. After the Powell amendment passed, they joined with southern Democrats opposed to the Supreme Court's decision in *Brown* v. *Board of Education of Topeka* to defeat the amended bill on final passage. Some legislators spoke in favor of federal aid and integration and yet voted against the amended bill in the final vote. And twelve of the 97 Republicans, later in the session, voted against civil rights legislation, an action inconsistent with their support of the Powell amendment but consistent with an attempt to defeat federal aid with a contrived paradox.

We should not interpret these remarks to mean that only majority rule is susceptible to manipulation. Consider the preferences in Figure 2.15 and the Borda procedure, which under sincere voting produces the social ordering $c P x I y P d P e$: c receives 18 points, x and y receive 17, d 12, and e 11. But suppose that person 1 reports the preference order $x P_1 e P_1 d P_1 c P_1 y$. In this instance, alternative x receives 17 points as before, but y and c each receive 16 points, while d receives 12, and e 15. Thus, x, which is 1's first choice, becomes the social choice.

Lest we infer from these examples that only voting institutions are susceptible to misrepresentation, suppose we ask people how much they would be willing to pay to construct some public project, such as a municipal auditorium. After we collect all responses, the city council

	Person			
1	2	3	4	5
x	y	y	c	x
c	c	c	d	y
d	e	x	x	e
y	d	e	e	d
e	x	d	y	c

2.15 Misrepresentation with the Borda count.

adds up the willingness of people to pay. If this total exceeds costs, then the auditorium is built, and each person's tax equals the amount that he said he would be willing to pay, divided by the total willingness to pay. The problem is that if the community is sufficiently large, then each person knows that his contribution affects the sum only slightly, and that no contribution is likely to be critical to the final outcome. Thus, by indicating that the project is worth nothing to him, a person might reason that although his action affects the project's prospects infinitesimally, it reduces his tax to zero. Everyone might reason this way, and everyone might believe that everyone else reasons this way. But even if a person believes that everyone thinks alike, there is no incentive to reveal preferences truthfully unless a person is willing to fund the project alone. Similarly, suppose we ask that people state the intensity of their preference for the auditorium, knowing that funding for such a project relies on the existing tax structure. Persons favoring an auditorium but in low tax brackets, then, should overstate their preferences, and persons in high tax brackets, who see little value in a municipal auditorium, are wise to understate their valuations.

These examples suggest that individual preferences need not correspond to what people tell us or actually vote. That negotiators of international treaties and labor contracts misrepresent their preferences as a matter of strategy comes as no surprise, but why should we believe that roll call votes in a legislature are any less strategic and are any more representative of true preferences? In fact, *strategic misrepresentation* makes politics interesting: Without it, voting entails little more than the solicitation and dry recording of preferences, and the outcomes follow in a cold deterministic way, save for the whims of nature.

The preceding examples raise several questions. First, are any institutions guaranteed to solicit the truthful revelation of preferences? Just as we made certain by using Arrow's theorem and its extensions that the Condorcet paradox is not a curious anomaly, we must do the

Person

1	2	3
a	a	b
d	c	a
b	d	c
c	b	d

2.16 Condorcet winner with a cycle.

same here for preference revelation. If no such institution exists so that the possibility of misrepresentation is universal, then we must address a second question: How can we plug individual preferences into alternative institutions to predict outcomes and to compare institutional performance? A subsidiary question is: If more than one person recognizes the value of misrepresentation, might the potential for strategic action become unmercifully complex as everyone tries to assess his optimal choice, anticipating that everyone else will do the same?

We require answers to the second and third questions, of course, only if the answer to the first is NO, which actually *is* our answer. But as with Arrow's theorem, we must proceed rigorously and generally to make certain that we overlook no institutions. We begin by interpreting G as a social choice function that selects outcomes in O, rather than by assuming that G orders all outcomes to produce a social preference order, and we assume that it is *resolute* because it chooses a unique element of O as the social outcome. There are several reasons for turning to social choice functions at this point. If we suppose that G produces a complete ordering of the outcomes in O for society, then we may observe orders that, although intransitive, nevertheless imply a specific outcome as best for society. The example in Figure 2.16 shows that not all intransitive social preferences are important in predicting final outcomes. In that example, a defeats b, c, and d, even though b, c, and d cycle. Thus, if we are concerned with how people might manipulate final outcomes by misrepresentation, it is sufficient to know what outcome emerges victorious from among a specific set of individual preference profiles. And since we require less of social choice functions than of social preference functions, perhaps social choice functions provide an escape from Arrow's theorem. Unfortunately, this hope proves futile. A third fundamental theorem, by Gibbard and Satterthwaite, raises the possibility that strategic voting is advantageous to someone for *any* institution or procedure.

To state their result formally, we say that an institution is *manipul-*

able if, for some set of outcomes and individual preferences over that set, there is some person who can benefit by misrepresenting his preferences. That is,

> *Manipulable*: The social choice function $G(\mathbf{R}, O)$ is manipulable if there exists a preference profile, $\mathbf{R} = (R_1, \ldots, R_n)$, for the group, and at least one person i such that for some preference order R_i',
>
> $G(R_1, \ldots, R_{i-1}, R_i', R_{i+1}, \ldots, R_n, O) \; P_i \; G(\mathbf{R}, O).$

If we assume now that G is resolute (that G always chooses a unique outcome), then the following theorem can be established:

> 2.3 (Gibbard-Satterthwaite): *If O contains more than two elements, and if G is nonmanipulable and yields a single outcome as the social choice, then G is a dictatorship.*

Admittedly, there is some disagreement as to whether the assumption that G is resolute is restrictive since it precludes institutions that might yield ties. Nevertheless, Thomas Schwartz proves, with a slight modification of assumptions, that even nonresolute institutions (and, thus, any nondictatorial institution of interest) can be manipulated for some preference profile. Assuming that everyone's preferences over O are strict, Schwartz has adapted the proof of this result to provide a simple proof of a theorem that is for all practical purposes equivalent to Theorem 2.3. This proof establishes manpulability as a corollary to Arrow's theorem, and thereby it shows the power of Arrow's axiomatic perspective.

We begin with the following assumptions, which we intend to show are inconsistent:

1. For every finite, nonempty set O, and for every preference profile $\mathbf{R}$, the social choice function $G(\mathbf{R}, O)$ is resolute and selects a unique element of O. Thus, $G(\mathbf{R}, \{x, y\}) = x$ if and only if $x \; P \; y$.
2. O contains at least three elements.
3. Nondictatorship (same as Arrow's axiom A5).
4. If everyone prefers every member of $O - \{y\}$ to y, then $G(\mathbf{R}, O) = G(\mathbf{R}, O\text{-}\{y\})$. That is, if y is at the bottom of everyone's preference order, deleting y will not change the social choice.
5. $G(\mathbf{R}, O)$ is nonmanipulable.

To show that these five assumptions are inconsistent, we establish that together they imply Arrow's five axioms, which we know are inconsistent. Of course, since Assumption 1 implies unrestricted domain (A2), since 2 is the hypothesis of Arrow's theorem, and since 3 is identical to

A5, it is sufficient to show that these assumptions imply Axioms A1, A3, and A4.

First, notice that the Pareto principle, A3, follows easily from Assumption 4 if we let O be the two-element set $\{x, y\}$. That is, if $O = \{x, y\}$ and if $x\ P_i\ y$ for all i, then 4 requires that $x\ P\ y$, which, from Assumption 1, means that $G(\mathbf{R}, \{x, y\}) = G(\mathbf{R}, \{x\}) = x$. This is simply a restatement of Axiom A3.

Second, to show that Arrow's independence condition, A4, is implied by these assumptions, we need only establish that if $\mathbf{R}$ and $\mathbf{R}'$ differ outside of O only for one person, i, then $G(\mathbf{R}, O) = G(\mathbf{R}', O)$; the implication then follows by repeated application of this conclusion. Note that unless this equality holds, person i prefers either the outcome $G(\mathbf{R}, O)$ to the outcome $G(\mathbf{R}', O)$ or vice-versa (since all preferences are strict and a person cannot be indifferent between two outcomes). But this violates Assumption 5, in which case it follows that i cannot change his preferences outside of O to affect the social choice.

Finally, to establish that A1 is implied by these assumptions, notice that since all preferences are strict, the only violations of transitivity that we need to consider are cycles of the form $x\ P\ y\ P\ z\ P\ x$. Our method is to assume that there is a cycle, and then to show that this cycle is inconsistent with Assumptions 1–5; this inconsistency establishes that these assumptions imply Arrow's first axiom. Suppose, then, that $x\ P\ y$ $P\ z\ P\ x$, and, since by Assumption 1 G must select one outcome from the set $\{x, y, z\}$, let $G(\mathbf{R}, \{x, y, z\}) = x$. Suppose further that we move the alternative that x defeats, y, to the bottom of person 1's preference order; the result is the new preference profile $\mathbf{R}'$. Then, unless $G(\mathbf{R}', \{x, y, z\}) = G(\mathbf{R}, \{x, y, z\}) = x$, Assumption 5 is not satisfied (notice that our argument here is the same as in the previous paragraph). Hence, $G(\mathbf{R}', \{x, y, z\}) = x$. By reiterating this argument for all n persons, we conclude if $\mathbf{R}^*$ corresponds to a preference profile in which y is at the bottom of everyone's preference order, then $G(\mathbf{R}^*, \{x, y, z\}) = x$. But then, by Assumption 4, this means that $G(\mathbf{R}^*, \{x, z\}) = x$. And since we already know that the independence axiom is satisfied, then $G(\mathbf{R}, \{x, z\}) = x$, or, equivalently, that $x\ P\ z$. But this means that no cycle is possible: the five assumptions *are* consistent with Axiom A1.

This argument establishes that Assumptions 1–5 are consistent with all of Arrow's axioms. But Arrow's theorem is that these axioms are inconsistent; hence, Assumptions 1–5 must be inconsistent. This means that any social choice rule or institution that satisfies the first four assumptions necessarily violates the fifth, nonmanipulability. Alternatively, if the social choice rule satisfies Assumptions 1, 2, 4, and 5, then it must be a dictatorship.

We emphasize now that Theorem 2.3 and its extensions do not imply that someone can manipulate a procedure or an institution in *every* instance. Rather, they imply that at least one person can manipulate a nondictatorial institution for some arrangement of individual preferences. Thus, although someone can manipulate the Borda count for the preferences found in Figure 2.15, other preferences can preclude manipulation (as, for instance, if everyone holds the same order). Arrow's theorem works the same way: Not all preferences yield intransitive social orders, but for every institution that satisfies his axioms, there exist some individual preferences such that the social order is intransitive.

Theorem 2.3 yields two profound consequences for the study of politics. First, regardless of the institution under consideration, the preferences that people *reveal* need not be the preferences that they *truly hold*. Thus, our earlier examples of strategic misrepresentation are not artifacts of Borda counts, opinion polls, or sequential voting with fixed agendas. Strategic misrepresentation may be individually profitable in any resolute institution under appropriately configured individual preferences. No institution and no social process is immune from the complications of strategic maneuver. A theory of politics based on the assumption that people reveal their preferences honestly is a theory that cannot encompass all of reality.

The second implication of this result concerns the design of political institutions. Put simply, before we can judge the performance of any proposed institution or voting procedure, we must assess the impact of strategic misrepresentation. If we do not take account of how people manipulate outcomes by strategic revelation of preferences, then those institutions may perform quite differently than predicted or intended. The arguments of both critics and defenders of the electoral college illustrate this point. A favorite pastime is to use previous election returns to compute who would have won a presidential election had different election rules prevailed. Thus, some have asserted, if the rules awarded candidates a state's electoral votes in proportion to a candidate's percentage of the popular vote in that state, rather than on the basis of winner-take-all electoral votes, Nixon would have defeated Kennedy in 1960. But this prediction assumes that procedural changes would not affect candidates' election strategies and voters' choices. Under a different system fewer voters might have voted for unpledged or third party electors, for example, or candidates might have campaigned more in some states and less in others, thereby making such an analysis meaningless.

For a second example, consider the debate over the adequacy of plurality rule. The concern here is that if more than two candidates compete for the same office, plurality rule does not always select the Condorcet winner, as in Farquharson's example. New York's 1962 senatorial election, in which the conservative James Buckley defeated two more liberal candidates, Charles Goodell and Richard Ottinger, provides an illustration of this possibility. Public opinion polls and the like revealed that in a pairwise vote, Goodell would have defeated both opponents. But, in a three-way contest, Goodell found himself ideologically sandwiched between his two opponents, and thereby lost. As an implementable procedure, however, exhaustive pairwise voting to find and elect a Condorcet winner is impractical in mass elections, and since it seems reasonable to require that democratic voting procedures select Condorcet winners if they exist, scholars have proposed a variety of voting schemes as alternatives to plurality rule. But the problem with most analyses of these alternatives is that they take inadequate account of the possibility that voters might misrepresent their preferences and candidates might change their campaigns under different rules. Although reforms might or might not perform better than plurality rule, the analysis remains incomplete unless it recognizes the consequences of Theorem 2.3.

2.5   Vote trading

The form of strategic misrepresentation of preferences most familiar to political scientists is *vote trading*, an agreement between two (or more) legislators for mutual support, even though it requires each to vote contrary to his real preferences on some legislation. Vote trading has fascinated scholars for years, and several questions about it have found answers. What, for example, are the necessary and sufficient conditions for vote trading to occur, to be individually rational? What happens if everyone vote-trades? Does such trading necessarily lead to or away from Pareto-optimal outcomes? If a Condorcet winner exists, will vote trading lead to it? What kinds of outcomes will vote trading avoid? We can answer some of these questions in this chapter. Because vote trading requires coordination among two or more persons, however, we must postpone definitive answers to some of them until we discuss cooperative games and coalitions. Here, we can at least identify some necessary and sufficient conditions for this form of strategic misrepresentation of preferences to occur.

Suppose a legislature considers a sequence of m bills, each of which can either pass or fail. Thus, the set of outcomes, O, consists of the 2^m possible permutations of pass–fail, ranging from no bills passing, to some of them passing, to all of them passing. We can now identify two

Legislator:

	1	2	3
Preference order:	(0, 0)	(0, 1)	(1, 0)
	(1, 1)	(1, 1)	(1, 1)
	(1, 0)	(0, 0)	(0, 0)
	(0, 1)	(1, 0)	(0, 1)

2.17 Nonseparable preferences.

alternative representations of preferences over these outcomes. First, we can let legislators have ordinal utility functions over O without restrictions on their form. Alternatively, we can examine the special case in which preferences over O are *separable*. Formally,

> *Separable preferences*: Let $o \in O$ be denoted $o = (x_1, x_2, \ldots, x_j, \ldots, x_m)$, where x_j is the characteristic of o on the jth dimension. A person's preferences are separable across the m dimensions if the utility function summarizing his preferences across all outcomes in O can be expressed as
>
> $$u(o) = u_1(x_1) + u_2(x_2) + \cdots + u_m(x_m).$$

For an equivalent statement of separability, suppose that for any person i,

$$o_1 = (x_1, \ldots, x_j, \ldots, x_m) \, P_i \, (x_1, \ldots, x_j', \ldots, x_m) = o_2, \quad (2.4a)$$

where x_k, $k = 1, \ldots, m$, equals either 0 or 1, depending on whether the kth bill fails or passes. Then if two dispositions of some set of m bills, such as o_1 and o_2, differ only in their disposition of the jth bill, and if $o_1 \, P \, o_2$, then

$$o_3 = (x_1'', \ldots, x_j, \ldots, x_m'') \, P_i \, (x_1'', \ldots, x_j', \ldots, x_m'') = o_4.$$
$$(2.4b)$$

That is, regardless of how we set the disposition of the other m-1 bills, the person always prefers the same disposition of the jth bill. Hence, in the context of legislative decision making, separability represents preferences in which bills have nothing to do with each other. Thus, separability might model preferences over an authorization for a dam in Utah and a bill for federal aid to secondary schools, but it is almost certainly violated if the bills are germane amendments to a single project. For example, if bill 1 builds half a bridge, and bill 2 builds the

				Outcome				
Legislator	$(0, 0, 0)$	$(1, 0, 0)$	$(0, 1, 0)$	$(0, 0, 1)$	$(1, 1, 0)$	$(1, 0, 1)$	$(0, 1, 1)$	$(1, 1, 1)$
1	0	2	4	−5	6	−3	−1	1
2	0	4	−5	2	−1	6	−3	1
3	0	−1	−3	5	−4	4	2	1

2.18 Separable preferences with three bills.

other half, then preferences over the two bills can hardly be separable.

We are concerned with this restriction on preferences because the concept of vote trading may be ambiguous if preferences are not separable. To see this, consider a three-member legislature with the ordinal preferences over the disposition of two bills shown in Figure 2.17. The entry "(1, 0)," for example, indicates that the first bill passes, whereas the second bill fails. To confirm that these preferences are not separable, note that legislator 1's ordinal utility must be $u(0, 0) > u(1, 1) > u(1, 0) > u(0, 1)$. Hence, if preferences are separable,

$$u_1(0) + u_2(0) > u_1(1) + u_2(1) > u_1(1) + u_2(0) > u_1(0) + u_2(1).$$

But with appropriate cancellations, this series of inequalities yields the contradictory conclusion that $u_2(1)$ is both greater than and less than $u_2(0)$, and with a different set of cancellations, it yields the contradiction that $u_1(1)$ is greater and less than $u_1(0)$.

The crucial feature of nonseparable preferences in this case is that it is impossible for us to judge how legislator 1 will vote on bill 1 unless we know his beliefs about the likelihood that bill 2 will pass. If he believes that bill 2 will pass with certainty, then he should vote for 1, but if he believes that bill 2 will fail, then he should vote against bill 1. Analysis of the nonseparable case, then, requires consideration of all bills simultaneously. We can look at coalitions seeking to pass one package of bills over another, but we cannot speak of legislators "voting against their preferences on some bill," since those preferences have meaning only in the full context of the entire agenda.

For the case of separability, let us consider the three-legislator, three-bill example in Figure 2.18. We change the format of our presentation somewhat in this figure by giving ordinal utility numbers to outcomes and by assuming that the utility associated with a bill's defeat is zero. To see that these preferences are separable, note, for example, that the utility of legislator 2 for the outcome (1, 1, 0) equals his utility from the first bill's passage, 4, plus the utility from the second bill's passage, −5, plus the utility from the third bill's defeat, 0, for a total of −1.

Legislators 1 and 2 prefer to pass bill 1, legislators 2 and 3 prefer to fail bill 2, and legislators 2 and 3 prefer to pass bill 3. Hence, if everyone truthfully reveals their preferences in a sequential vote, the final outcome is (1, 0, 1). Interestingly, though, this outcome is a Condorcet winner. For example, it defeats (1, 1, 0) with the support of legislators 2 and 3; it defeats (0, 1, 1) with the support of legislators 2 and 3; it defeats (0, 0, 1) with the support of legislators 1 and 2; and so forth. And it is no accident that truthful revelation of preferences yields the Condorcet outcome here, as Theorem 2.4, first proved by Thomas Schwartz, makes clear:

2.4.   (Schwartz): *If O corresponds to all combinations of pass-fail over a set of bills, if preferences over O are separable, and if a Condorcet winner exists, sincere voting yields that winner.*

The proof of this result consists of showing that if vote trading is required to achieve some outcome – if that outcome occurs only if one person or more misrepresents his true preference – then it cannot be a Condorcet winner. Thus, Condorcet winners, if they exist, will win if everyone votes sincerely. To see this result, let O be the set of all possible outcomes, where $\mathbf{o} = (x_1, \ldots, x_m)$ is an element of O and x_j, $j = 1$ to m, is 1 if bill j passes and 0, otherwise. Next, let $\mathbf{o}' = (x'_1, \ldots, x'_m)$ be the sincere vote outcome. In Figure 2.18 the sincere outcome is $(x'_1 = 1, x'_2 = 0, x'_3 = 1)$. Notice that if we change the vote on any single bill only, then $\mathbf{o}'$ defeats this new outcome. That is, if P is the majority preference relation, then,

$$\text{sincere outcome} = (x'_1, \ldots, x'_j, \ldots, x'_m) \; P \; (x'_1, \ldots, x''_j, \ldots, x''_m),$$
(2.5)

where $x'_j \neq x''_j$. Referring again to the example, the sincere outcome (1, 0, 1) defeats (0, 0, 1), because a majority favors bill 1, because preferences are separable, and because the disposition of bills 2 and 3 remains identical in the two outcomes. In accordance with expressions (2.4a) and (2.4b), expression (2.5) is what separability of individual preference implies – if a majority prefer (1, 0, 1) to (1, 1, 1), then that same majority will prefer (0, 0, 0) to (0, 1, 0). Now let $\mathbf{o} = (x_1, \ldots, x_j, \ldots, x_m)$ be an arbitrary outcome other than the one that sincere voting produces. To complete the proof of 2.4, we need only show that some other outcome dominates $\mathbf{o}$ in a majority vote, so $\mathbf{o}$, which can only be arrived at by vote trading, cannot be a Condorcet winner. First, suppose that the sincere outcome, $\mathbf{o}'$, and $\mathbf{o}$ differ on the disposition of the jth

Legislator	(0, 0, 0)	(1, 0, 0)	(0, 1, 0)	(0, 0, 1)	(1, 1, 0)	(1, 0, 1)	(0, 1, 1)	(1, 1, 1)
1	0	2	4	−5	6	−3	−1	1
2	0	4	−5	2	−1	6	−3	1
3	0	−5	2	4	−3	−1	6	1

2.19 Vote trading without a Condorcet winner.

bill since they must differ on some bill. Following expression (2.5), since a majority prefer the disposition of this bill provided by o' to the disposition provided by o, then, from separability,

sincere outcome $= (x'_1, \ldots, x'_j, \ldots, x'_m) \ P \ (x'_1, \ldots, x_j, \ldots, x'_m)$.

But again from the assumption that all individual preferences are separable, if a majority hold the preceding preference, then that same majority must yield the social preference,

$$(x_1, \ldots, x'_j, \ldots, x_m) \ P \ (x_1, \ldots, x_j, \ldots, x_m),$$

so some other motion dominates o. This completes the proof of Theorem 2.4.

Thus, if a Condorcet winner exists, misrepresentation is counterproductive for a majority of legislators. Although legislators might bring about some other outcome by trading votes, the simple "trade" of no trade at all dominates in majority rule. Theorem 2.4, however, does not preclude the value of vote trading for a majority in every instance. Suppose that a Condorcet winner does not exist, and consider the example in Figure 2.19, which differs from the one in Figure 2.18 only by a slight modification of legislator 3's preferences. As before, all preferences are separable and the sincere voting outcome finds all three bills passing, in which case all three legislators receive a payoff of 1. Observe, however, that any pair of legislators prefer to vote trade, to pass only one bill. Legislators 1 and 2 can collude to defeat bills 2 and 3 − 1 votes to defeat bill 2, even though his sincere preference is for passage, whereas 2 votes to defeat bill 3, even though his sincere preference is for passage as well. After this trade, the outcome (1, 0, 0) prevails, which legislator 1 and 2 both prefer to the sincere outcome. But now legislators 1 and 3 can trade to pass bills 1 and 2. From there, other trades improve the payoffs of two legislators, and so forth. In fact, an examination of the utilities in Figure 2.19 reveals that there is no Condorcet winner, and all outcomes cycle.

Admittedly, this sort of misrepresentation is not the kind that Theorem 2.3 envisions. That theorem asserts that for any institution

Legislator	Bill 1	Bill 2	Bill 3
1	10	−2	−4
2	−2	10	−4
3	−8	4	−2
4	4	−8	2
5	−5	−5	−5

2.20 Vote trading with a Condorcet winner.

there is some preference configuration for which sincere revelation is unwise from an *individual* perspective. The preceding example, however, assumes some minimal coordination between two or more persons. We can best study vote trading, then, when we analyze cooperative games.

Before we conclude, however, we ought to clear up one potential misinterpretation of Theorem 2.4. It is sometimes asserted that vote trading is profitable only if a Condorcet winner does not exist. But this is not the implication of that theorem, as the example in Figure 2.20 shows. With respect to that figure, suppose preferences are separable and that the payoff each legislator receives if a bill fails is zero, so that the outcome (0, 0, 0) is a Condorcet winner. Nevertheless, notice that legislators 1 and 2 have an incentive to trade votes on bills 1 and 2 so that both bills pass. However, 3 and 4 now should trade on these same two bills so as to nullify the agreement between legislators 1 and 2. In fact, because the outcome of the trade between 1 and 2 is different from the Condorcet winner, there necessarily exists a majority of legislators (3, 4, and 5 in the example) who will prefer to negate any such trade. What Theorem 2.4 implies, then, is that if a Condorcet winner exists, the only outcome that will be supported by a majority of legislators is the same outcome that would have prevailed if no vote trading ever took place. Thus, the correct interpretation of this theorem is that if all legislators possess perfect information about each other's preferences, if there are no impediments to trading (that is, if communication and coordination are costless among legislators), then vote trading is pointless – some majority will prefer and presumably act to negate trades.

2.6   Summary

This chapter reviews four theorems, the first three of which are central to understanding the theoretical concerns of politics. The first theorem, Arrow's result (and its immediate predecessors), has at least two interpretations. One is that we can justify normative utterances by referring to "the public interest" or to "society's welfare" only if they follow from analyses of individual self-interest. Since there is no guarantee that social preferences are transitive, there is no guarantee that anything stands highest in the social preference order, and, thus, no guarantee that there is an unambiguous best choice for society. A second interpretation is that groups differ fundamentally from people. Although the attribution of goals to legislatures, interest groups, political parties, and nations is often a convenient journalistic shorthand, the decision-making structure that Chapter 1 develops applies only to individual human beings. Groups are not legitimate primitive theoretical concepts, and an understanding of collective action must derive from an understanding of procedures and institutions and of how people choose in such environments.

The second theorem, 2.2, opens the door wider for complexity in politics. Focusing on majority rule in a spatial context, it shows that, unless a Condorcet winner exists, social preferences are wholly cyclic. There are two interpretations of this theorem as well. First, we can view it simply as a description of the social preference relation in a particular circumstance. The derivation of spatial preferences in Chapter 1 suggests that unless majority rule mechanisms that take cognizance of budgetary limits yield a Condorcet winner, social preferences cycle over the entire space. In addition, it establishes the implications of one institutional structure, a naive committee using simple majority rule chaired by a smart agenda setter. We suspect, however, that few interesting committees correspond to this description. Legislatures, for example, rarely use simple majority rule in their deliberations. Instead, bills must pass through a maze of committees before reaching the floor, at which point parliamentary rules themselves constrain a speaker and other potential agenda setters. But Theorem 2.2 is hardly irrelevant to politics. On the contrary, because this theorem reveals an agenda setter's great power, it means that debates and strategic maneuvers to secure one set of rules rather than another may be the decisive battleground over public policy. And although Theorem 2.2 concerns a single procedural variable – the agenda, in a specific context, majority rule – it should leave us with the suspicion that people can manipulate other procedures as well.

The third theorem, 2.3, like Arrow's, does not describe what will happen, only what is possible, but what is possible is supremely important. Although Theorem 2.2 demonstrates the potential power of a chairman, Gibbard and Satterthwaite's theorem suggests that committee members are not powerless. Aside from replacing the agenda setter or adopting an alternative procedure, members can also misrepresent their preferences. Theorem 2.3 states that no institution, including pairwise voting with an agenda, is immune from the possibility that people will prefer to misrepresent their true preferences. Thus, Theorem 2.2 implies that we cannot rule out as a possibility strategic action when we look at a committee chairman in a majority rule institution. But Theorem 2.3 implies that we can never rule out as a possibility strategic action by ordinary committee members in *any* nondictatorial institution.

We emphasize that these three theorems do not offer predictions about what will happen in politics, but only about what is possible. But what we have learned is that manipulation and strategy are probably pervasive in politics. First, Farquharson's example is a warning that even with Condorcet winners and transitive social preferences, procedures can be selected judiciously to yield other outcomes. Secondly, even after a specific institution or procedure is selected, we cannot simply plug individual preferences into such procedures and anticipate that an outcome will emerge as a trivial aggregation of those preferences. Theorem 2.2, although limited to majority rule agendas, suggests that control of the detailed application of a procedure can manipulate outcomes further. And Theorem 2.3 suggests that politics and the way in which institutions transform individual preferences into a social choice are more complicated still: We cannot rule out the possibility that everyone in society has, from time to time, the incentive to reveal their preferences strategically. Thus, we must find a way to understand this strategic complexity.

As in Chapter 1, we now present a series of concepts with which the beginning reader should be familiar.

binary procedure	plurality rule	social preference order
majority rule	Condorcet paradox	cycle
axiomatic method	social choice function	social preference
resolute	collective rationality	function
Pareto principle	citizen sovereignty	unrestricted domain
independence	decisive	nonperversity
acyclic	agenda	nondictatorship
Borda count	finite	Pareto-dominated
contract curve	Pareto-optimal	top-cycle set
sincere voting	manipulable	Condorcet winner
separable preferences	strategic misrepresentation	vote trading

CHAPTER 3

Basic theory of noncooperative games

non-binding agreements

(no communication)?

no assump. of collective rationality.

By strategically revealing their preferences, people can manipulate the outcomes of institutions, which may range from formal voting mechanisms to informal social processes. But the consequences of one's actions depend on what others do, and, thus, the preferences people reveal should depend on what preferences they think others will reveal. We now know that this interdependence is fundamental to politics, and one way to model it is to interpret the states of nature that a decision maker confronts so that those states include estimates of other people's probable choices. But if a person conditions his own choices on what he believes others will do, then he should assume that others do the same. Thus, we cannot assume generally that states of nature link alternatives to outcomes in some mechanical way. A science of politics thereby requires a theory about how people condition their decisions on the decisions of others when they believe that others are doing the same. To complicate matters further, people's motives may not be identical. Hence, whether we are talking about voting, candidates in election campaigns, coalitions, or bureaucratic choice, people's fates are both interdependent and conflictual, and any assumption about a benign environment of other persons, who assume the role of nature, is untenable. The hallmark of the reasoning we are about to expound takes both interdependence and conflict into account and assumes that people choose to maximize utility, recognizing their joint dependence. Hence, they take the calculations and dissimilar goals of others into account, and expect everyone else to do the same.

   This chapter establishes the foundations of that reasoning, the theory of noncooperative games, which, in its traditional interpretation, assumes that although fates are interdependent, people cannot coordinate their choices. Section 3.1 outlines the essential character of a noncooperative game and then specifies more formally the basic structure of games. Succeeding sections define two representations of noncooperative situations, the extensive (game tree) and the strategic (normal) form. Later sections concern the strategies (individual choices) that people might choose because those strategies possess certain equilibrium properties. We consider the nature of equilibria, some of the conditions under which equilibria exist, and the sense in which we can say that equilibria "solve" games, by which we mean "predict choices and outcomes."

Much of this and succeeding chapters, then, concerns the concept of an equilibrium and refinements of its definition. This concept is central to our theory. An equilibrium is a statement, given the interdependence of fates, about the actions that people choose and the outcomes that prevail. The properties of equilibria vary, depending on the situation being analyzed and the assumptions that we use to model the situation. Sometimes the equilibrium concept makes imprecise predictions and at other times it narrows predictions to a unique outcome. And in certain situations no equilibrium may exist at all. But even nonexistence is a clue to what actions and outcomes we can anticipate. Finally, the concept of an equilibrium can take a definition such as a Condorcet winner, which by itself simply defines properties that a particular outcome may or may not satisfy, and convert it into a prediction about events, or it can tell us that such winners will not prevail in certain institutions. An equilibrium, then, corresponds to the empirical regularities that our models predict. Concepts of equilibrium are the link between our abstract models and the empirical world that we are trying to understand.

3.1   Noncooperative games: an example

We begin with the concept of a "noncooperative game," which a simple example illustrates: Suppose that a legislature must vote on whether to raise its members' salaries. Things being what they are, each legislator prefers any outcome that raises his salary to one that does not. Also, with an eye to the next election, each member would rather vote *against* rather than *for* the raise, subject to the constraint, of course, that the salary raise meets approval. These utility numbers are consistent with such preferences:

$$u_i(o_1 = \text{raise passes}, i \text{ votes } against) = 2$$
$$u_i(o_2 = \text{raise passes}, i \text{ votes } for) = 1$$
$$u_i(o_3 = \text{raise fails}, i \text{ votes } against) = 0$$
$$u_i(o_4 = \text{raise fails}, i \text{ votes } for) = -1$$

We can model this situation as a *game*, since the legislators' fates are interdependent. Assuming majority rule, the final outcome and each legislator's eventual welfare depend not only on how he votes but also on how his colleagues vote. Second, the optimal choice for a legislator depends on other legislators' choices. If a legislator believes that his vote is decisive, he should vote for the salary increase, but if he believes that the outcome is a foregone conclusion either for passage or for defeat, he should vote against the raise (since $u_i(o_1) > u_i(o_2)$, and since $u_i(o_3) > u_i(o_4)$).

The structure of this game, however, does not tell us whether it illustrates the meaning of noncooperative, so we must rely on what we know about other aspects of the situation. Can legislators discuss their votes beforehand and reach agreements to coordinate ballots? Might legislators agree that only members from safe districts will be required to vote for the raise? Has the minority party publicly announced its opposition to the raise, thereby forcing the members of the majority party to support or oppose the measure? Do mechanisms exist for enforcing collusive agreements, which may be especially important with secret ballots?

Answering such questions may be difficult. Thus, any choice between a noncooperative and cooperative formulation of a situation may be debatable. Indeed, much recent theorizing about coalition formation proposes adapting noncooperative concepts to cooperative situations. Learning to walk before we run, however, we define a *noncooperative game* as a situation in which people must choose without the opportunity for explicitly coordinating their actions. Stated differently, if no two persons can establish a contractual arrangement whereby their decisions are unambiguously linked, then a game is noncooperative. For example, if the answer to all of our previous questions about a legislature is no, the corresponding game is noncooperative. (People might want to cooperate in a noncooperative game, and in many such games strong incentives for cooperation exist; but if the game is noncooperative, they cannot do so.)

3.2 Games in extensive form: the game tree

Game theory uses three kinds of descriptions to model situations: (1) the extensive, or game-tree form; (2) the normal, or strategic form; and (3) the characteristic-function form. The last form, (3), concerns cooperative games; hence, we ignore this descriptive mode for the present. Here, we are concerned only with the extensive and normal forms, and we begin with the extensive form.

The *extensive form* is an elementary description of the situation that we are modeling, and it contains this information: Whose choice (move) is it at any particular point in time? What alternative actions are available to each person at any particular move? What does each player know about other persons' prior choices? What are the alternative states

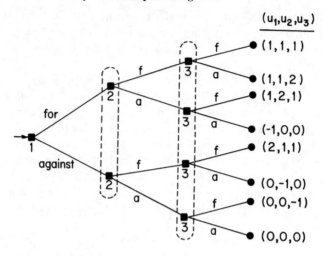

3.1 Game tree for secret ballot voting on pay raise.

of nature and their likelihood? And what are each person's preferences (utilities) over outcomes? To illustrate, consider the example of the legislature voting on a salary increase, and, to simplify discussion, suppose that the situation concerns only three legislators, 1, 2, and 3 and that voting is by secret ballot. We also ignore the possibility that collusion may have occurred beforehand and we model the situation with the assumption that each legislator must choose between voting for or against the raise, with no opportunity to discuss the vote first with anyone.

Figure 3.1 depicts this situation with what we commonly refer to as the *game tree* representation of the extensive form. An arrow indicates the start of the game. It points to a _node,_ which stands for the choice facing legislator 1. Thus, 1 moves first, choosing to vote for or against, without knowing the acts that his colleagues select later. Each of 1's alternatives connects to a second node indicating 2's move. The dotted envelope in the figure enclosing legislator 2's two choice points indicates that 2 makes his decision without knowing how 1 has voted, and therefore he does not know whether he is at the bottom or top choice point. Similarly, each of 2's alternative acts from his two choice points connects to a choice point for legislator 3. Again, a dotted envelope indicates that 3 also chooses in ignorance of the other players' decisions. We call these dotted envelopes *information sets,* and they

imperfect information
incomplete information

you don't know the whole tree

you know the tree but not where you are

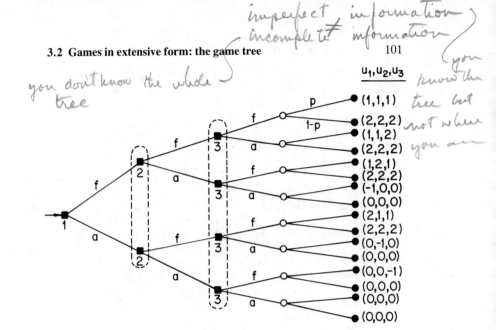

3.2 Game tree for secret ballot voting on pay raise with unknown constituent sensitivity.

play an important role in game-theoretic modeling. Not only do we use them to represent what people know or can recall about past moves, but also, as this example shows, we use them to model a situation involving *simultaneous moves*. Although a game-tree representation forces a sequential representation for individual choices, judicious selection of information sets permits us to model simultaneous choices. Hence, for the example in Figure 3.1, in which all three legislators vote at the same time or by secret ballot, it does not matter whether we assume that legislator 1, 2, or 3 goes first. Completing the description of the tree, we find that the eventual outcome depends on how the three legislators vote, so the last choice (3's) connects to a solid dot indicating the end of the game. The utilities beside the tree's end points represent the game's outcome for each player, and correspond to the particular sequence of actions leading to that point.

We could expand this example to incorporate the possibility that each legislator is unsure about how constituents will react to a vote for a salary increase. Constituents taken as a whole, for example, may regard the issue of legislative salaries as a critical determinant of their vote in the next election, or they may be insensitive or unaware of the issue owing, for instance, to little press coverage. If they are insensitive to the issue, they neither reward nor punish their legislators for how they vote. Figure 3.2 incorporates this possibility by assigning nature a final move in the game, in which nature "chooses" between a sensitive and an insensitive electorate with probabilities p and 1-p, respectively.

With Figures 3.1 and 3.2 in mind, we can now summarize the essential components of a game in extensive form:

> *The extensive form* of a game consists of a tree having six features describing the relations among the alternative actions and outcomes:
>
> 1. The game tree has a finite set of nodes or points, some of which ($\bullet$) are *terminal nodes*, and indicate end points or outcomes of the game; each of the rest represents a *choice node*, point of choice by a designated player ($\blacksquare$), or by nature (o).
> 2. From each choice node there is a branch assigned to each alternative (action or state of nature) available at that point.
> 3. Nature makes each choice by chance from a specified lottery over the alternative states.
> 4. The tree indicates the sequence of choices thus: Its base consists of a single choice point (indicated by an arrow), which represents the game's starting point (first move). Each of the branches from the starting point connects to the next point in the game if a player or nature chooses that alternative on the first move. If this point is an end point, the game ends; otherwise this is the next choice point in the game, and so on.
> 5. We use *information sets* in the game tree to indicate the uncertainty of the player who is to move next about the earlier choices that the other players in the game have made. If the information set contains only one choice point, then the player who makes a decision at this point knows the location of the choice in the tree. But if the information set contains more than one point, then the player knows that the game has reached one of the points in this information set, but he does not know which one. To ensure logical consistency we must place several restrictions on information sets: Each choice point in the same information set must apply to the same player and for the same move. Each point in the set, moreover, must have the identical set of alternatives, lest variations in the alternatives inform the player of his position in the information set.
> 6. The remaining conditions pertain more to the players themselves than to the rules of the game, as the extensive form summarizes them. The extensive form includes the players' utility functions defined over the tree's end points (outcomes). Also, each player knows, and expects the other players to know, all details of the situation that the game presents, each person knows that the others know that he knows the tree, and so forth. In the jargon of game theory, then, the game tree is *common knowledge*.

Although Figures 3.1 and 3.2 illustrate the components of an extensive form game, the example in Figure 3.3 illustrates the variety that this form can represent. This figure models, albeit trivially, the choices of three presidential candidates, in which candidates 1 and 2 must first compete for their party's nomination and 3 is an incumbent certain to secure nomination. Candidates 1 and 2 must first choose between a_1, a campaign tactic for the primaries that appeals to the party faithful but not to the general electorate, and a_2, a tactic that appeals to the general electorate but not to the party faithful. With respect to the general election, the tactic b_1 is directed at party identifiers, whereas the tactic b_2 casts a broad appeal to all potential voters. This game tree reveals that if 1 and 2 make the same initial decision, then chance decides who competes against the incumbent. Hence, a game tree can permit intermediate moves by chance, including moves that determine which players remain in the game. Second, if 1 and 2 make different initial decisions, then the candidate who appeals to the party faithful is nominated and chooses a general-election tactic. Thus, the choice of one player can affect who moves next, the alternatives that are available in that move, and even which players remain in the game. Finally, although the outcomes (the probabilities that the incumbent will be reelected) are somewhat arbitrary in this example, if the challenger and the incumbent choose the same tactic *and* if the challenger uses the same tactic as in the primaries, then the election is a toss-up. But if the challenger shifts tactics, he loses credibility with some voters and the advantage goes to the incumbent. Hence, game trees can also model the strategic advantages of certain players over others and demonstrate that the consequences of actions can depend on the paths taken through the tree.

The extensive form can represent precise details about a situation. Even simple parlor games, such as poker or tic-tac-toe, however, may be too intricate to describe in detail. One solution is to abandon details that appear unimportant or that we can otherwise temporarily suppress, and focus instead on only the more general features of the situation. Another solution is to limit attention to just a portion of the game tree. For example, one might develop a game tree for just the next four or five moves in a particular game. Of course, every extensive form necessarily must be incomplete, because the only truly complete form is one that models all of future history. In any event, the extensive form of a game provides a foundation for developing the basic principles of game theory. The analyst who understands these principles can decide if he can relax some of the theory's conditions without damaging the basic

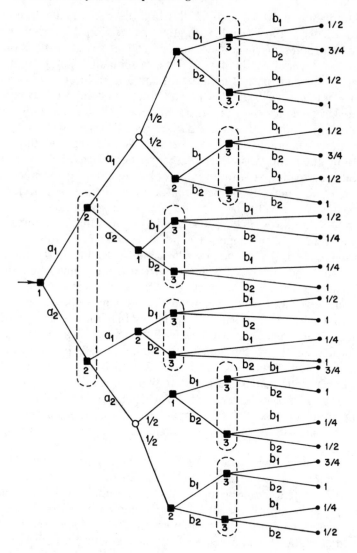

3.3 Presidential campaign game tree.

rationale. For example, our description of the extensive form explicitly requires a finite number of nodes in the game tree, which means that the number of players, moves, actions, and states of nature in the tree must all be finite. This need not be the case, however, since we make such an

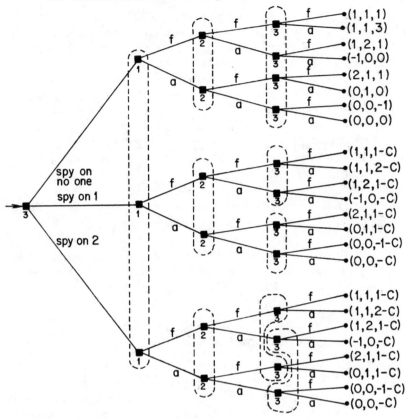

3.4 Secret ballot and spying by voter 3.

assumption here only to facilitate the presentation. Later, we devote considerable attention to games in which people enjoy infinitely many strategies, such as positions on issues, monies to be allocated in the pursuit of some activity, and the timing of a particular act.

One final comment is essential. We assume for the present that a person's information set summarizes whatever uncertainty that person has about a situation. Thus, all persons know the properties of a game's extensive form, including the utility functions of other persons (at least up to some probability distribution), their alternatives, and the characteristics of their information sets. This assumption is unrealistic in many applications, such as cases in which some legislators do not know the preferences of other legislators or of their constituents. Later we review relaxations of this assumption, but here we suppose that a situation's extensive form is common knowledge.

These remarks do not mean that we cannot study information in politics. Information sets in the extensive form are an important variable in modeling. To illustrate, suppose that in a secret ballot vote, legislator 3 alone can learn, before he casts his ballot, how 1 or 2 votes, but at a cost of C units of utility. Suppose further that 2, but not 1, can learn whether 3 has spied and on whom. The extensive form in Figure 3.4 summarizes these assumptions. Thus, 1 does not know 3's initial choice, but 2 learns, without learning 1's vote, whether he is himself in the top, middle, or lower branch. The top branch is identical to Figure 3.1, since here 3 chooses not to spy. In the middle branch, 3 learns 1's vote, and in the lower branch, 3 learns 2's vote. We can imagine more complicated possibilities, but the point is that we can model various informational asymmetries even with the assumption that the extensive form is common knowledge.

3.3   Strategy and the normal form

A situation's extensive-form representation can be an essential component of modeling. But even with suitable abstractions, it can remain unmercifully complex and defy analysis. Imagine complicating the game in Figure 3.3 by permitting candidates to choose running mates and to adjust their tactics during a campaign in order to allocate resources between different forms of advertising, in which those resources depend on the policies that they advocate and the running mate that they chose earlier. Fortunately, we can transform all finite games in extensive form into a special game tree that is especially tractable for analysis. In this kind of tree there are no chance nodes; each person, i, chooses only one action, and he makes this choice before knowing the choices of the other players.

To construct such an extensive form, we must introduce one more piece of notation. In Chapter 1 we let S denote states of nature and s_j a particular state. Now these states explicitly include the choices of others, and so we let s_i denote person i's *strategy* and S_i the set of all strategies for i. Temporarily, we can think of a strategy as an alternative drawn from the set A, as the particular action that i selects. We can represent person i's utility from an end point in the tree, then, with the

Stackelberg follower — 2d player will choose best
response.

See principal-agent.
S. leader   S. follower

Person 3 votes

		for		against	
Person 2 votes:		for	against	for	against
Person 1 votes:	for	1, 1, 1	1, 2, 1	1, 1, 2	−1, 0, 0
	against	2, 1, 1	0, 0, −1	0, −1, 0	0, 0, 0

3.5 Normal form of secret ballot game.

utility function $u_i(s_1, s_2, \ldots, s_i, \ldots, s_n)$ defined over the strategies that the n persons in the game choose.

Figure 3.1 matches this situation, and its content can be represented, without loss of generality, with the matrix in Figure 3.5. In this figure, person 3's choice between voting for and against the pay raise determines whether the outcome is in the left- or right-hand matrix, 2's choice determines whether the outcome is in the first or second column of the matrix that 3 chooses, and person 1's choice selects the row. The entry in each cell is the vector of utilities for the three legislators if they jointly (but independently) choose that outcome. In keeping with convention, the first entry in each cell denotes 1's utility, the second denotes 2's utility, and so on. Figure 3.5 is an elementary example of a game in strategic or *normal* form.

> *Noncooperative game in normal form*: Each person i selects a strategy s_i, from the set of strategies, S_i, available to i before knowing the choice that any other person makes. The utility function on the joint actions that all n persons take, $u_i(s_1, s_2, \ldots, s_i, \ldots, s_n)$, gives each player's preferences.

Any game in normal form, then, corresponds to a simple kind of game tree in which each person has just one information set containing his alternatives and the game tree has no chance nodes. Thus, the game tree in Figure 3.1 is tailor-made for a normal-form representation, since there the players choose simultaneously and must make only one choice. But what of other more complicated situations in which the players do not choose simultaneously? At this point it is useful to modify the secret ballot example slightly to show not only how the normal form can be used to represent situations in which not all choices are simultaneous, but also to illustrate how seemingly innocuous changes in a situation's extensive form can complicate that normal form. Later, we show how such complexity can be treated.

Suppose, then, that we modify our voting scenario to incorporate roll call voting, in which legislator 1 votes first, then legislator 2, and finally 3. Also, suppose that each legislator is aware of how those before him voted. In this situation, people do not choose simultaneously, and modeling the situation in normal form, which presupposes simultaneous choice, requires that we draw a more careful distinction between actions and strategies. Briefly, a strategy is a complete plan for playing a game that takes account of all possible contingencies, including the choices of other people. That is,

> A *strategy* is a rule for a person that, on any particular move, selects just one act from each information set.

In effect, once a person selects a strategy, he could leave this rule with a referee (or computer) to apply in making choices of particular acts for him at any point of the game.

For the moment we confine our attention to one kind of strategy, a pure strategy, that does not select actions probabilistically. Specifically,

> A *pure strategy* always selects the same sequence of acts for a person on any play of a game if we hold constant the choices of the other persons and the prevailing states of nature.

The concept of a pure strategy is fundamental and its meaning will become clearer as we proceed. Until we explicitly introduce other possibilities, all strategies that we discuss here are pure strategies.

Returning to the example of a legislative salary raise, note that Figure 3.6 shows the extensive form if voting is by roll call. Hence, this figure differs from 3.1 (secret ballot) only in that it suppresses all information sets to indicate that whenever a legislator chooses, he knows how others before him voted. Although the extensive forms in Figures 3.1 and 3.6 are similar, their normal forms are substantially different. With secret ballot voting a legislator cannot implement plans that are contingent on the actions of his colleagues; by assumption, he is ignorant of their actions until after he chooses. With roll call voting, though, contingency plans may be wise, especially if one votes late in the sequence. For example, given the assumptions about preference incorporated into this game, no legislator should vote *for* the salary increase if, on the basis of what he learns about those voting before him, sufficient votes exist to pass the measure. More precisely, if the legislators vote in the sequence 1, 2, 3, legislator 1 has, as before, but two strategies: a_1, vote *for*; and a_2, vote *against*. Legislator 2, however, must now choose among four strategies. Since he knows how 1 voted, he should use this information

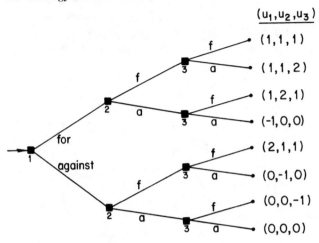

3.6 Roll call voting on a legislative pay raise.

and plan his choices accordingly. He can take into consideration how 1 voted with one of the following strategies:

b_1: If 1 votes *for*, vote *for*; but if 1 votes *against*, vote *for*.
b_2: If 1 votes *for*, vote *for*; but if 1 votes *against*, vote *against*.
b_3: If 1 votes *for*, vote *against*; but if 1 votes *against*, vote *for*.
b_4: If 1 votes *for*, vote *against*; but if 1 votes *against*, vote *against*.

Each of these strategies selects an act for each of 2's information sets, and each thereby specifies a choice for every alternative contingency that can confront 2 in this game. Each strategy in the set $S_2 = \{b_1, b_2, b_3, b_4\}$ is mutually exclusive, and S_2 is exhaustive, in that it includes every possible pure strategy for 2.

Legislator 3 must allow for contingencies that the combination of the other *two* legislators' votes determines, and hence his set of strategies is even more complicated than 2's. For example, one such strategy reads:

if 1 votes *for* and 2 votes *for*, vote *against*; but
if 1 votes *for* and 2 votes *against*, vote *against*; but
if 1 votes *against* and 2 votes *for*, vote *for*; but
if 1 votes *against* and 2 votes *against*, vote *for*.

Thus, legislator 3 faces four contingencies: the two alternative ways in which legislator 1 can vote times the two ways legislator 2 can vote. For each contingency, 3 can vote *for* or *against* the pay raise, which yields 2^4

= 16 different pure strategies. To enumerate all of these strategies let, (ff/f), say, read "if 1 votes *for* and if 2 votes *for*, then vote *for*" while (af/a) reads "if 1 votes *against* and 2 votes *for*, then vote *against*." So 3's 16 strategies are thus

$$
\begin{array}{lcccc}
c_1: & [(ff/f), & (fa/f), & (af/f), & (aa/f)] \\
c_2: & [\quad '' & '' & '' & (aa/a)] \\
c_3: & [\quad '' & '' & (af/a), & (aa/f)] \\
c_4: & [\quad '' & '' & '' & (aa/a)] \\
c_5: & [\quad '' & (fa/a), & (af/f), & (aa/f)] \\
c_6: & [\quad '' & '' & '' & (aa/a)] \\
c_7: & [\quad '' & '' & (af/a), & (aa/f)] \\
c_8: & [\quad '' & '' & '' & (aa/a)] \\
c_9: & [(ff/a), & (fa/f), & (af/f), & (aa/f)] \\
c_{10}: & [\quad '' & '' & '' & (aa/a)] \\
c_{11}: & [\quad '' & '' & (af/a), & (aa/f)] \\
c_{12}: & [\quad '' & '' & '' & (aa/a)] \\
c_{13}: & [\quad '' & (fa/a), & (af/f) & (aa/f)] \\
c_{14}: & [\quad '' & '' & '' & (aa/a)] \\
c_{15}: & [\quad '' & '' & (af/a), & (aa/f)] \\
c_{16}: & [\quad '' & '' & '' & (aa/a)]
\end{array}
$$

Figure 3.7 presents the normal form of the roll call voting game. This matrix contains one row for each of legislator 3's 16 pure strategies, and one column for each combination of 1 and 2's pure strategies. We emphasize that specifying a pure strategy for each legislator implies a unique sequence through the game tree in Figure 3.6, a unique outcome, and hence a unique vector of utilities. Consequently, each cell in the normal-form matrix in Figure 3.7 corresponds to an end point and utility vector of the corresponding game tree. For example, suppose the legislators choose the strategy 3-tuple (a_1, b_2, c_6). The strategy a_1 specifies that legislator 1, who must vote first, votes *for*. The strategy b_2 specifies that legislator 2 votes *for* if 1 votes *for* and *against* if one vote *against*. Since, in a roll-call, 2 knows how 1 has voted, 2 votes *for*. Finally, the strategy c_6 specifies that if legislators 1 and 2 both vote *for*, then legislator 3 votes *for* as well. Hence, the strategy 3-tuple (a_1, b_2, c_6) leads to the passage of the pay raise with all three legislators voting for its approval. Since each legislator values the passage of the raise when he votes for it at 1 unit of utility, we enter the utility 3-tuple $(1, 1, 1)$ in the corresponding cell of Figure 3.7.

Although the choices of *acts* in the extensive-form game of Figure 3.6 are not simultaneous, the choices of *strategies* in Figure 3.7 are. Hence, the derivation of Figure 3.7 converts a dynamic sequential situation into

	Player 1							
	a_1				a_2			
	Player 2				Player 2			
Player 3	b_1	b_2	b_3	b_4	b_1	b_2	b_3	b_4
c_1	1, 1, 1	1, 1, 1	1, 2, 1	1, 2, 1	1, 1, 2	−1, 0, 0	1, 1, 2	−1, 0, 0
c_2	1, 1, 1,	1, 1, 1	1, 2, 1	1, 2, 1	1, 1, 2	0, 0, 0	1, 1, 2	0, 0, 0
c_3	1, 1, 1	1, 1, 1	1, 2, 1	1, 2, 1	0, −1, 0	−1, 0, 0	0, −1, 0	−1, 0, 0
c_4	1, 1, 1	1, 1, 1	1, 2, 1	1, 2, 1	0, −1, 0	0, 0, 0	0, −1, 0	0, 0, 0
c_5	1, 1, 1	1, 1, 1	0, 0, −1	0, 0, −1	1, 1, 2	−1, 0, 0	1, 1, 2	−1, 0, 0
c_6	1, 1, 1	1, 1, 1	0, 0, −1	0, 0, −1	1, 1, 2	0, 0, 0	1, 1, 2	0, 0, 0
c_7	1, 1, 1	1, 1, 1	0, 0, −1	0, 0, −1	0, −1, 0	−1, 0, 0	0, −1, 0	−1, 0, 0
c_8	1, 1, 1	1, 1, 1	0, 0, −1	0, 0, −1	0, −1, 0	0, 0, 0	0, −1, 0	0, 0, 0
c_9	2, 1, 1	2, 1, 1	1, 2, 1	1, 2, 1	1, 1, 2	−1, 0, 0	1, 1, 2	−1, 0, 0
c_{10}	2, 1, 1	2, 1, 1	1, 2, 1	1, 2, 1	1, 1, 2	0, 0, 0	1, 1, 2	0, 0, 0
c_{11}	2, 1, 1	2, 1, 1	1, 2, 1	1, 2, 1	0, −1, 0	−1, 0, 0	0, −1, 0	−1, 0, 0
c_{12}	2, 1, 1	2, 1, 1	1, 2, 1	1, 2, 1	0, −1, 0	0, 0, 0	0, −1, 0	0, 0, 0
c_{13}	2, 1, 1	2, 1, 1	0, 0, −1	0, 0, −1	1, 1, 2	−1, 0, 0	1, 1, 2	−1, 0, 0
c_{14}	2, 1, 1	2, 1, 1	0, 0, −1	0, 0, −1	1, 1, 2	0, 0, 0	1, 1, 2	0, 0, 0
c_{15}	2, 1, 1	2, 1, 1	0, 0, −1	0, 0, −1	0, −1, 0	−1, 0, 0	0, −1, 0	−1, 0, 0
c_{16}	2, 1, 1	2, 1, 1	0, 0, −1	0, 0, −1	0, −1, 0	0, 0, 0	0, −1, 0	0, 0, 0

3.7 Roll call voting for salary increase in normal form (entries list payoff to legislators 3, 2, and 1, respectively).

a static model, but one that contains certain analytical advantages. The normal form, however, inherits one property from the extensive form: It presumes that all persons know the full properties found in it, including everyone else's payoffs and strategies. Thus, as with the extensive form, all information found in the normal-form game is common knowledge.

This assumption of common knowledge is critical to a correct interpretation of the normal form. Some readers, when first encountering game theory, find it troubling that the theory requires people to choose strategies simultaneously, because some decision-making flexibility appears to be lost. But a normal-form representation does not suppose that people cannot adjust their *actions* as the game proceeds. Since the players have common knowledge about the game's structure (payoffs, moves, information sets, and so forth), they can learn nothing new except what others have chosen in the past. Thus, people can condition their decisions only on these choices and not on anything else. Since the strategies in the normal form take account of all possible contingent actions, they place no restriction on how people might adjust their actions.

This is not to say that the preceding development of the normal form is wholly general. If people are uncertain about the preferences of others or about what others know about preferences – if preferences are not common knowledge – then perhaps they can learn something about these preferences or about what others know from the choices they observe. In this event, the normal form may be an inappropriate vehicle for representing a situation. In fact, game theorists are now beginning to research such areas in a concerted way, and the subsequent chapters containing specific examples explore some research possibilities (see Chapters 4, 5, 6, and 10). Here, though, we assume that people are fully informed about a game's structure, that they know that others are also equally informed, that they know that others know that they are informed, and so on.

3.4   Normal form of infinite games

Although the preceding discussion emphasizes finite games with finite trees, we can use the normal form to represent games in which some or all players enjoy an infinite number of strategies. For example, consider two presidential candidates who must each decide how to allocate their time campaigning in the 50 states. Let t_{ij} denote the amount of time that candidate i spends in state j, such that

$$\sum_{j=1}^{50} t_{ij} = T_o \text{ for } i = 1, 2,$$

where T_o is each candidate's time-budget constraint. Hence, each candidate's strategy, s_i is a vector $s_i = (t_{i1}, t_{i2}, \ldots, t_{i50})$ that specifies how much of T_o candidate i allocates to each state, and each s_i is an element of

$$S_i = \{s_i \mid \sum_{j=1}^{50} t_{ij} = T_o, T_o \geqslant t_{ij} \geqslant 0 \text{ for all } j\}. \tag{3.1}$$

S_i, then, is the set of all strategies that satisfies the time-budget constraint.

Completing the example, let the probability that candidate i carries state j, p_{ij}, equal the proportion of time that candidate i spends in state j relative to his opponent's allocation of time in that state. Thus,

$$P_{ij} = \frac{t_{ij}}{t_{1j} + t_{2j}}.$$

Finally, let e_j be the electoral college weight of state j. Hence, if we suppose that candidates maximize their expected electoral vote, then we can represent candidate i's payoff as a function of the allocation strategies that he and his opponent choose thus:

$$u_i(s_1, s_2) = \sum_{j=1}^{50} \frac{e_j t_{ij}}{t_{1j} + t_{2j}}, \tag{3.2}$$

where s_1 is in S_1 and s_2 is in S_2. Expressions (3.1) and (3.2), which define strategies and payoffs, describe a simple two-person resource-allocation game between the two candidates.

For another example, suppose that the election concerns the candidates' positions on the issue of how much weight, w, each ought to give to defense spending compared with social-welfare spending, as well as a choice of an overall tax rate, r. Denoting the candidates 1 and 2, let $s_i = (r_i, w_i)$, $i = 1, 2$, be their strategies. Their respective strategy spaces, presumably, are infinite and correspond to the budget simplex that Figure 1.14 describes. With this specification of the strategy spaces, the game's description becomes complete with a specification of payoff functions. The simplest possibility is to assume that

$$u_i(s_1, s_2) = \begin{cases} 1 & \text{if candidate } i \text{ wins} \\ 0 & \text{if the election is a tie} \\ -1 & \text{if candidate } i \text{ loses.} \end{cases}$$

What is missing here, however, is the link between these numbers and the candidates' strategies. This link requires a model of how citizens vote. Again, the simplest possibility is to suppose that voters have preferences like the ones in Figure 1.14 and, in particular, that simple Euclidean distance describes voters' indifference contours, as defined in expression (1.2). In this case, voter j prefers candidate 1 over 2 if the distance between j's ideal point and s_1 is less than the distance between his ideal point and s_2. If $u_j(s_i)$ is the utility that voter j associates with candidate i's position on the two issues, the election game's payoff function is

$$u_1(\mathbf{s}_1, \mathbf{s}_2) = \begin{Bmatrix} 1 \\ 0 \\ -1 \end{Bmatrix} \text{ if num } \{u_j(\mathbf{s}_1) > u_j(\mathbf{s}_2)\} \begin{Bmatrix} > \\ = \\ < \end{Bmatrix} \text{ num } \{u_j(\mathbf{s}_2) > u_j(\mathbf{s}_1)\},$$

(3.3)

where "num" means "the number of." Thus, candidate 1's payoff is $+1$ if the number of voters who prefer his strategy over his opponent's is greater than the number who prefer his opponent's strategy. We can formulate candidate 2's payoff identically.

This completes the description of a rather simple game-theoretic model of a two-candidate election. If we limit strategies to a finite number of possibilities, of course, the game is finite and we can represent it in tabular form. But if the candidates can adopt any position on the simplex, then a tabular form is cumbersome, and a formal description of the strategy spaces along with the preceding specification of a payoff function in expression (3.3) describes the game. Thus, regardless of whether a game is finite or infinite, its normal-form representation consists of the following: N, a set of players; S_i, a set of strategies for each player $i \in N$; and a function $u_i(s_1, s_2, \ldots, s_n)$ that specifies, for each player $i \in N$, the payoff or utility that the player associates with the outcome that prevails when the strategy n-tuple $(s_1, s_2, \ldots, s_n)$ is chosen, where s_i is a particular strategy in S_i. (Individual strategies can be both a specific variable, such as when they denote the time some act will be chosen, or they can be a vector. As with outcomes, if there is no reason to require that a strategy be a vector, we use the notation s_i to denote a strategy for player i; but if a strategy is necessarily a vector, we use the boldface notation $\mathbf{s}_i$.)

3.5   Best-response strategies and domination

A description or abstract representation is of little value unless we can use it to predict what strategies people choose. The particular difficulty that we must now confront is that a person's best (utility-maximizing) choice commonly depends on what others do, and that what is best for others depends on what he does. The perspective of the normal form highlights the nature of this complex decision problem. Recall that in such games all persons choose strategies simultaneously. Thus, no one has the opportunity to see first the strategies that others choose. They may observe specific choices of alternatives as the extensive form of a game unfolds, but their choice of a strategy has already determined their conditional responses. And as we note earlier, this decision-making problem is not the fault of our representation, since the normal form summarizes all information that the players possess and, thus, all the information that they can use to condition their decisions. How, then, do people choose in an environment of mutual dependency?

This is the central problem of game theory. But before we go too far, the first possibility that we should explore is that one strategy is unambiguously better than all others and thereby implies an unambiguous best choice. To formalize the notion of unambiguously better we first represent a particular set of choices of pure strategies by the set of all persons, except i, by the vector of n-1 strategies,

$$\mathbf{s}_{-i} = (s_1, \ldots, s_{i-1}, s_{i+1}, \ldots, s_n).$$

Thus, $\mathbf{s}_{-i}$ identifies a choice of strategy for all persons except i. Using this notation, we can now express the choice of strategies by *all* persons, which we wrote previously as $\mathbf{s} = \{s_1, \ldots, s_i, \ldots, s_n\}$, as $\mathbf{s} = (s_i, \mathbf{s}_{-i})$. Then,

> *Best-response pure strategy*: Person i's strategy s_i^* is a best response against a specific choice of strategies by the remaining players in the game, $\mathbf{s}_{-i}$, if and only if
>
> $$u_i(s_i^*, \mathbf{s}_{-i}) \geq u_i(s_i, \mathbf{s}_{-i}),$$
>
> for every s_i in S_i.

It is reasonable to suppose that people choose strategies that are best responses to the strategies of others. That each person must choose before others reveal their choices, however, serves to complicate this problem. Nevertheless, it may be possible to eliminate certain strategies from consideration, and, ideally, to find a single strategy that is a best response against every possible choice by the other players. To articulate these ideas, we define some simple terms relating to the notion of domination of strategies. Letting S_{-i} denote all possible strategies that all persons except i can choose, and letting s_i^* be a particular strategy for person i and s_i be any other strategy for i, then,

> *Domination*: s_i^* dominates s_i if and only if
>
> $$u_i(s_i^*, \mathbf{s}_{-i}) \geq u_i(s_i, \mathbf{s}_{-i}),$$
>
> for every $\mathbf{s}_{-i}$, and where strict inequality holds for at least one strategy n-1 tuple in S_{-i}. Further, s_i^* *strictly dominates* s_i if strict inequality holds for every strategy in S_{-i}. Finally, s_i^* is *undominated* if no strategy strictly or weakly dominates it.

	b_1	b_2
a_1	3, 2	0, 4
a_2	5, 1	1, 1

A

	b_1	b_2
a_1	3, 3	2, 0
a_2	0, 1	2, 5

B

	b_1	b_2
a_1	1, 0	0, 1
a_2	0, 2	2, 0

C

3.8 Three games in normal form.

To avoid confusion between best-response and dominant strategies, notice that the strategy s_i^* is a best response against $\mathbf{s}_{-i}$ if s_i^* maximizes i's utility when everyone else chooses $\mathbf{s}_{-i}$. Suppose, on the other hand, that we pretend that person i has only two strategies: s_i and s_i^*. Then the strategy s_i^* dominates s_i if, regardless of what the other players in the game choose, s_i^* is always at least as good and sometimes better for i than is $s_i - s_i^*$ is at least as good a response as s_i under *all* contingencies, and is sometimes a better response.

The three examples in Figure 3.8 illustrate the definition of domination. (The notation developed for $\mathbf{s}$ is convenient for general definitions, especially if the number of players is unknown. In most examples, however, it is more convenient to use a's for person 1's strategies, b's for 2's strategies, and so on.) In Game A, the strategy a_2 strictly dominates a_1 since $5 > 3$ and $1 > 0$. That is, regardless of what 2 chooses, person 1 is strictly better off with his second strategy. For person 2, b_2 dominates (but not strictly) b_1 since $4 > 2$ and $1 = 1$. In Game B, a_1 dominates a_2 since $3 > 0$ and $2 = 2$, but neither of player 2's strategies dominates the other – the strategy that 2 should choose depends on what 1 chooses. In Game C, no strategy dominates another for either player. In all these games we limit ourselves to pure strategies, single plays, and no communication between the players.

The concept of domination can be especially useful for analyzing a game in normal form, because it is reasonable to suppose both that a person will not choose a dominated strategy, and that all other persons in the game know this. (Recall our earlier assumption, that all players are aware of the game's extensive form, and that they each know that every other player has this knowledge.) Hence, a player gains nothing by choosing a dominated strategy, but he might lose something.

To appreciate the application of domination, consider again the game in Figure 3.7. Notice that the strategy c_{10} dominates (not strictly) all other strategies available to legislator 3. Hence, we can assume that this is the only strategy that legislator 3 will consider. Knowing this, legislator 2 can safely eliminate all strategies but b_3: If 1 chooses a_1, only

2's fourth choice gives him as much as he gets from choosing b_3, whereas if 1 chooses a_2, only 2's first choice gives him as much as b_3. And as long as 3 chooses c_{10}, no strategy gives 2 more than b_3. Knowing that 3 will choose c_{10} and that 2 will choose b_3, legislator 1, then, has a simple choice: a_1 yields him a payoff of 1, and a_2 yields him 2, so he chooses his second strategy and votes *against*. The concept of domination, then, reduces a $16 \times 4 \times 2 = 128$ cell game to the unique prediction that the players will choose (c_{10}, b_3, a_2).

Looking now at this strategy 3-tuple to see what outcome it implies, note that b_3 requires legislator 2 to vote the opposite of 1. Since 1 votes *against*, 2 votes *for* the raise. Finally, c_{10} requires 3 to vote *for* if 1 and 2 vote *against* and *for*, respectively. In the final outcome, then, the pay raise passes, with only legislator 1 voting *against*, so that legislator 1 receives a payoff of 2, and legislators 2 and 3 each receive a payoff of 1. In this instance of roll call voting, it pays to vote first, to vote *against* the raise, and to force the remaining legislators to vote *for* the raise. One prediction, then, is that if the Speaker or majority leader has a name that begins with the letter A or B, and if voting is by alphabetical order, then he will prefer a roll call vote; otherwise, he would probably choose a secret ballot (or a bipartisan commission on salaries).

This example illustrates the potential power of the concept of domination and how we can use it to simplify a game in normal form. But we cannot simplify every game in this manner, as the games in Figures 3.5 and 3.8c illustrate. Nevertheless, when analyzing a game, we should first examine the situation for dominated strategies and the opportunity to simplify any subsequent analysis.

3.6   Pure-strategy equilibria

Suppose that we cannot reduce a game to an unambiguous choice for each person by successive applications of the concept of domination. One possibility, of course, is that a person can treat others as nature and, as in a simple risky-choice decision problem, attach probabilities to their alternative choices. But this approach would discard the potentially useful fact that the other persons in the game are not nature, that they, too, can think, and that they confront the same choice dilemma as he does. For example, consider the two-person game in Figure 3.9, and suppose initially that person 1 thinks that 2 will choose b_1, since that gives 2 the greatest potential payoff. Believing this, 1 should choose a_1. Knowing, however, that 2 is also cognizant of the properties of this game, 1 might reason thus: "2 thinks that I will choose a_1, since that is my best response to the strategy b_1. But if 2 thinks that I will choose a_1,

Person 2:

		b_1	b_2	b_3
	a_1	8, 4	4, 0	2, 7
Person 1:	a_2	6, 5	5, 6	5, 3
	a_3	3, 9	4, 0	6, 1

3.9 A pure-strategy equilibrium at (a_2, b_2).

he should choose b_3 instead. Thus, I should not choose a_1, but a_3, which is a better response to b_3 than is a_1. Of course, after 2 thinks this through and concludes that I will choose a_3, he will respond with b_1, in which case I should reaffirm my original decision to choose a_1. But then, if 2 thinks further he will respond with b_3. . . ." In short, person 1 (and 2) faces an infinite *he-thinks-that-I-think regress*.

To resolve such infinite regresses, suppose there is a particular vector of strategies for the players of an n-person noncooperative game, $\mathbf{s}^* = (s_1^*, s_2^*, \ldots, s_n^*)$, such that no player has an incentive to change his strategy in this vector if he believes that everyone else will choose his corresponding strategy in the vector. Put differently, suppose that s_i^* is a best-response strategy to $\mathbf{s}_{-i}^*$ for all i. We call such a vector an equilibrium point of the noncooperative game. More formally,

Pure-strategy (Nash) equilibrium: The vector of pure strategies $\mathbf{s}^* = (s_1^*, \ldots, s_i^*, \ldots, s_n^*)$ is an equilibrium n-tuple if and only if, for *every* player i in the game,

$$u_i(s_i^*, \mathbf{s}_{-i}^*) \geq u_i(s_i, \mathbf{s}_{-i}^*),$$

for every $s_i \in S_i$.

Thus, if $\mathbf{s}^*$ is an equilibrium, and if each person i expects the others to adopt $\mathbf{s}_{-i}^*$, then i can do no better by choosing some strategy other than s_i^*, and he may do worse.

The concept of a Nash equilibrium n-tuple is perhaps *the* most important idea in noncooperative game theory. Although later we consider various modifications and refinements of this definition, it remains the most important instrument for converting descriptions of situations into predictions about people's choices. Whether we are analyzing candidates' election strategies, the causes of war, agenda manipulation in legislatures, or the actions of interest groups, predictions about events reduce to a search for and description of equilibria.

Put simply, equilibrium strategies are the things that we predict about people.

We can appreciate the value of this concept by referring again to Figure 3.9. If person 1 expects 2 to choose b_2, then his best response is a_2. Moreover, if 1 believes that 2 thinks he will choose a_2, then 2 will choose b_2. The strategy pair (a_2, b_2) is an equilibrium pair and terminates the infinite he-thinks-that-I-think-that-he-thinks regress. Before we conclude that this concept "solves" the theoretical issue of predicting choice, however, we must consider these questions about equilibria: (1) Do all games have equilibria? (2) Are equilibria unique? (3) What reason do we have to suppose that people will arrive at an equilibrium? Much of the theory of noncooperative games concerns these questions, and a considerable part of the application of that theory to politics focuses on them as well.

The answers to these questions are not trivial. For instance, consider Figure 3.9 again, and notice that if person 1's payoff from the strategy pair (a_2, b_2) is lowered from 5 to 3, then (a_2, b_2) is no longer an equilibrium. Person 1 would unilaterally shift from a_2 to either of his other two strategies. Indeed, no pair of strategies is an equilibrium. Thus, to answer the first question, games need not have equilibria, at least not in pure strategies. This leaves us the dual task of identifying those games that have such points and of formulating alternative solutions for those that do not.

If we look now at the game in Figure 3.5, we find four equilibria: the three permutations of votes that correspond to outcomes in which one person receives a payoff of 2 while the other two persons each receive 1, and the outcome resulting when all three legislators vote *against* the raise. Thus, the answer to our second question is, "Not necessarily: A game may have zero, one, or many equilibria." This example makes an answer to our third question problematical. Although the he-thinks regress terminates for legislator 1 if he believes that his colleagues will split their vote (he then votes *for*), if all three share this belief, then they all vote *for*, but this outcome is not an equilibrium.

In spite of the disquieting answers to these three questions, they do not negate the usefulness of the equilibrium concept. Instead, they yield new questions: Under what conditions will a game necessarily have an equilibrium? If equilibria are not unique, can we eliminate certain equilibria as implausible choices? And under what conditions are nonunique equilibria equivalent from the perspective of a game's players? The answers to these questions help us develop answers to our earlier question about the rationale for supposing that people choose equilibrium strategies.

3.7   Some conditions for existence of pure-strategy equilibria

This section discusses some basic theorems in the theory of noncooperative games concerning the existence of pure-strategy equilibria. Each
theorem guarantees that any game satisfying certain conditions has at
least one pure-strategy equilibrium n-tuple of strategies. These conditions, however, are *sufficient* but not *necessary*. Hence, although many
games that the theorems do not cover have equilibria, we can use these
results and extensions of the concept of strategy to develop equilibria
for other games.

Games of perfect information

In some games, such as chess, tic-tac-toe, and roll call voting, the rules
permit each person to know, whenever it is his turn to act, precisely
what choices all others as well as he made on earlier moves. Thus, at
every decision point a person knows exactly where he is in the game's
extensive form. Formally.

A *game of perfect information* is a game in which every
information set in its extensive form contains just one choice
point.

In a game of perfect information each player makes each choice with the
full knowledge of the exact path in the game tree that earlier choices in
the play follow. The example in Figure 3.1 of voting with a secret ballot
does not meet this definition, since legislators 2 and 3's information sets
contain more than two choice points. Roll call voting, however, satisfies
the perfect-information condition. The following result summarizes the
importance of this class of games:

3.1.   *Every finite noncooperative game of perfect information has at
least one pure-strategy equilibrium n-tuple.*

Thus, chess (with a stopping rule to prevent stalemates) and tic-tac-toe
have pure-strategy equilibria. This feature of tic-tac-toe, as well as its
simplicity, makes it possible to program a computer that never loses.
The extensive form of chess, however, is so complex that at present no
computer can store and manipulate this form. Although we know that a
best strategy exists, we cannot find it; this state of affairs, of course,
maintains interest in this game.

Roll call voting has a pure-strategy equilibrium, as Theorem 3.1
requires. If we reexamine the game in Figure 3.6, moreover, we can
gain a sense of the theorem's proof. Specifically, if we work backward
up the extensive-form tree of this game (called the *backward reduction*

method), we can deduce directly that legislator 1 should vote *against*, forcing his two colleagues to vote *for*. Beginning with legislator 3, observe, first, that since he knows how 1 and 2 voted, his decision task is simple: Vote *against* if legislators 1 and 2 both vote *against* (voting *against* yields 3 a payoff of 0, while voting *for* yields -1) or if legislators 1 and 2 both vote *for* (to avoid a payoff of 1 rather than 2); otherwise vote *for* (to secure a payoff of 1 rather than 0). Thus, under the appropriate contingencies, we can eliminate inappropriate choices for 3. Legislator 2, by the assumption that all persons are fully cognizant of a game's extensive form and everyone knows this, should be able to infer that 3 will make these choices. That is, legislator 2 knows that 3 is a fully rational and informed fellow, in which case 2 also knows 3's strategy. Hence, 2 can reduce the extensive form game in Figure 3.6 to the form in Figure 3.10. But then a similar argument applies for legislator 1: If 1 votes *for*, then 2's voting *for* is certain to induce 3 to vote *against*, while an *against* vote by 2 at this node induces 3 to vote *for*. Clearly, legislator 2 prefers to vote *against* since he prefers that 3 rather than himself incur the ire of constituents. Hence, in the event that 1 votes *for*, 2 will vote *against*. But if 1 votes *against*, then 2 is certain to vote *for*, and 3 will therefore vote *for*, since 3 prefers the pay raise that he voted for to no pay raise at all. Thus, we can reduce the extensive form further to that portrayed in Figure 3.11. This form, as we see, presents legislator 1 with a trivial choice: He prefers *against* to *for*. So by working backward up the game's extensive form, we can confirm the intuitively reasonable idea that legislator 1, who chooses first and knows that 2 and 3 will know how he voted, should vote *against* and force 2 and 3 into voting *for* the raise, given their preference for a pay raise, even if it costs them some votes in the next election.

We can work up the extensive form of any game of perfect information in a similar manner to arrive at an equilibrium, and the general representation of this process proves Theorem 3.1. (The only ambiguities that might arise with a game such as chess are that stalemates cannot go on forever and we must ensure that the game is finite.)

Some simple voting games

The method of working backward up an extensive-form game tree can be applied to other games besides those of perfect information. Consider again the scenario of three legislators whose preferences yield an intransitive social order over three alternatives under majority rule. Thus, suppose that legislative preferences over the alternatives x, y, and z are as recorded here, in which case x defeats y, y defeats z, and z defeats x:

Legislator

	1	2	3
1st	x	z	y
2d	y	x	z
3d	z	y	x

Recall that if all three legislators vote their preferences *sincerely*, then an agenda setter can dictate the outcome. For example, if x and y are voted on first, then x wins; but when paired against z, x loses, so z is the social choice. Alternatively, if y and z are paired first, then y wins, and x emerges as the eventual social choice, since x defeats y in the second vote. Complicating this analysis, however, is the assertion of Theorem 2.3, that no institution is immune, under some configuration of individual preferences, from strategic misrepresentation of those preferences. Specifically, legislator 1 can partially negate the effects of the agenda "x vs. y, winner against z," which we denote by (x, y, z), by voting for y rather than x on the first ballot. Hence, y defeats x and z, so that by this strategic ploy, 1 secures his second choice instead of his last.

The problem with this discussion is that even if the agenda is predetermined, all three legislators may choose to vote *sophisticatedly*. That is, although Theorem 2.3 tells us that legislators might have incentives to misrepresent their preferences in their voting, it does not tell us how they respond to each other and to this fact. Legislator 1's ploy might provoke 2 to vote other than sincerely, so that 1 might in turn vote sincerely, so that 3 might, and so on. Since we want to know whether there is an end to this reasoning, the concept of an equilibrium strategy should enter the discussion. Specifically, if we view the possible preferences that a legislator might reveal in a voting sequence as a strategy, an equilibrium tuple is an end to this sequence of responses. To identify the final outcome from a particular agenda, then, requires a game-theoretic analysis.

We might begin by displaying the entire game in extensive form, including even a choice of agendas by one of the three legislators. Our tree would quickly become unwieldy, however, because each legislator has two choices on the first ballot and two on the second, which, in conjunction with three agendas, yield a tree with $3 \times 2^3 \times 2^3 = 192$ terminal nodes. We can use the concept of dominance, though, to simplify the analysis and to solve for equilibria.

First, notice that on the final vote, no matter which alternatives are paired, each legislator should vote for the alternative that he most prefers, because voting on the final ballot for the preferred alternative

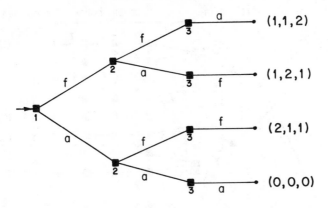

3.10 Second reduction.

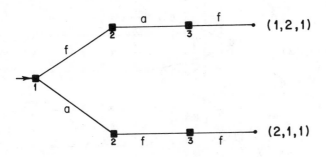

3.11 Last reduction.

can only increase the chances of that alternative passing. For example, if x is paired against z on the final ballot, then legislators 2 and 3 should vote for z, and 1 should vote for x. Similarly, if y and z are paired on the final ballot, then y prevails, but if x and y are thus paired, x prevails. Voting sincerely on the second (final) ballot is a dominant strategy.

This finding permits a considerable simplification of the extensive form. In particular, we can draw the tree in Figure 3.12, knowing, for example, that if the agenda (x, y, z) prevails, and if a majority votes for x on the first ballot, then the eventual winner is z, because x loses to z on the second ballot.

An examination of this tree reveals, for example, that if the agenda is (x, y, z), then no matter how legislators 1 and 2 vote, legislator 3, who

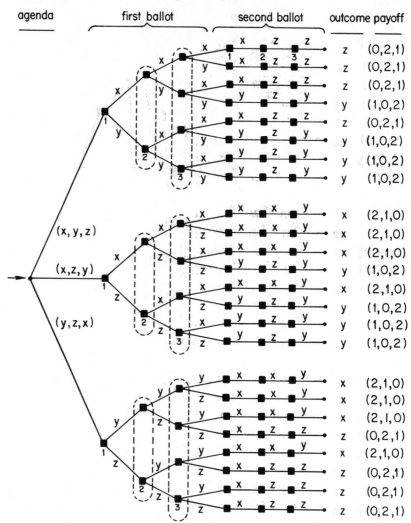

3.12 Agenda setting among three alternatives.

prefers y to z, has a dominant strategy, voting for y. Although this vote does not ensure that y will be the outcome, it is sometimes decisive in favor of y, but never for z. Thus, we can reduce this game again to that shown in Figure 3.13. Now, if the agenda is (x, y, z), 1 and 2 have dominant strategies. For 1, voting for y is always at least as good as voting for x, and sometimes it is better, since 1 prefers y to z; for 2, who prefers z to y, the dominant strategy is to vote for x. Thus, legislators 1

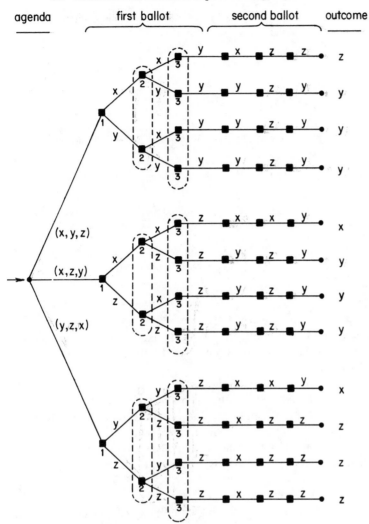

3.13 Partially reduced tree for agenda setting among three
alternatives.

and 3 vote for y, whereas 2 votes for x on the first ballot. On the second
ballot, all three legislators vote sincerely, so y beats z.

Hence, if all legislators fully assess the game tree for the agenda $(x, y,$
$z)$, legislators 2 and 3 vote sincerely, whereas legislator 1 votes contrary
to his immediate preference on the first ballot, and y prevails. A similar
analysis shows that the agenda (x, z, y), with legislator 2 voting in a

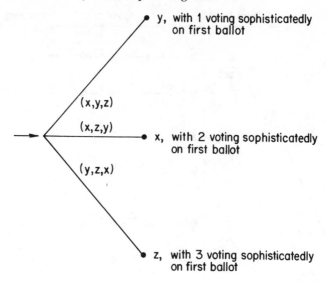

$$y, \text{ with 1 voting sophisticatedly on first ballot}$$

(x,y,z)
(x,z,y)

$$x, \text{ with 2 voting sophisticatedly on first ballot}$$

(y,z,x)

$$z, \text{ with 3 voting sophisticatedly on first ballot}$$

3.14 Fully reduced tree for agenda setting among three alternatives.

sophisticated manner on the first ballot, yields x as the eventual winner, and the agenda (y, z, x) with sophisticated voting by 3, yields z as the eventual winner.

This example illustrates that, even if the players do not enjoy perfect information, working backward up a game tree can assist our analysis of their decisions. In this example, we can reduce a tree with 192 terminal nodes to the simple tree in Figure 3.14, and we can deduce the nonobvious fact that if any legislator in the example is the agenda setter, then he should first pair his most preferred alternative against the alternative that defeats it and rely on one of his colleagues to vote in a sophisticated way, to avoid his last choice.

This discussion does not conclude the analysis of sophisticated voting. Our example is but a prelude to a more comprehensive discussion in Chapter 6. Several caveats, however, are in order. First, recall that the discussion of misrepresentation and of Theorem 2.3 in Chapter 2 concerns only logical possibilities; the discussion offers no predictions about events. Although we learned with Theorem 2.3 that misrepresentation of preferences is always possible, we cannot make predictions because we did not model specific institutions and we did not have a theory about how people respond to each other in an environment of mutual dependence. The concept of an equilibrium provides the requisite theoretical tool, and by using it in conjunction with the notion of dominated strategies, the preceding discussion yields some predictions, albeit in an especially simple circumstance. Referring again to Figure 3.14, we can now predict, given an amendment agenda, the identity of the final outcome and of the legislator who will misrepresent his preference. Hence, the concept of an equilibrium, by ensuring that we have accommodated the motives and strategic alternatives of all persons in a particular circumstance, renders a logical possibility a prediction.

The second caveat is a comment on the analysis of the normal form of the preceding example, which we have largely ignored owing to its complexity: Each legislator has eight strategies and the normal form has $8 \times 8 \times 8 = 512$ cells, so an analysis of the game in extensive form is easiest. But the existence of a complicated normal form does not always indicate that an analysis of the extensive form is easier. The normal form for the game in Figure 3.4, were we to construct it, would contain $2 \times 8 \times 10 = 160$ cells. But analyzing the extensive form is difficult, owing to the more complicated and asymmetric nature of the players' information sets. We cannot work backward up the tree in any straightforward way. To analyze this game requires constructing its normal form and identifying its equilibrium strategies, of which there are many. Sometimes, then, the normal form becomes an analytic necessity.

We should ask whether analyzing one form is equivalent to analyzing the other. We wish that a simple answer to this question were possible, but unfortunately it is not. Generally, the equilibria that we find in a normal-form game will include all equilibria that we could derive from an analysis of its extensive form. In some instances, though, the normal form contains many more equilibria. In the roll call voting example, if *everyone* votes for the same alternative on each ballot, then the corresponding strategies are an equilibrium; no one has a positive incentive to shift to some other strategy because no individual legislator can change an outcome. Obviously, some of these equilibria are unreasonable predictions about outcomes and choices since they correspond to people choosing dominated strategies and voting for their last choice. Later, this chapter gives a general method to eliminate some of these equilibria that is based on the extensive form that the normal form summarizes. Game-theoretic analysis involves an interplay between the normal and extensive forms, each of which has its advantages. As we proceed through this volume, the reader will gain an increasing sense of when one or the other form is especially appropriate.

Concave–convex games and the fundamental theorem

Up to now this section has examined only finite games with pure-strategy equilibria. But the definition of a Nash equilibrium does not suppose that strategies are finite, and thus we can apply it to any kind of noncooperative game, including those in which some or all of the S_i's are infinite. Infinite games warrant special attention for two reasons. First, if we can give strategies a geometric representation (for example, as a budget simplex), and if we can summarize payoffs by simple mathematical functions [for example, expression (3.2)], then we can often use calculus to solve a game, and we can establish general conditions under which pure-strategy equilibria necessarily exist. Second, we can establish a particular theorem that proves, with a slight reconceptualization of strategies, that all finite games have at least one equilibrium.

To establish the groundwork for this theorem, consider an n-person, noncooperative game in normal form, in which each person has an infinite and uncountable number of strategies. As before, a strategy for person i, s_i, must be an element of the set S_i, but in the kind of game that we shall consider, we suppose that we can represent s_i as a vector of m components, $s_i = (s_{i1}, s_{i2}, \ldots, s_{im})$. Thus, as in the earlier examples of infinite games, strategies might concern the time that a candidate devotes to a state, so that s_{ij} denotes the time that candidate i allocates to state j. Alternatively, s_{i1} can denote candidate i's proposed tax rate, and s_{i2} the proportion of the government's budget that he proposes to spend on defense. In general, and given some other assumptions, we can think of s_i as a point in an m-dimensional coordinate space.

Although further mathematical generalization is possible, the examples and applications that concern us in this book permit us to assume that the dimensions of s_i are the real numbers and that s_i is a subset of m-dimensional Euclidean space. Thus, although we abstractly assume that for every person i, S_i is *convex, closed, and bounded,* Chapter 1 shows that this condition is satisfied if a budget constrains actions (strategies in this instance), so that a budget simplex represents the set of all possible actions. These preliminaries permit definition of an important kind of infinite, noncooperative game:

> A *concave game in normal form* is a noncooperative game such that (1) the strategy set of each player i, S_i, is a convex, closed, and bounded subset of m-dimensional Euclidean space, and (2) for every player i and for every strategy n-tuple $s = (s_1, \ldots, s_n)$, the utility function $u_i(s)$ is both *continuous* in s and *concave* in the ith player's strategy, s_i, holding the other players' strategies, s_{-i}, fixed.

Although this definition is difficult to grasp for anyone unfamiliar with mathematics, we already have an example of such a game. Since budget simplexes correspond to convex, closed, and bounded sets, the earlier example of campaign resources that candidates allocate to states under the electoral college, in which t_{ij} represents the time that candidate i allocates to the jth state, is a game that satisfies the first part of this definition. In lieu of a formal definition, condition (2), continuity, requires essentially that $u_i(\mathbf{s})$ change smoothly as the elements of $\mathbf{s}$ vary. The requirement that u_i be concave in the ith player's strategy for all i means that if we hold the strategies of the other n-1 players fixed at any set of values, and if we graph the utility function of the remaining player, i, against any component of $\mathbf{s}_i$, then that graph looks like the risk-averse utility function in Figure 1.15. In the electoral college example, expression (3.2) is a concave function of candidate i's strategy. Holding $\mathbf{s}_2$ constant, if we graph $u_1(\mathbf{s}_1, \mathbf{s}_2)$ against t_{1j} for any j, we are, in effect, graphing the function $e_j t_{1j}/(t_{1j} + t_{2j})$ plus a constant (since the remaining terms do not depend on t_{1j}). The resulting curve, starting at $(0, 0)$, increases as t_{1j} increases, but it increases at a decreasing rate. This example illustrates a concave function. Since this argument holds for any j and for $i = 1$ and 2, the electoral college game is a concave game.

This definition is central to a fundamental theorem.

> 3.2. *Every concave game in normal form has at least one equilibrium in pure strategies.*

Before proving this theorem, we want to emphasize the implication that if the payoff functions for resource allocation games are concave and continuous, then at least one pure-strategy equilibrium exists. This does not mean, of course, that concave and continuous payoff functions make sense. We must decide whether or not they do so in the context of a particular model. But to the extent that we can argue that Nash equilibria are good predictions about choices, we now know that we can predict an outcome in resource-allocation games with such payoff functions.

As we shall see shortly, however, Theorem 3.2 has broader applications, and it is central to the mathematical development of noncooperative game theory. Its proof, moreover, is instructive because it uses techniques that are common to many other proofs in formal theory. Unfortunately, it requires that we use some results that are unfamiliar to many readers. So instead of providing a fully general proof, we derive an equilibrium for a particular two-person game in a way that illustrates the more general argument.

Suppose that two persons must each choose a number in the closed interval $[0, 1]$, where x is the number that 1 chooses, and y is the number that 2 chooses. We assume that their utilities are as follows:

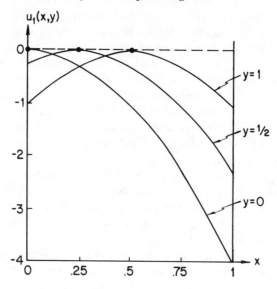

3.15 The payoff function $-(y - 2x)^2$.

$$u_1(x, y) = -(y - 2x)^2 \text{ and } u_2(x, y) = -(x - 2y)^2.$$

If we now differentiate u_1 with respect to x and set the resulting first derivative equal to zero, we learn that the necessary condition for a maximum is $x = y/2$ (the second derivative is negative at this point, and therefore confirms that $y/2$ is a maximum). Thus person 1 prefers to choose a number exactly equal to one-half that chosen by 2; by a similar argument, 2 prefers to choose a number equal to one-half that chosen by 1. Modest contemplation of this problem reveals its solution: If, for example, person 1 chooses $\frac{1}{2}$, then 2's best response is $\frac{1}{4}$, in which case 1's best response is $\frac{1}{8}$, and 2's best response is $\frac{1}{16}$, and so on. The unique equilibrium to this process occurs if both persons choose zero. That is, 1's best response to zero is zero, and 2's best response to 1's best response is also zero. Notice what is occurring here: If we let the function $r_1(s_2)$ identify 1's best response to 2's strategy, and $r_2(s_1)$ be 2's best response to 1's strategy, then a Nash equilibrium is a pair of strategies, (s_1^*, s_2^*), such that $r_1(s_2^*) = s_1^*$ and $r_2(s_1^*) = s_2^*$.

We now use this insight to prove what we already know, namely, that the game in question has an equilibrium. First, to see that Theorem 3.2 applies, consider Figure 3.15, which graphs u_1 as a function of x for various values of y. The figure shows that this utility function is concave in x, regardless of what value we assign to y. Thus, since the interval

[0, 1], which is each person's strategy space, is convex, closed, and bounded, Theorem 3.2 asserts that the game has an equilibrium. To show that this theorem's assertion is correct, consider the function $f = u_1 + u_2$, except that we substitute the variable r_1 for x in u_1 and r_2 for y in u_2. These new variables are also constrained to the interval [0, 1]:

$$f((x, y), (r_1, r_2)) = -(y - 2r_1)^2 - (x - 2r_2)^2.$$

This may seem an odd function, but notice that if, for a fixed x and y, we maximize f with respect to r_1, we are in effect maximizing $u_1(r_1, y) = -(y - 2r_1)^2$, since this is the only term that depends on r_1. Hence, finding the value of r_1 that maximizes f is equivalent to finding 1's best response to y, which we denote by r_1^*. Similarly, the solution to the maximization of f with respect to r_2, denoted r_2^*, finds 2's best response to 1's strategy, x. The values of r_1^* and r_2^* depend on x and y, of course, and in this example if $r_1^* = y/2$ and $r_2^* = x/2$, $f = 0$, whereas for all other values of r_1^* and r_2^*, f is negative. Hence, f is maximized if $r_1^* = y/2$ and if $r_2^* = x/2$. We can represent this fact by a new function, $g(x, y) = (r_1^* = y/2, r_2^* = x/2)$ that identifies the best responses to y and x, respectively. For example, $g(\frac{1}{2}, \frac{1}{2}) = (\frac{1}{4}, \frac{1}{4})$, while $g(\frac{1}{8}, \frac{1}{10}) = (\frac{1}{20}, \frac{1}{16})$.

We know from the definition of an equilibrium, however, that if (x^*, y^*) is an equilibrium pair, then x^* is a best response to y^*, and y^* is a best response to x^*. Hence, if a game has an equilibrium, there must exist two strategies such that $g(x^*, y^*) = (x^*, y^*)$. There is exactly one such point in our example (owing to the fact that u_1 and u_2 are strictly concave functions of x and y, respectively), namely, $(0, 0)$, so $(0, 0)$ is an equilibrium pair. To confirm this result directly, note that if we set $y = 0$ in 1's utility function, we get $u_1(x, 0) = -(0 - 2x)^2 = -4x^2$, in which case 1 has no incentive to choose any number other than zero. Similarly for 2, if we set $x = 0$ in 2's utility function, we get $u_2(0, y) = -(0 - 2y)^2 = -4y^2$, which is also maximized for $y = 0$. Hence, if either person chooses 0, then the other has no incentive to choose any number other than zero himself. That is, $g(0, 0) = (0, 0)$.

The generalization of this example yields a proof of Theorem 3.2, but the derivation of $(0, 0)$ as the equilibrium disguises a crucial step in the argument. We easily found a value of x and y such that the values of r_1 and r_2 that maximize f are x and y themselves. Put differently, $(0, 0)$ is a *fixed point* of $g(x, y)$, a point such that when we substitute it into the arguments of g, g returns it as its value. The crucial step in the proof of Theorem 3.2 is to show that the function g for any concave n-person game has at least one fixed point. There is no reason for us to enter into

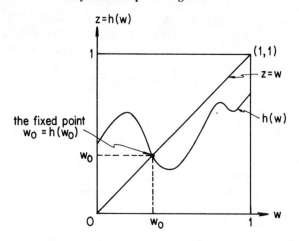

3.16 Illustration of a fixed point.

the mathematical complexities needed to prove this fact. The assumptions that strategy spaces are convex and that utility functions are concave and continuous make it possible to appeal to another mathematical theorem called the Kakutani fixed-point theorem. This theorem is unwieldy to state but easy to illustrate.

Note first, that the domain and range of the function $g(x, y)$ in our example are identical. That is, g takes points in a two-dimensional strategy space of the two players and maps them back into the same space (or a subset of that space) by identifying a new pair of admissible strategies. To represent this function geometrically, we need four dimensions, but this requirement exceeds our conceptual capabilities or graphic skills. To see the logic of a fixed point theorem, then, consider instead a function $z = h(w)$ that takes values of w in the interval $[0, 1]$ and assigns them to values of z, which also is constrained to the interval $[0, 1]$ or to a subset of this interval. Referring to Figure 3.16, we see that the permissible ranges of w and z require that we consider only functions that go from the left to the right side of the box (from $w = 0$ to $w = 1$) and never go above or below the box (z never exceeds 1 or is less than 0). The diagonal line in Figure 3.16, drawn from the point $(0, 0)$ to the point $(1, 1)$, represents the expression $w = z$. To construct $h(w)$, we begin at $w = 0$ and z equal to anything in the range $[0, 1]$, and, without lifting our pencil or pen from the page [so $h(w)$ is continuous], we draw a curve to the other side of the box (to $w = 1$). Notice now that $h(w)$ *must* do one of three things: cross the diagonal at some point; start at $(0, 0)$, and stay below the diagonal line everywhere; or stay above the diagonal line but be connected to it at $(1, 1)$. Thus, the equation $h(w) = w$ is satisfied at least once: when $h(w)$ crosses the diagonal, when $h(w)$ and w are both 0, or when $h(w)$ and w are both 1. Such points are the fixed points of the function $h(w)$ in the interval $[0, 1]$.

Returning to the interpretation of Theorem 3.2, we emphasize that this theorem presents only a sufficient condition for the existence of equilibria and not a necessary condition. Hence, many games have equilibria in pure strategies even though the players' utility functions are neither continuous nor concave. Several examples in the next chapter illustrate this claim. As the next section shows, however, this theorem also ensures the existence of equilibria for every noncooperative game in which the players have a finite number of strategies.

3.8   Mixed strategies

It is not necessarily obvious that Theorem 3.2 is relevant to games with a finite number of strategies, but by rethinking the concept of a strategy, we can make this theorem relevant. Throughout the preceding discussion we say little about the process which players use to select strategies, aside from presuming that each participant evaluates the game's properties rationally. But reflect on this process: Before acting, each person, i, assigns each strategy a probability number, p_j^i, such that the probabilities across all of i's strategies sum to one. After the player assigns these probabilities, he conducts a lottery (spins a spinner, rolls a die, etc.) that is consistent with the assigned probabilities and he chooses the strategy that wins the lottery.

This may seem to be a strange way to choose strategies, since it is so random. And who, after all, wishes to leave his fate to chance or randomness? But this scheme does not preclude assigning a probability of 1 to a strategy and of 0 to the rest. Thus, the old method of analyzing a game and choosing a pure strategy simply becomes a special case of choosing by lottery. Choosing by lottery, furthermore, is not equivalent to succumbing to randomness or chance: Each person selects his own lottery and adjusts probabilities as he sees fit. Later we provide a better motivation for choosing strategies by lottery. Here, we intend to show simply that lotteries, in conjunction with Theorem 3.2, solve all finite noncooperative games, at least mathematically.

To begin, we formally define

> A *mixed strategy* for person i is a lottery $(p_1, p_2, \ldots)$ over i's pure strategies such that $1 \geq p_j \geq 0$ for all j, all probabilities sum to one, and i chooses s_{ij} with probability p_j. Denote the set of i's mixed strategies by the convex, closed, and bounded set S_i.

Person 2

	b_1	b_2
Person 1: a_1	4, 1	1, 4
a_2	2, 3	3, 2

3.17 A game without pure-strategy equilibrium.

The fact that S_i is convex, closed, and bounded follows from the requirement that the p_j's sum to 1. This requirement is like a budget constraint in that it states that, no matter what probabilities are assigned to the different strategies, the sum of the probabilities assigned cannot exceed 1. Hence, a geometric representation of the set of all admissible lotteries over three outcomes or alternatives looks exactly like the budget simplex in Chapter 1, which we already know corresponds to a convex, closed, and bounded set. This fact becomes important later when we apply Theorem 3.2

The concept of a mixed strategy permits us to define the mixed extension of any finite noncooperative game in normal form: Letting **p**'s denote mixed strategies for person 1, **q**'s denote mixed strategies for 2, ..., and **r**'s denote mixed strategies for person n,

> *Mixed extension*: The mixed extension of a finite noncooperative game in normal form consists of the n strategy spaces S_i, $i = 1, 2, \ldots, n$, and the n expected utility functions, $u_i(\mathbf{p}, \mathbf{q}, \ldots, \mathbf{r})$, where, for example,
>
> $$u_1(\mathbf{p}, \mathbf{q}, \ldots, \mathbf{r}) = \Sigma_j \Sigma_k \cdots \Sigma_m \, p_j q_k \cdots r_m \, u_1(s_{1j}, s_{2k}, \ldots, s_{nm}),$$
>
> where **p** is a lottery in S_1, **q** is in S_2, and so on.

To illustrate this definition, consider the game in Figure 3.17, and observe that this game does not have an equilibrium in pure strategies. To form its mixed extension, let person 1's strategies consist of all lotteries of the form $\mathbf{p} = (p_1, p_2) = (p_1, 1 - p_1)$, where p_1 is the probability that 1 chooses a_1 and p_2, that 1 chooses a_2. Similarly, $\mathbf{q} = (q_1, 1 - q_1)$ represents a particular strategy for person 2. Person 1's payoff function, then, is the expected utility expression,

$$u_1(\mathbf{p}, \mathbf{q}) = 4p_1 q_1 + 1p_1(1 - q_1) + 2(1 - p_1)q_1 + 3(1 - p_1)(1 - q_1)$$
$$= 4p_1 q_1 + 3 - 2p_1 - q_1. \tag{3.4}$$

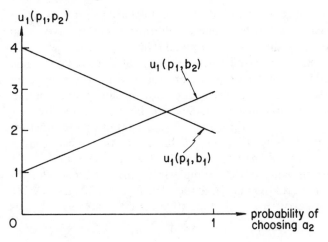

3.18 Expected payoffs to player 1.

Similarly for person 2,

$$u_2(\mathbf{p}, \mathbf{q}) = 1p_1q_1 + 4p_1(1 - q_1) + 3(1 - p_1)q_1 + 2(1 - p_1)(1 - q_1)$$
$$= -4p_1q_1 + 2p_1 + q_1 + 2, \qquad (3.5)$$

which completes the description of the game's mixed extension.

The concept of a mixed extension is important, now, because of the following fundamental theorem:

3.3. *Every mixed extension of a finite, n-person, noncooperative game has at least one equilibrium n-tuple in either pure or mixed strategies.*

To see how this theorem relates to Theorem 3.2 and that it is actually a corollary of that theorem, consider again the mixed extension of the game in Figure 3.17. Using expression (3.4), Figure 3.18 graphs 1's expected utility as a function of the mixed strategies for 1 (the continuum of lotteries between a_1 and a_2) after we constrain 2 to choose either b_1 or b_2 with certainty. Note first, that no matter which strategy 2 chooses, person 1's expected payoff, graphed as a function of his strategy p_1, is a straight line (*continuous* and *linear*). But a linear function is both concave and convex. Hence, we can assert that 1's expected utility is a continuous and concave function of the lottery that he chooses and that therefore it satisfies condition (2) of Theorem 3.2. Since the set of mixed strategies available to player 1 can be represented by the closed and convex interval [0, 1] in Figure 3.18, condition (1) of this theorem is also satisfied. An identical argument holds for person 2, so this game has at least one equilibrium point in mixed strategies.

More generally, for all n-person finite games if we extend each person's strategy space to include the set of all mixed strategies over S_i, the conditions of Theorem 3.2 are satisfied for the new game. In this game, the set of strategies, S_i, are convex, closed, and bounded (recall our previous argument that they can be represented by budget simplexes), and expected payoffs are represented by the function $u_i(\mathbf{p}, \mathbf{q}, \ldots, \mathbf{r})$, which is a continuous and linear function of each player's mixed strategies. Thus, Theorem 3.3 follows from Theorem 3.2.

Referring again to the example in Figure 3.17, consider the pair of mixed strategies $[\mathbf{p} = (\frac{1}{4}, \frac{3}{4}), \mathbf{q} = (\frac{1}{2}, \frac{1}{2})]$. If we fix 2's strategy at $(\frac{1}{2}, \frac{1}{2})$, 1's expected payoff from all alternative mixed strategies becomes

$$u_1(\mathbf{p}, (\tfrac{1}{2}, \tfrac{1}{2})) = 4(\tfrac{1}{2})p_1 + 3 - 2p_1 - \tfrac{1}{2} = 2.5.$$

Thus, if 2 chooses the lottery $(\frac{1}{2}, \frac{1}{2})$, person 1 has no positive incentive to change from any one strategy to another. His expected payoff from all strategies equals 2.5, so, in particular, he has no incentive to shift from $(\frac{1}{4}, \frac{3}{4})$. Similarly for player 2, if we fix 1 at $(\frac{1}{4}, \frac{3}{4})$, 2's expected utility is

$$u_2((\tfrac{1}{4}, \tfrac{3}{4}), \mathbf{q}) = -4(\tfrac{1}{4})q_1 + 2(\tfrac{1}{4}) + q_1 + 2 = 2.5.$$

Hence, like 1, if 2 believes that 1 will choose $(\frac{1}{4}, \frac{3}{4})$, then he has no incentive to shift from any mixed strategy to any other; in particular, he has no incentive to shift from $(\frac{1}{2}, \frac{1}{2})$. This establishes that $((\frac{1}{4}, \frac{3}{4}), (\frac{1}{2}, \frac{1}{2}))$ is an equilibrium pair of (mixed) strategies that guarantees both players a payoff of 2.5.

To appreciate the advantage of these strategies compared with others, suppose that person 1 fails to choose his equilibrium strategy and instead matches 2's strategy, $(\frac{1}{2}, \frac{1}{2})$. If 2 stays at $(\frac{1}{2}, \frac{1}{2})$, both persons continue to receive payoffs of 2.5. But there is now no reason to suppose that 2 will not shift to some other strategy. If we substitute $(\frac{1}{2}, \frac{1}{2})$ for $\mathbf{p} = (\frac{1}{4}, \frac{3}{4})$ in expression (3.2), we get $u_2(\frac{1}{2}, \frac{1}{2}), \mathbf{q} = 3 - q_1$, which is maximized if $q_1 = 0$ – if 2 chooses b_2 with certainty. If 2 chooses b_2 and 1 chooses the mixed strategy $(\frac{1}{2}, \frac{1}{2})$, 1's payoff equals 2.0. Thus, by deviating from the equilibrium mixed strategy $(\frac{1}{4}, \frac{3}{4})$, 1 has become vulnerable to 2's choices, so that if 2 responds appropriately to 1's deviation, 1's payoff declines. A similar argument shows that if 2 deviates from his equilibrium mixed strategy, then he becomes vulnerable to 1's choices.

There are two interpretations of the importance of Theorem 3.3. The first is that it actually provides a useful model of people's choices, and that mixed strategies solve all finite games. In certain contexts this seems reasonable. For example, if several times in a football game a quarterback faces third down and three yards to go for a first down, he probably should not always pass, because the defense will see the simple pattern of his strategy and play accordingly. But if he mixes his plays, he must avoid simply alternating between passing and running, for the defense might deduce this pattern as well. A mixed strategy of some sort, such as one that uses a random mechanism, is probably best, and Theorem 3.3 suggests that the better quarterbacks, those who win more games, ceteris paribus, are those who somehow infer the appropriate mixed strategy. In other contexts, however, it seems more reasonable to interpret Theorem 3.3 simply as a mathematical trick that, although providing analytic closure to an abstract problem, does not offer a useful model of choice. Equilibrium mixed strategies can be exceedingly difficult to compute. For sufficiently realistic games they may be practically impossible to compute. So it seems unreasonable to suppose that by induction or otherwise, people learn to act as if they make such calculations. Also, the costs of computation may exceed any gains.

The relevance of this theorem to politics, then, depends on the situation under study. As we consider alternative models, we will reevaluate the meaning of mixed strategies. Often, reevaluation requires development of alternative theoretical ideas, whereas at other times we suggest that there are fundamental indeterminacies in politics in games for which no equilibrium solution appears reasonable.

3.9 Perfect equilibria

Often, however, we meet a different problem than the one that Theorem 3.3 tries to solve – namely, games with too many equilibria. In committee-voting games, especially, we find many highly implausible equilibria. Suppose that all members of a three-person legislature unanimously rank some alternative – say, y – last in their preferences. Nevertheless, as we indicated earlier, to have all three legislators voting for y is an equilibrium. Assuming majority rule, if everyone votes for y, no one has a *positive* incentive to change his vote, since this cannot change the outcome. Naturally, we do not anticipate anyone voting for y and thus we might properly regard y as a bogus equilibrium.

The problem is to find a way to eliminate bogus equilibria without destroying the existence of equilibria that make sense. To illustrate what we require, consider the extensive form in Figure 3.19 and its normal-form equivalent in Figure 3.20. To convince the reader that this is not a

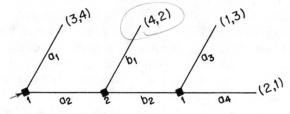

3.19 Extensive form for invasion game.

Player 2:

	s_{21}	s_{22}
s_{11}	3, 4	3, 4
Player 1: s_{12}	4, 2	1, 3
s_{13}	4, 2	2, 1

3.20 Normal form for invasion and retaliation.

wholly contrived example, imagine the following substantive interpretation of each player's actions in the extensive form:

Player 1: USSR	Player 2: USA
a_1: don't invade	b_1: don't fire missiles
a_2: invade Western Europe	b_2: fire missiles at USSR
a_3: don't fire missiles	
a_4: fire missiles at USA	

The strategies in the game's normal form are

s_{11}: don't invade

s_{12}: invade Western Europe and, if the United States retaliates, respond in kind

s_{13}: invade Western Europe and, if the United States retaliates, do not respond

s_{21}: if the USSR invades Western Europe, do not fire missiles

s_{22}: if the USSR invades Western Europe, fire missiles in response

Examining the normal form first, we find that there are two equilibrium pairs: (s_{13}, s_{21}) and (s_{11}, s_{22}). But looking at the extensive form shows us that the second of these equilibria is bogus. Working backward

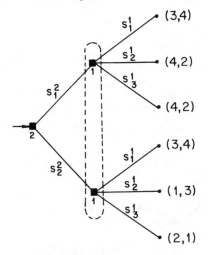

3.21 An alternative extensive form.

up the tree, notice that at 1's last move, a_4 dominates a_3. Thus, 2's choice between b_1 and b_2 is actually a choice between the payoff vectors (4, 2) and (2, 1). Naturally, 2 prefers (4, 2) and thus chooses b_1. This implies that 1's initial choice is between (3, 4) and (4, 2), so 1 chooses a_2 to obtain (4, 2). Therefore, (a_2, b_1) is the only reasonable equilibrium of this game.

The problem that this example uncovers is that *in moving from the extensive form to the normal form, we may lose information about a game's strategic character.* For example, the extensive form in Figure 3.21 yields the same normal form, Figure 3.20, as does the tree in Figure 3.19. But these trees are vastly dissimilar. In Figure 3.21 we cannot eliminate any strategy by working backward up the tree, so that in this instance both equilibrium outcomes, (3, 4) and (4, 2), make sense. Thus, if a normal-form game has multiple equilibria, and if we know its corresponding extensive form, then we must check that form to assess the plausibility of the different equilibria.

Plainly, we require a general method for checking for and eliminating implausible or bogus equilibria. To establish such a method, we first define the concept of a subgame. If we use the notation **G** to denote an extensive-form game, then

> **G**′ is a *subgame* of **G** if **G**′ is an extensive-form game beginning at some node of **G**, which contains all subsequent nodes of **G**, and in which all information sets in **G** of all players remain intact.

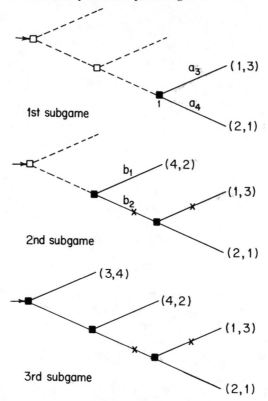

1st subgame

2nd subgame

3rd subgame

3.22 Subgames of the invasion game.

Thus, the game in Figure 3.19 has three subgames: the extensive form that begins at 1's second decision point, the form that begins at 2's decision point, and the full form (see Figure 3.22). The game in Figure 3.21, however, has only one subgame, the full game itself, since any division of this game necessarily divides player 1's information set.

We introduce the concept of a subgame because it permits us to define the following refinement of an equilibrium:

> *A subgame perfect equilibrium* of **G** is an equilibrium *n*-tuple that induces an equilibrium on every subgame of **G**.

Although this definition applies to mixed- as well as to pure-strategy equilibria, the applications in this book are limited to pure strategies. Hence, to illustrate a subgame perfect equilibrium, consider again the game in Figure 3.19 and observe that the following sequence of choices leads to one of the two pure-strategy equilibria that appear in the game's normal form,

Sequence (s_{11}, s_{22})		Sequence (s_{13}, s_{21})	
1	a_1, b_1, a_3	5	a_2, b_1, a_3
2	a_1, b_1, a_4	6	a_2, b_1, a_4
3	a_1, b_2, a_3		
4	a_1, b_2, a_4		

Referring to Figure 3.22 and looking at the first subgame, which begins at 1's second choice point, we find that there exists only one equilibrium: Player 1 chooses the dominant strategy, a_4. This fact eliminates sequences 1, 3, and 5 as inducing subgame perfect equilibria. Looking next at the second subgame beginning with 2's only decision point, we see that if 1 chooses his dominant strategy, a_4, 2's dominant strategy for this subgame is b_1. Thus, sequence 4 cannot induce a subgame perfect equilibrium, since it requires player 2 to choose b_2. Finally, looking at the game beginning with 1's first choice point, we find that if 1 and 2 choose their dominant strategies in subsequent moves, then 1's dominant choice is a_2. This fact eliminates sequence 2. Hence, (s_{13}, s_{21}) is the unique subgame perfect equilibrium.

It seems sensible to assume that only subgame perfect equilibria offer reasonable predictions. Checking whether an equilibrium is subgame perfect is equivalent to working back up the tree, and assuming that people can work back up a tree is equivalent to presuming simply that people can plan ahead for likely events. For the game in Figure 3.19, for example, player 2 knows that if he chooses b_2, 1 will choose a_4; here the outcome will be (2, 1). Thus, using the foresight that knowledge of the game's structure permits, player 2 can eliminate b_2 in favor of b_1.

This concept requires refinement, however, since we can imagine games with multiple subgame perfect equilibria, some of which are counterintuitive. For example, in a legislature voting by secret ballot between two alternatives, x and y, the only subgame is the full extensive form. But then "all voting for y" is a subgame perfect equilibrium even if the legislature unanimously prefers x to y. One resolution of this problem is to suppose that each player believes that opponents play with a *shaky hand*, that there is always some small probability that people will deviate from any strategy. Although we cannot do justice to his elegant development, Reinhardt Selten formalizes this perspective by supposing that every person in an n-person, noncooperative game believes that at any opponent's choice node there is a nonzero probability that each act will be chosen. We call such strategies, which assign some probability to every possible branch in a game tree, *totally mixed behavioral strategies*. A best response for i, then, is an s_{ij} such that if $\mathbf{b}_{-i} = (\mathbf{b}_1, \ldots \mathbf{b}_{i-1}, \mathbf{b}_{i+1}, \ldots, \mathbf{b}_n)$ is a totally mixed behavioral strategy for the other n-1 players, then the strategy s_{ij} maximizes i's expected utility when he plays against b_{-i}. Somewhat imprecisely, a totally mixed behavioral strategy n-tuple is a *perfect equilibrium* for any player i and strategy s_{ij}, if we find that if s_{ij} is not a best response to b_{-i}, then the behavioral strategy b_i should place infinitesimal weight on s_{ij}. Actually, the precise definition of a perfect equilibrium is formulated in terms of limits to these probabilities, but this definition suits our purposes. Selten shows, moreover, that *every finite normal-form game and every extensive-form game in which every player can remember his past moves has at least one perfect equilibrium.*

For an example of how this shaky-hand perspective eliminates unrealistic equilibria, consider again the situation in which everyone in a three-person committee prefers y less than x, but in which, if all members vote for y, they are at a pure-strategy Nash equilibrium, which is also a subgame perfect equilibrium. Suppose that each receives a payoff of 0 from y and 1 from x. If every member of the committee [evaluating the equilibrium outcome $(0, 0, 0)$ that results if they all vote for y] believes that there is some small probability, say e, that any particular member of the committee will not vote for y, then voting for y is not a best-response strategy. For example, consider the expected utility that one voter gets from sticking with y, $E(y)$, compared with his payoff from shifting to x, $E(x)$. The probability that the other two members will stick with y is $(1 - e)(1 - e)$; the probability that either one will stick with y and the other shift to x is $2e(1 - e)$, and the probability that both will shift to x is e^2. Thus, with majority rule,

$$E(y) = 0(1 - e)(1 - e) + 0(2e(1 - e)) + e^2,$$

since in the first two terms, y continues to win, while

$$E(x) = 0(1 - e)(1 - e) + (2e(1 - e)) + e^2,$$

since in the case of the second term, the voter is decisive for x. Clearly, $E(x) > E(y)$ provided that $e > 0$. Hence, regardless of how small a chance a voter places on the possibility that the other members will make an error, he should place an infinitesimal probability on voting for y. If everyone acts thus, voting for y disappears in the limit as a predicted outcome.

3.10   Summary

Chapter 2 told us that political institutions are unlike simple black boxes into which we plug individual preferences and out of which emerges, in some deterministic way, a social preference or choice. Instead, institutions admit strategic maneuvering. Noncooperative game theory seeks to unravel the complexities of this maneuvering if communication and coordination are either impossible or excessively costly. Although the concepts of the extensive and normal form provide a representation of interdependent choice situations, we require the concept of an equilibrium to resolve "he-thinks-that-I-think" regresses.

An equilibrium is a strategy n-tuple, such that each person's strategy is a best response to the choices of the other n-1 players' choices that the equilibrium implies. In equilibrium, no person has any incentive to shift unilaterally to some other strategy, and thus this concept permits predictions about people's choices.

Two critical issues arise with respect to equilibria: existence and uniqueness. First, if an equilibrium does not exist, then we remain without a prediction. Theorem 3.2 and its corollary for finite, noncooperative games, Theorem 3.3, are important, then, to the extent that they resolve the issue of existence for general classes of games. Uniqueness is the other side of the coin. Many games, especially those that model voting in legislatures, have too many equilibria. Thus, we introduce the concepts of subgame perfect and perfect equilibria. These concepts seek to eliminate those equilibria that arise because we lose information as we move from the extensive to the normal-form, or because certain strategy n-tuples, although they satisfy a mathematically precise definition, make no sense.

This chapter does not end our explorations of the properties of equilibria. We can judge the value of theorems such as 3.2 and 3.3 as well as refinements of the Nash equilibrium concept only after we examine specific applications. Hence, beginning with the next chapter, we turn to special classes of noncooperative games that model some very general institutions and processes. Before proceeding, though, readers should be certain that they are familiar with the following concepts:

extensive form	game tree	node
information set	simultaneous moves	terminal node
choice node	common knowledge	strategy
normal form	pure strategy	infinite game
best-response strategy	domination	strictly dominates
undominated	infinite regress	pure-strategy
perfect information	backward reduction	equilibrium
sophisticated voting	concave game	sincere voting
mixed strategy	mixed extension	fixed point
subgame	behavioral strategy	linear
perfect equilibrium	subgame perfect equilibrium	shaky hand

CHAPTER 4

Elections and two-person zero-sum games

Chapter 3 has provided us with the general structure of the theory of noncooperative games, but we cannot gain a sense of the theory's relevance to politics simply by admiring its internal logic. And although it is tempting to consider examples, we must be careful. Too often the apparent applications of game theory to politics take the simplistic form, "Let player 1 be the U.S. with two strategies and let player 2 be the USSR with two strategies...." Such examples illuminate the general structure of games and reveal the strategic commonality of substantively distinct situations. But a theory of politics requires that we move beyond simple "stories" to modeling specific institutions and processes in a sustained way.

This chapter reviews one such effort, the modeling of plurality-rule elections, which compels us to focus on a particular class of non-cooperative games, zero-sum games, the most widely studied and applied class in games theory. Chapter 3 contains all the conceptual ideas that we require to construct and analyze a zero-sum game. Nevertheless, because the zero-sum assumption imposes specific restrictions on individual utility functions, we can prove some important additional results about equilibria for games that utilize this assumption. After reviewing those results, we turn to the specific application of two-candidate elections. There, we examine alternative specifications of candidate objectives, conditions for the existence of equilibria in single and multi-issue contests, elections that concern income redistribution, elections in which the information of various participants is incomplete, and finally, multicandidate elections. Here, then, we begin to see the development of a consistent body of theoretical knowledge about an important institution.

4.1 Zero-sum games

A zero-sum game is no different from the games described in Chapter 3 except that preferences must satisfy one additional property. Namely,

> A noncooperative game is *zero sum* if the payoffs across all players sum to zero for *every* strategy n-tuple.

Thus, if $n = 2$, then $u_1(s_1, s_2) + u_2(s_1, s_2) = 0$ for all $s_1 \in S_1$ and $s_2 \in S_2$. We can also turn any *constant-sum* game, one in which the payoff summed across all persons equals a constant, k, for *all* strategies, into a zero-sum game. We need only take advantage of the fact that utility functions are invariant up to positive linear transformations. Hence, we can make a constant-sum game zero sum, without altering its strategic character, by subtracting the constant k from one player's utility function. For example, suppose that each of two election candidates measures his utility by the number of votes he receives. Thus $u_1 = V_1$ and $u_2 = V_2$, where V_i denotes the votes received by candidate i. Suppose also that, regardless of the candidates' strategies, turnout is constant. That is, $V_1 + V_2 = k$. If we subtract k from, say, candidate 1's utility, we arrive at an equivalent utility function, $u_1 = V_1 - (V_1 + V_2) = -V_2$, and the election game is now zero sum: Candidate 1 prefers to minimize 2's vote, whereas 2 prefers to maximize his own vote.

Before we study election games, we should appreciate why we anticipate being able to use the zero-sum assumption to model various political processes. Suppose that we conduct a series of experiments to ascertain the utility that each candidate associates with candidate 1's vote plurality, Pl_1, in a two-candidate plurality-rule election. Candidate 1's utility should increase and 2's decrease as Pl_1 increases. But even after we take suitable account of measurement error and so forth, it is unlikely that the candidates' utilities would sum *exactly* to zero for all values of Pl_1. Thus, we might conclude that we cannot model the election as zero sum and that any election is so unlikely to satisfy this condition that such games, although theoretically elegant, remain substantively uninteresting.

This conclusion is premature. We should view the zero-sum condition as a useful *approximation*, in which the benefits of its use often outweigh any distortion of reality. This notion of a useful approximation holds especially if we cannot measure utility directly but can describe outcomes in some objective quantitative way, and if it is reasonable to suppose that both persons in a game order the outcomes inversely. In elections, for example, we can assume that for candidates who are concerned solely with winning, $u_i = +1$ if Pl_i is positive, $u_i = -1$ if i's plurality is negative, and $u_i = 0$ if the outcome is a tie. We might then compare the resulting analysis to one in which candidates care about the magnitudes of their pluralities. That is, we could let $u_i = Pl_i$. Both assumptions about candidate motives yield zero-sum games. As we become more sophisticated, we can consider more complex specifications of utility. The result of such variations is a comparative analysis of objectives that may explain why two candidates in identical circumstances act differently.

In addition to two-candidate elections, other situations that we might model with the zero-sum assumption include: unconditional wars, in which no one cares much about the costs of conflict but in which everyone is concerned solely with who "wins" and who "loses"; a debate over income-redistribution policies in which none of the antagonists cares about the welfare of anyone else but in which each seeks solely to maximize his own income; and international trade negotiations, in which the parties view various trade barriers as conferring benefits on one nation at the expense of the workers in some other. Although we might agree that none of these situations is ever purely zero sum – wars are costly to fight, and alternative redistribution and trade policies have macro-economic effects on all economies – the zero-sum assumption nevertheless is often a valuable first approximation of the motives of key actors.

4.2   Interchangeability and equivalence

Aside from their substantive application, two-person zero-sum games warrant special attention because equilibria for them necessarily satisfy two conditions, interchangeability and equivalence. Although the concept of an equilibrium forms the basis for our models' predictions, the strength or believability of these predictions varies as a function of the properties of different equilibria. Some equilibria, such as those in Figure 3.7, correspond to the conjunction of dominant strategies and are especially believable. Others, such as those requiring calculation of complex mixed strategies, represent credible predictions only if we are willing to stretch the "as if" principle beyond reason. Hence, the mere existence of an equilibrium is no guarantee that we can formulate a credible prediction about choice and outcomes. The properties of interchangeability and equivalence lend credence to equilibria as predictions.

To define these terms and to appreciate their value, we first restate the definition of a Nash equilibrium for a two-person zero-sum game. Letting A and B represent person 1 and 2's strategy sets, respectively, with $a \in A$ and $b \in B$, then (a^*, b^*) is an equilibrium pair for a two-person zero-sum game if and only if

$$u_1(a^*, b^*) \geq u_1(a, b^*) \text{ for all } a \in A$$
$$u_2(a^*, b^*) \geq u_2(a^*, b) \text{ for all } b \in B.$$

But if the game is zero sum, then $u_1 = -u_2$, in which case we can write the second inequality as

$$-u_1(a^*, b^*) \geq -u_1(a^*, b) \text{ for all } b \in B.$$

Person 2

		b_1	b_2	b_3
Person 1:	a_1	3, -3	5, -5	4, -4
	a_2	-5, 5	9, -9	6, -6

4.1 Zero-sum game with a pure-strategy equilibrium.

After we eliminate the minus signs and combine this inequality with the first, we obtain the following restatement of the definition of a Nash equilibrium:

$$u_1(a^*, b) \geq u_1(a^*, b^*) \geq u_1(a, b^*) \tag{4.1}$$

for all $a \in A$ and $b \in B$.

This inequality states that if both persons choose equilibrium strategies, then any deviation by one person decreases (or at least does not increase) his payoff and increases (or at least does not decrease) his opponent's payoff. To illustrate inequality (4.1), consider the game in Figure 4.1 and note, first, that (a_1, b_1) is the unique Nash equilibrium: 1's payoff drops from 3 to -5 if he shifts unilaterally to a_2 and 2's payoff drops to -5 or -4 if he shifts to b_2 or b_3, respectively. But since the game is zero sum, if 1 shifts unilaterally from a_1 to a_2, then 2's payoff increases from -3 to 5; and if 2 shifts from b_1 to b_2 or b_3, then 1's payoff increases from 3 to 5 or 4, respectively.

With these definitions, we can now consider why two-person, zero-sum games receive special attention. Recall from Chapter 3 that if noncooperative equilibria are not unique, then the "he thinks" regress can persist since people may require some coordination of choices to attain an equilibrium. Suppose, though, that equilibrium strategies satisfy this property:

> Equilibrium strategies are *interchangeable* if, when (s_i^*, s_{-i}^*) and (s_i', s_{-i}') are equilibrium n-tuples, then (s_i^*, s_{-i}') and (s_i', s_{-i}^*) are equilibrium n-tuples as well for all i.

Thus, for two-person games in particular, equilibrium strategies are interchangeable if, when (a^*, b^*) and (a', b') are equilibrium pairs, then (a^*, b') and (a', b^*) are also equilibrium pairs. Hence, interchangeability guarantees that, whenever players identify their equilibrium strategies, the situation requires no implicit coordination to ensure that some equilibrium prevails. Each player need only choose one of his equilibrium strategies.

	b_1	b_2
a_1	1, −1	y, $-y$
a_2	z, $-z$	x, $-x$

4.2 Proving equivalence and interchangeability.

A second valuable property for equilibrium n-tuples is that they be equivalent, in the sense that no player holds a preference for one over another. Formally,

Equivalence: Two equilibrium n-tuples, $\mathbf{s}^* = (s_1^*, \ldots, s_n^*)$ and $\mathbf{s}' = (s_1', \ldots, s_n')$, are equivalent if $u_i(\mathbf{s}^*) = u_i(\mathbf{s}')$ for all i.

Thus, if a game's equilibria are equivalent, then all persons are indifferent as to which equilibrium prevails.

The following theorem establishes the importance of two-person zero-sum games as a special case:

4.1.   *For any two-person zero-sum game, all equilibria are equivalent and interchangeable.*

The incompletely specified finite game in Figure 4.2 illustrates the proof of this theorem. Assume for this game that payoffs are zero sum and that (a_1, b_1) and (a_2, b_2) are equilibrium pairs. To establish interchangeability, we want to show that (a_1, b_2) and (a_2, b_1) are equilibria, and, to establish equivalence, that the payoff entries in all four cells are $(1, -1)$. From the definition of an equilibrium, and the condition that (a_1, b_1) and (a_2, b_2) must both satisfy expression (4.1), we have

$$1 \geq z \text{ and } -1 \geq -y \text{ and } x \geq y \text{ and } -x \geq -z.$$

If we eliminate minus signs and combine, these inequalities become

$$1 \geq z \geq x \geq y \geq 1,$$

which holds only if $x = y = z = 1$. But then all four cells of the game are identical and all are equilibria. Thus, all equilibria are equivalent and interchangeable.

Interchangeability and equivalence are important properties. Equivalence asserts that the players are indifferent as to which equilibrium prevails whereas interchangeability implies that if each player finds two or more strategies involved in equilibria, no implicit or explicit coordination is required between the players to ensure that some equilibrium prevails. Later chapters show that equilibria for nonzero-sum games need not satisfy these properties. Therefore it may be difficult to assert in this instance that equilibria "solve" them, that they render believable predictions. For two-person zero-sum games, however, interchangeability and equivalence ensure believability.

4.3 Maxmin and minmax

Although people will not deviate unilaterally from equilibria, and although a person should adopt an equilibrium strategy if he believes that his opponent will do the same, we must nevertheless ask why anyone should adopt such a strategy in the first place, or believe that an opponent would do so. We can answer this query in part by supposing that if people approach a zero-sum game from the perspective of minimizing their maximum possible losses, they will necessarily arrive at an equilibrium. This approach also establishes a simple procedure for ascertaining whether a finite game has an equilibrium in pure strategies or whether its solution requires mixed strategies.

In a normal-form game, people choose their strategies simultaneously, and thus they are uninformed about an opponent's choice. But suppose that both persons in a two-person zero-sum game assume the worst, that no matter what they do, their opponent will choose a strategy that minimizes their payoffs. Given this pessimistic outlook, each person should choose a strategy that maximizes his minimum payoff, or equivalently, that minimizes the maximum damage that an opponent can inflict. Person 1, then, should first examine each strategy that he might choose and calculate its *security level*, the minimum payoff that a strategy affords if person 2 chooses the most damaging response. Thus, for each a_j, person 1 calculates

$$\min_{b_k} u_1(a_j, b_k) = \text{security level of } a_j.$$

Second, person 1 identifies the strategy that maximizes his security level. That is, he calculates v_1, where

$$v_1 = \max_{a_j} \min_{b_k} u_1(a_j, b_k).$$

Looking at Figure 4.3a, which reproduces the game in Figure 4.1 and which we already know has an equilibrium at (a_1, b_1), we find that the security value of a_1 is 3, and that of a_2 is -5. Hence, a_1 maximizes 1's security level, and $v_1 = 3$. Suppose that person 2 makes a similar calculation. That is,

$$v_2 = \max_{b_k} \min_{a_j} u_2(a_j, b_k).$$

	b_1	b_2	b_3	1's security level			b_1	b_2	b_3	1's security level
a_1	3	5	4	3		a_1	4	-1	3	-1
a_2	-5	9	6	-5		a_2	-8	7	0	-8

2's security
level   -3   -9   -6   -4   -7   -3
                  (a)   (b)

4.3 A game with and a game without a pure-strategy equilibrium.

Referring again to Figure 4.3a and keeping in mind that 2's payoffs are the negative of those shown, we see that b_1 maximizes 2's security level and $v_2 = -3$. Hence, if each person adopts his *maxmin* strategy, a_1 for 1 and b_1 for 2, they arrive at the equilibrium.

In general, *this procedure leads both persons to a pure-strategy equilibrium if such an equilibrium exists* (and to a mixed-strategy equilibrium, if we substitute the notation for such strategies appropriately in our discussion). Thus, it is no accident that in the example, (a_1, b_1) is an equilibrium and that the process of maximizing security levels isolates this strategy pair. To see this result formally, note that if (a^*, b^*) is an equilibrium, then the security level for person 1 of a^* is $u_1(a^*, b^*)$; that is,

$$\min_{b_k} u_1(a^*, b_k) = u_1(a^*, b^*),$$

since, if person 2 deviates from b^* while 1 chooses a^*, then 2 cannot increase his own payoff and cannot thereby decrease 1's payoff. Accordingly, b^* is 2's best response to a^*. Furthermore, for person 1 the security level of any strategy cannot be greater than that provided by a^*, since by choosing b^* person 2 can always hold 1 down to $u_1(a_j, b^*) \leq u_1(a^*, b^*)$. Our next inequality states the obvious fact that the minimum number from some set of numbers cannot exceed any particular number; that is,

$$\min_{b_k} u_1(a_j, b_k) \leq u_1(a_j, b^*).$$

Combining the preceding equalities and inequalities, we arrive at

$$\min_{b_k} u_1(a_j, b_k) \leq u_1(a_j, b^*) \leq u_1(a^*, b^*) = \min_{b_k} u_1(a^*, b_k),$$

or simply

$$\min_{b_k} u_1(a_j, b_k) \leq \min_{b_k} u_1(a^*, b_k).$$

That is, the minimum payoff that person 1 receives from a^* is at least as great as the minimum payoff that he receives from any other strategy in A. This minimum is $u_1(a^*, b^*)$, since by an equivalent argument, b^* maximizes person 2's minimum payoff. Thus,

$$\max_{a_j} \min_{b_k} u_1(a_j, b_k) = u_1(a^*, b^*) = v_1,$$

$$\max_{b_k} \min_{a_j} u_2(a_j, b_k) = u_2(a^*, b^*) = v_2.$$

From the zero-sum condition that $u_1 = -u_2$, these last two expressions combine to yield

$$v_1 = -v_2. \tag{4.2}$$

Hence, because they act under the presumption that an opponent makes an appropriate response, both persons arrive at an equilibrium.

This pessimistic approach to choosing strategies might seem unwarranted, especially if a player presumes that his opponent is stupid or poorly informed about the game's structure. In reality, of course, some people can comprehend strategic complexities better than others. But we must keep in mind the rules for formulating a game in normal form. First, we should try to build into a game's extensive form a description of whatever is relevant about a person's information, including his knowledge about possible actions and the payoffs of others. That is, whenever possible we should design the game tree so that it takes into account any relevant information asymmetries. Recall the example of a secret ballot in which person 3 can choose to spy on 1 or 2 (see Figure 3.4). There we model the assumption that voter 2 knows more about 3's choice than does 1 directly by this game's extensive form.

Second, since we should suppose that we can derive the normal-form game from some appropriate extensive form, we should not assume that people are ignorant of some aspect of the game as presented. If one person is less informed than the other, this fact, already represented in the game's extensive form, should find reflection in its normal form. If we again insist on incorporating information asymmetries into our analysis of that form, then we may be making the error of some sort of double counting. Admittedly, it is often inconvenient to detail a situation's extensive form (our analysis of elections later in this chapter illustrates this inconvenience), and more convenient to assume that the information contained in a game's normal form represents the knowledge available to the analyst but not necessarily to the participants. Later, we consider games of incomplete information. Here, though, we try to establish the fundamentals of game theory, and the task is

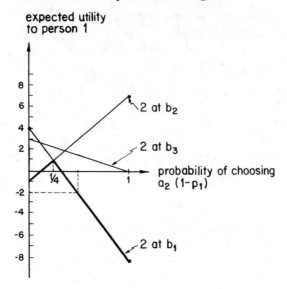

4.4 Security levels of player 1 with mixed strategies.

easier if we assume that the normal form is common knowledge to all participants.

Aside from engaging in a debate as to whether the preceding pessimistic perspective justifies equilibria as predictions about outcomes, we also want to show that the application of this perspective helps us find out quickly whether a zero-sum game requires mixed strategies for a solution. In game (a) in Figure 4.3, a_1 and b_1 maximize person 1 and 2's security levels, respectively. And in accordance with expression (4.2), $v_1 = -v_2$. Hence, (a_1, b_1) is an equilibrium pair. In game (b), on the other hand, although a_1 and b_3 maximize 1 and 2's security levels, respectively, these levels are not the negative of each other. Since any equilibrium pair satisfies expression (4.2), it must be the case that no pure-strategy equilibrium exists. And since Theorem 3.3 tells us that every finite noncooperative game has at least one equilibrium in pure or mixed strategies, it follows that the game in Figure 4.3b has an equilibrium only in mixed strategies.

Let us consider this second game in more detail, since mixed strategies yield a geometric interpretation of the preceding formal argument. Consider Figure 4.4, which graphs u_1 as a function of 1's mixed strategies for each of 2's three pure strategies. Now consider the darkened portions of the two lines corresponding to player 2 choosing b_1 and b_2. This line graphs the minimum expected payoff to 1 for all of 1's mixed strategies when played against any of 2's pure strategies.

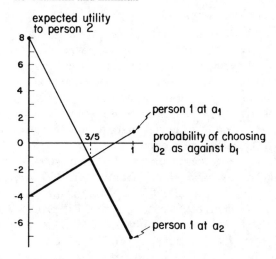

4.5 Security levels of player 2 with mixed strategies.

Equivalently, the darkened line segment shows the minimum payoff to which player 2 can hold 1 if 2 chooses an appropriate response. For example, if 1 chooses the mixed strategy $(\frac{1}{2}, \frac{1}{2})$, and if 2 plays b_1, then 1 gets $4(\frac{1}{2}) - 8(\frac{1}{2}) = -2$; if 2 plays b_2, then 1 gets 3; and if 2 plays b_3, then 1 gets $\frac{3}{2}$. Thus, 1's worst payoff from $(\frac{1}{2}, \frac{1}{2})$ is -2, which is what the darkened line in 4.4 indicates. Looking at the point on this line that is a maximum, this figure shows that 1 maximizes his security level with the mixed strategy $(\frac{3}{4}, \frac{1}{4})$, which is the mixed strategy corresponding to the intersection of the lines b_1 and b_2. [The equations for these lines are $u_1(\mathbf{p}, b_1) = 4 - 12(1 - p_1)$ and $u_1(\mathbf{p}, b_2) = -1 + 8(1 - p_1)$, respectively, and these two equations are equal – the two lines cross – at $1 - p_1 = \frac{1}{4}$.] Thus, $(\frac{3}{4}, \frac{1}{4})$ maximizes 1's security level. Once player 1 is at $(\frac{3}{4}, \frac{1}{4})$, it matters little to him whether 2 chooses a pure strategy or a lottery between b_1 and b_2, because $(\frac{3}{4}, \frac{1}{4})$ yields the same expected payoff in either case.

Figure 4.4 also reveals that b_3 is not a good response against any of player 1's mixed strategies. No matter what pure or mixed strategy 1 chooses, either b_1 or b_2 holds 1 down to a lower expected payoff than does b_3. Thus, we can safely eliminate b_3 from consideration when calculating the mixed strategy that maximizes player 2's security level. To compute that strategy, we repeat the analysis just performed for 1 by graphing player 2's expected payoffs as a function of 2's mixed strategies. Figure 4.5 shows accordingly that $(\frac{2}{5}, \frac{3}{5}, 0)$ maximizes 2's security level.

The preceding discussion identifies the strategies that maximize each player's security level. To confirm that $((\frac{3}{4}, \frac{1}{4}), (\frac{2}{5}, \frac{3}{5}, 0))$ is an equilibrium pair, notice from Figure 4.3b that if player 1 chooses $(\frac{3}{4}, \frac{1}{4})$, then regardless of 2's strategy, 1's expected payoff is 1, whereas 2's expected payoff is -1. Thus, player 2 has no incentive to choose any strategy over another, and in particular, we can say that any strategy that places zero weight on b_3 – the strategy $(\frac{2}{5}, \frac{3}{5}, 0)$ in particular – is a best response against $(\frac{3}{4}, \frac{1}{4})$. Similarly, $(\frac{3}{4}, \frac{1}{4})$ is a best response against $(\frac{2}{5}, \frac{3}{5}, 0)$. Hence, since equilibria correspond to strategies that are best responses against each other, $((\frac{3}{4}, \frac{1}{4}), (\frac{2}{5}, \frac{3}{5}, 0))$ is an equilibrium pair.

At this point it may seem reasonable to ask, "If any strategy is a best response against $(\frac{3}{4}, \frac{1}{4})$ – excluding those that might yield b_3 – why should player 2 choose $(\frac{2}{5}, \frac{3}{5}, 0)$ in particular?" The answer is that although 2 is indifferent over a considerable range of strategies, *provided that 1 chooses* $(\frac{3}{4}, \frac{1}{4})$, only $(\frac{2}{5}, \frac{3}{5}, 0)$ leaves 2 invulnerable. That is, only this mixed strategy ensures that 1 has no opportunity to reduce 2's payoff further by shifting from $(\frac{3}{4}, \frac{1}{4})$. This is the meaning of equilibrium.

4.4   Concave games and resource allocations in elections

Turning now to games that are explicitly continuous instead of those that we make continuous by mixed strategies, suppose that a game satisfies the conditions of Theorem 3.2 and suppose that each person's payoff or utility function is a concave function of his strategies. If the game is also zero sum, then each person's utility function is *convex* in his opponent's strategies. This result holds because the negative of a concave function is convex [if $f(x)$ is concave in x, then $-f(x)$ is convex in x]. For a zero-sum game, $u_1 = -u_2$, and if $u_1(a, b)$ is concave in a and $u_2(a, b)$ is concave in b, then $u_1(a, b) = -u_2(a, b)$ is convex in b. We can now use the following theorem, which is a corollary of Theorem 3.2.

> 4.2.   *For two-person zero-sum games, if the sets of strategies available to both players, A and B, are both convex, closed, and bounded, and if $u_1(a, b)$ is concave in a and convex in b, as well as continuous in a and b for all $a \in A$ and $b \in B$, then the game has an equilibrium in pure strategies.*

And *if we substitute the words "strictly concave" and "strictly convex" for "concave" and "convex," respectively, then the equilibrium is unique.*

Figure 4.6 graphs the payoff function for a typical concave–convex zero-sum game. Notice that the general form of $u_1(a, b)$ resembles a saddle, which is a consequence of being concave everywhere in a and convex everywhere in b. (Thus, Theorem 4.2 is often called the *saddle*

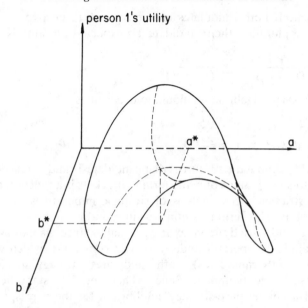

person 1's utility

4.6 Saddle point.

point theorem.) At the point (a^*, b^*), u_1 necessarily decreases as a increases or decreases from a^*, while u_1 necessarily increases as b increases or decreases from b^*. Hence, (a^*, b^*) is an equilibrium and is the unique point with this property.

Theorem 4.2 establishes a sufficient condition for the existence of an equilibrium. We can now use some properties of the maximization and minimization of continuous concave–convex functions to find such an equilibrium in specific games. Since equilibrium strategies are best responses to each other, if (a^*, b^*) is an equilibrium pair, then a^* maximizes $u_1(a, b^*)$ and b^* maximizes $u_2(a^*, b)$, or equivalently, b^* minimizes $u_1(a^*, b)$. To minimize and maximize a continuous function, we take derivatives of that function, set the derivates equal to zero, and solve simultaneously for a^* and b^* (a unique solution then results in the case of strictly concave–convex utility functions). A more sophisticated analysis, of course, should make certain that utility is not maximized at the boundaries of either person's strategy set (we refer to such solutions as corner solutions), but we ignore such complications here.

To illustrate this procedure applied to a particular game, consider again the example from Chapter 3, of two candidates who compete by allocating resources to the 50 states. Recall that the probability that a candidate carries a state equals the relative proportion of resources he allocates to it. If both candidates maximize their respective expected electoral vote plurality, then candidate 1's expected plurality is

$$\sum_{j=1}^{50} \frac{e_j x_j}{(x_j + y_j)} - \sum_{j=1}^{50} \frac{e_j y_j}{(x_j + y_j)}, \qquad (4.3)$$

which, after some algebraic manipulation, becomes

$$2\left[\sum_{j=1}^{50} \frac{e_j x_j}{(x_j + y_j)}\right] - \sum_{j=1}^{50} e_j,$$

where x_j and y_j denote the resources that candidates 1 and 2, respectively, allocate to state j. Since only the term in brackets is a function of the candidates' strategies, the analysis of strategic imperatives can focus exclusively on it. This term is continuous in x_j and y_j, and it is (strictly) concave in x_j and (strictly) convex in y_j. Finally, since we can assume that each candidate operates under a budget constraint (which we let equal X_o for both candidates), both candidates' strategy spaces are convex, closed, and bounded. Hence, Theorem 4.2 establishes the existence of a unique pure-strategy equilibrium for the corresponding game.

To find the equilibrium, we first set up the constrained function that candidate 1 seeks to maximize (recall our use of the Lagrange multiplier, H, in Chapter 1),

$$\sum_{j=1}^{50} \frac{e_j x_j}{(x_j + y_j)} - H\left[\sum_{j=1}^{50} x_j - X_o\right].$$

Differentiating with respect to x_j and setting the result equal to zero,

$$\frac{e_j y_j}{(x_j + y_j)^2} - H = 0 \quad \text{for } j = 1, \ldots, 50. \qquad (4.4)$$

Thus, focusing on two arbitrary states, say j and k, after solving for H, we find that

$$\frac{e_j y_j}{(x_j + y_j)^2} = \frac{e_k y_k}{(x_k + y_k)^2} \qquad (4.5)$$

An equivalent analysis for candidate 2 yields

$$\frac{e_j x_j}{(x_j + y_j)^2} = \frac{e_k x_k}{(x_k + y_k)^2} \qquad (4.6)$$

Rearranging the terms in expression (4.5) gives $(x_k + y_k)^2/(x_j + y_j)^2 =$

$e_k y_k / e_j y_j$, while similar manipulations of (4.6) establish that the ratio of the same two squared terms equals $e_k x_k / e_j x_j$. Hence,

$$\frac{e_k y_k}{e_j y_j} = \frac{e_k x_k}{e_j x_j}$$

Equivalently,

$$\frac{y_j}{y_k} = \frac{x_j}{x_k}.$$

Thus, the ratio of resources that a candidate allocates between any two states should be the same for the candidates. In equilibrium, if candidate 2 allocates twice the level of resources to state j that he does to state k, then candidate 1 should do the same thing. And since the candidates have identical budgets, this means that in equilibrium the two candidates should choose identical strategies. That is, $x_j = y_j$ for all j.

As a final step, if we substitute the fact that x and y are equal in any state, then equation (4.6) reduces to

$$\frac{e_j}{e_k} = \frac{y_j}{y_k}$$

or simply that, in equilibrium, candidate 2 allocates his resources in proportion to each state's electoral vote weight. The same argument shows that this conclusion holds for candidate 1 as well.

We should not interpret this analysis as a general conclusion about campaigning under the electoral college. First, it ignores the dynamic aspects of campaigning and the fact that candidates often adjust their campaign tactics as the election approaches. Because the analysis supposes that candidates make a single allocation decision that must serve them for all time, the preceding normal form perhaps oversimplifies an election's true extensive form by ignoring the dynamic nature of campaigns, which permits continual adjustment of allocations. Second, expression (4.3) imposes a special assumption about the relationship between resources and the probability of carrying a state. Although that assumption guarantees the concavity of payoff functions and thus the existence of an equilibrium, it is not necessarily the most plausible possibility. Nevertheless, this example does illustrate how theorizing about the electoral college might proceed. This model shows that there is no large- or small-state bias insofar as resources are concerned, and refinements might yield conclusions that we regard as satisfactory for policy prescriptions.

4.5   Symmetric games and candidate objectives

Although the preceding example might tax some readers' mathematical comprehension, computing equilibria is sometimes even more difficult. For one class of games, however, we can simplify the analysis considerably.

> *Symmetric two-person games*: A two-person game is symmetric if the strategy sets of both players are identical (if $A = B$), and if, when the players exchange strategies, they also exchange payoffs. That is, $u_1(a, b) = u_2(b, a)$ for all $a \in A$ and $b \in B$.

Thus, for two-person zero-sum games, because $u_1 = -u_2$, then $u_1(a, a) = u_2(a, a) = -u_1(a, a)$, or simply

$$u_i(a, a) = 0 \quad \text{for all } a \in A.$$

Hence, if (a^*, b^*) is an equilibrium pair, then

$$0 = u_1(a^*, a^*) \geqslant u_1(a^*, b^*) \geqslant u_1(b^*, b^*) = 0,$$

or equivalently, $v_1 = v_2 = u_i(a^*, b^*) = 0$. Thus, for symmetric, zero-sum, two-person games, the equilibrium payoffs to both players are zero. But, from the fact that equilibrium strategies in two-person zero-sum games are equivalent and interchangeable, this means that

$$u_1(a^*, b_k) \geqslant u_1(a^*, b^*) = u_1(a^*, a^*) = u_1(b^*, b^*) = u_1(b^*, a^*) \geqslant u_1(a^*, b_k),$$

so if (a^*, b^*) is an equilibrium pair, then (a^*, a^*), (b^*, a^*), and (b^*, b^*) are equilibrium pairs also. Hence, if a^* is an equilibrium strategy for one player, then it is an equilibrium strategy for the other player as well.

The electoral college example illustrates a two-person symmetric game in which both candidates hold identical resources and no candidate enjoys an advantage in any state. Hence, we could conclude directly (rather than deriving the result using calculus) that for at least one equilibrium, the candidates allocate their resources in an identical pattern. And since payoffs are strictly concave, we know that this is the unique equilibrium.

To illustrate these ideas about symmetric games with another example, consider the question of the equivalence of alternative candidate objectives in elections. Recall our earlier argument that often it is impossible to measure people's preferences directly, so that we must use

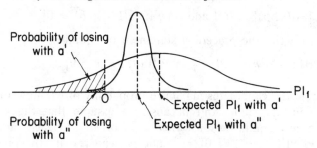

Probability of losing with a'

Probability of losing with a''

Expected Pl_1 with a'

Expected Pl_1 with a''

Pl_1

4.7 Probability of winning vs. expected plurality.

objective measures, such as votes or money as approximations for preference structures. Even if we limit discussion to objectives that depend solely on votes won or lost in two-candidate contests, however, we might use many different assumptions to represent a candidate's objectives. These include maximizing expected plurality, probability of winning, proportion of the vote, absolute vote, and so on. Unfortunately, we have no general means for deciding which is the more appropriate assumption.

Actually, we have stated this problem imprecisely, because we must define "equivalent objectives." Fortunately, game theory provides a definition: Two objectives are equivalent if they yield identical equilibria and thus identical predictions about outcomes. This definition shows us why it is important to conceptualize competitive processes in game-theoretic terms. Consider Figure 4.7, which graphs two probability distributions over candidate 1's plurality – one when candidate 1 adopts the strategy a' and the other for the strategy a'' (holding candidate 2's strategy fixed). Notice that although a' yields the greater expected plurality, the lower uncertainty associated with a'' yields a greater probability of winning. Thus, the two objectives of maximizing expected plurality and probability of winning appear distinct.

But suppose that the game is symmetric and that (a^*, b^*) is an equilibrium pair if the candidates maximize expected plurality. Because candidate 1's plurality, $\text{Pl}_1(a, b)$, is a random variable,

$$E[\text{Pl}_1(a^*, b)] \geq E[\text{Pl}_1(a^*, b^*)] = 0 \geq E[\text{Pl}_1(a, b^*)],$$

where E denotes expected value. But if the distribution of plurality is symmetric about its mean (for example, if it is normally distributed), then,

$$E[\text{Pl}_1(a, b)] \geq 0 \text{ if and only if } \Pr[\text{Pl}_1(a, b) \geq 0] \geq \tfrac{1}{2},$$

and

$E[\text{Pl}_1(a, b)] \le 0$ if and only if $\Pr[\text{Pl}_1(a, b) \ge 0] \le \frac{1}{2}$,

so we can rewrite the preceding statement of equilibria as

$$\Pr[\text{Pl}_1(a^*, b) \ge 0] \ge \frac{1}{2} \ge \Pr[\text{Pl}_1(a, b^*) \ge 0].$$

Of course, if the candidates maximize probability of winning, then the game is constant sum. And because we assume that the game is symmetric, any equilibria must yield a payoff of $\frac{1}{2}$ to each candidate. That is, $\Pr[\text{Pl}_1(a^*, b^*) \ge 0] = \frac{1}{2}$, and we can rewrite the preceding inequality as

$$\Pr[\text{Pl}_1(a^*, b) \ge 0] \ge \Pr[\text{Pl}_1(a^*, b^*) \ge 0] \ge \Pr[\text{Pl}_1(a, b^*) \ge 0],$$

which is simply the requirement for (a^*, b^*) to be an equilibrium pair under the hypothesis that candidates maximize probability of winning. Hence, any equilibrium under expected plurality maximization is an equilibrium under probability of winning maximization. We can easily reverse this argument to show that any equilibrium under probability of winning maximization is also an equilibrium under expected plurality maximization. Hence, for symmetric games and symmetric representations of uncertainty, the two objectives are equivalent.

Although we needed some special assumptions to derive this conclusion, it is important to emphasize what we do not assume. We make no specific assumptions, for example, about the substantive content of strategies or about whether the election is a simple direct-vote, majority-rule process or whether it is mediated by an electoral college. In this way we can begin to derive propositions about candidates' decisions that do not depend on specific details of electoral institutions.

4.6   Two-candidate elections with a single issue

Because objectives such as maximizing the probability of winning or maximizing plurality yield zero-sum payoffs to candidates, the study of two-candidate elections is one of the principal ways in which two-person zero-sum game theory is applied to politics. The discussion in Section 4.5, however, is simply a starting point, because it does not tell us what specific platforms candidates will adopt in response to the electorate's preferences. To form a complete model, we must establish the relationship between candidates' payoff functions and citizens' preferences. Here we proceed by assuming that the electorate is concerned with a single issue, called the *issue space*, while the next section generalizes the analysis to elections that concern several issues simultaneously.

Suppose that two candidates in a simple plurality-rule system are competing for office, and that the primary issue in the campaign is how the candidates weight social-service programs and military preparedness in government spending decisions. Although we can describe citizens' preferences in such elections by a two-dimensional commodity or coordinate space, the government's budget, which we suppose is fixed, constrains that space. Recall from Chapter 1 that if preferences satisfy some simple convexity and monotonicity conditions, then we have induced *single-peaked preferences* on the relevant budget constraint. Hence, if each candidate's election strategy is a position on this constraint, then single-peaked functions describe citizens' preferences over the constraint.

Consider a *symmetric, two-candidate, plurality-rule election*, by which we mean one that satisfies five conditions. First, all citizens vote. Second, all citizens know the candidates' positions on the issues, and partisian identification and the like play no role. All decisions are based on the issue. Thus, if candidates adopt identical strategies, the expected outcome is a tie. Third, the candidates' strategic opportunities are identical. Thus, neither candidate is constrained to the right or to the left on an issue, and both candidates can adopt any position on it. Fourth, candidates maximize plurality or some equivalent objective. If, for example, they maximize plurality and if all citizens vote, or if we suppose simply that a candidate's payoff is $+1, 0$, or, -1 – depending on whether he wins, ties, or loses – then the election game is zero sum. Finally, candidates know the form of citizens' preferences on the issue (that they are single peaked, symmetric, and so on), and they know the *preference distribution* of the electorate on the issue (they know whether ideal points bunch up to form a unimodal distribution, whether there are two opposing centers of opinion so that preferences are bimodal, or whatever). This means that for a specific pair of candidate strategies, each candidate knows his and the opponent's payoffs.

More formally, if a and b denote the strategies of candidates 1 and 2, respectively, where a and b are positions on the line denoting the relative weight each would give to defense appropriations, and if we denote the set of voters by $N = \{1, 2, \ldots, i, \ldots, n\}$, then 1's plurality is

$$\text{Pl}_1(a, b) = \text{num}\{i \in N \mid aP_ib\} - \text{num}\{i \in N \mid bP_ia\}, \quad (4.7)$$

where num denotes number of. An election equilibrium, then, is a pair of strategies, (a^*, b^*), satisfying, for all a and b:

$$\text{Pl}_1(a^*, b) \geq \text{Pl}_1(a^*, b^*) \geq \text{Pl}_1(a, b^*).$$

The previous section, however, tells us that maximizing expected plurality and probability of winning are equivalent if the election is symmetric and if the distribution of plurality is symmetric. This second symmetry requirement is obviously satisfied if there is no uncertainty (that is, if the distribution of Pl_i is a point-mass). Thus, we can also use the following objective function for candidates if there is nothing in the election model that biases outcomes in favor of one candidate or the other:

$$u_1(a, b) = \left\{ \begin{array}{c} 1 \\ 0 \\ -1 \end{array} \right\} \text{ if } Pl_1 \left\{ \begin{array}{c} > \\ = \\ < \end{array} \right\} 0. \tag{4.8}$$

We can now fill in this model with our assumptions about voter preferences, since these assumptions tell us, for specific candidate positions, how many voters prefer one candidate over the other. Although this is a highly abstract version of an election, we prove shortly that this model uncovers a general tendency in two-candidate, plurality-rule election systems that is an important result, called *the median voter theorem*:

4.3. *In symmetric two-candidate plurality-rule elections with only one issue on which voters have single-peaked preferences, in equilibrium both candidates adopt the median ideal preference.*

By *median ideal preference* we mean the point on the issue dimension, say, x^*, such that at least one-half the voters most prefer positions at or to the right of x^* and at least one-half most prefer positions at or to the left of x^*. In Figure 4.8, voter 3's ideal preference is the median ideal preference. (If the number of voters is odd, and if preference curves have no flat spots, then this median point is unique. But if the number of voters is even, or if ideal points are plateaus, then the median is generally a region rather than a unique point.)

Actually, we can restate this theorem in the terminology of Chapter 2. Proved in the 1950s by Duncan Black, the theorem defines a sufficient condition for a transitive social preference order under majority rule.

4.3′. (Black): *If all citizens' preferences are single peaked on a single dimension, then the median ideal preference is a Condorcet winner and the social preference order under simple majority rule is transitive, with the median standing highest in the order.*

Thus, in this *unidimensional* case, we have escaped the specter of Arrow's theorem as well as the possibilities for agenda manipulation that Theorem 2.2 describes. We accomplish this by violating Arrow's axiom A_2, unrestricted domain. The assumption of single peakedness rules out the kinds of individual preferences that generate intransitive social preferences.

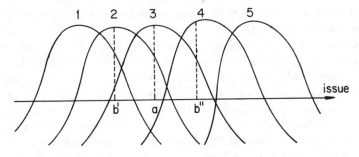

4.8 Single-peaked preferences.

To illustrate these theorems and their proof, consider the situation in Figure 4.8, which graphs the preferences of five voters and locates the first candidate's strategy, a, at the median ideal point, that of voter 3. Suppose that candidate 2 responds by choosing b' as his strategy. Then, while voters 1 and 2 vote for candidate 2, voters 3, 4, and 5 vote for 1. Hence 1's payoff is $+1$, while 2's is -1. Similarly, if candidate 2 adopts b'', then he secures the votes of voters 4 and 5, but again he loses the election because voters 1, 2, and 3 prefer candidate 1. Indeed, the best candidate 2 can do is to match his opponent at the median preference, in which case the election is a tie. This reasoning establishes that the median ideal preference is the Condorcet winner as well as the unique equilibrium to the zero-sum election game.

To establish the part of result 4.3′ that asserts that social preferences are transitive under majority rule, with the median ideal point standing highest in the social order, suppose that x, y, and z are three positions on the issue dimension and that xPy and yPz. Suppose also, without loss of generality, that x and y are both to the left of the median. Then if xPy, single peakedness implies that x must be closer to the median than y: that is, $d(x, a) < d(y, a)$, where d denotes distance. We now have two cases: (1) z is on the same side of the median as x and y and (2) z is on the opposite side of the median. For case (1), yPz implies $d(y, a) < d(z, a)$ so that $d(x, a) < d(z, a)$. From single peakedness again, all persons to the right of the median, a, prefer x at least as much as z, so, from the definition of the median, if x any z are paired in a majority vote, x receives at least half the votes. Thus, xRz. Indeed, single peakedness implies that all persons who prefer y to z also prefer x to z, so xPz. For case (2), consider the majority that prefers y to z. A simple paper and pencil exercise shows that single peakedness implies that alternative x, which lies between y and z, must be at least as good for these voters as z, in which case xPz, which establishes that the social preference order is not cyclic. And if we add a few details to this argument, we could show that intransitivities in the indifference relation are impossible as well.

Observe now that neither Theorem 4.3 nor 4.3′ assumes anything about how the electorate's ideal points are distributed on the issue. The theorems hold if ideal points are bunched up near the median (unimodally distributed) or if they are polarized between the two extremes of policy (bimodally distributed). The importance of these theorems, then, is that they establish a central tendency in two-candidate, plurality-rule elections at the electorate's median preference. Hence, if we believe that two candidates adopt distinct positions on important election issues, or that they should adopt such positions, then we must identify those features of reality that conflict with the preconditions of the theorems.

Before doing this, however, we should review the importance of Theorem 4.3′. It might seem that Theorem 4.3 is all that we require as a starting point to model elections. Uniqueness and the fact that equilibrium strategies maximize security levels give us some confidence that candidates choose such strategies. But attainment of an equilibrium in practice may still require trial and error on the part of a game's players, especially if the concept of an equilibrium is not a natural one to them. Theorem 4.3, then, gives us a prediction, but it may be a prediction that we prefer to augment with additional arguments for supposing that candidates actually choose such equilibrium strategies in elections. Theorem 4.3′ provides such an argument, since with transitivity we can imagine some simple electoral dynamics that lead to equilibria. If, for example, candidates sequentially adjust their platforms in response to each other, choosing strategies that beat an opponent's current position, spatial strategies necessarily move up the social preference order and eventually converge to the electorate's median preference.

Turning to extensions and limitations of these two theorems, we inquire first about their robustness with respect to the single-peakedness assumption. The unfortunate fact is that this assumption is critical. The nonsingle-peaked preferences that Figure 4.9 portrays, for example, correspond to the usual cycle $xPyPzPx$ with three voters, in which case the candidates have no pure-strategy equilibrium platforms. This does not mean that single peakedness is a necessary condition for transitive social orders or for the existence of an election equilibrium in pure strategies. Several alternative sufficient conditions for the existence of Condorcet winners and transitive orders appear in the literature. But these conditions lack the geometric simplicity of single peakedness, and they are not widely applied. Substantively, then, it is important to ask: When are we likely to witness single-peaked preferences, and when

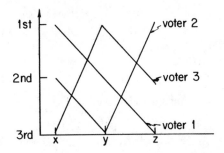

4.9 Nonsingle-peaked preferences.

should we anticipate multipeaked preference profiles? Although we cannot answer this question definitively, the most common answer grows out of experiences with multidimensional scaling of attitudes. There, researchers take the recovery of unidimensional single-peaked preferences from survey data to mean that voters use a common criterion, such as ideology, to evaluate alternatives (positions on the issue), and they might interpret multipeaked preferences to mean that voters use several criteria simultaneously. Hence, the existence of multipeaked preferences implies that the issue space is not unidimensional but multidimensional.

We emphasize again that Theorem 4.3 provides a limited model of elections. It assumes that all citizens vote even though candidates design strategies in many elections as much to affect turnout as voters' preferences between candidates. Further, Theorem 4.3 assumes that voters have *complete information* about the candidates' positions, an assumption that finds little empirical support. Third, the theorem assumes single peakedness, although multidimensional preferences seem a more likely possibility. In response to these and other criticisms a significant literature has developed on "spatial models of election competition." We cannot review that literature here, although shortly we review the problems that multidimensional competition and *incomplete information* occasion. Much of this literature, however, relies on the zero-sum formulation of election games, and the various attempts to render election models more realistic use theorems from game theory. These attempts incorporate voting participation as a function of candidate strategies, for example, with various assumptions relating the probability of voting to the utility that a citizen associates with each candidate's position. The functional relationships specified generally posit that these probabilities vary concavely with utility differences or with utility from a most preferred candidate, which, when combined with other appropriate assumptions, yields a game that satisfies the conditions of Theorem 3.2. Since we illustrate this approach in different contexts in subsequent chapters, we set it aside for the moment and turn now to elections that involve more than one issue.

4.7   Two-candidate multidimensional elections

The assumption that preferences are single peaked over a single issue avoids the instabilities associated with intransitive social preferences. Students of elections, however, find the assumption of a single issue unrealistic and unsuited for a general theory of elections. This section generalizes the median voter theorem, 4.3, but in so doing we find that an election equilibrium exists only under very restrictive conditions. To the extent that majority-rule elections entail more than one issue, then, the instabilities that intransitive social preferences create seem inescapable.

As in the unidimensional case, we begin with expression (4.7) or (4.8) as the specification of the candidates' objectives, and we identify voter preferences given fixed candidate positions. To describe the multidimensional context, we consider again the legislator from Chapter 1 who must choose not only a relative share for defense versus social-welfare spending, but a tax rate as well. There we induce multidimensional preferences on a budget-constraint simplex, so that a person's preferences form convex preference sets that we summarize by circular or elliptical indifference contours. Presently, we simplify discussion by considering only the case of circular indifference contours, which represent preferences by simple Euclidean distance from some ideal point (expression 1.1).

To characterize the relationships among candidates' payoffs, candidates' strategies, and the electorate's preferences, notice that if indifference contours are circles, and if a voter's utility declines monotonically with distance from his ideal point, then he should always vote for the candidate closest to his ideal. Thus, if two candidates are at distinct positions in the issue space, we can establish who votes for whom as follows: Draw a *perpendicular bisector*, a line that perpendicularly bisects the line connecting the two candidates' positions. All voters with ideal points on the same side of this bisector as candidate 1 prefer 1, and all voters with ideal points on the same side of the bisector as candidate 2 prefer 2. Voters with ideal points on the line are indifferent between the candidates. Thus, in Figure 4.10, which plots the ideal points of five voters (denoted x_1 through x_5) and two candidate positions (a for 1 and b for 2), voters 1, 2, and 3 prefer candidate 1, voter 4 prefers candidate 2, and voter 5 is indifferent between the two candidates.

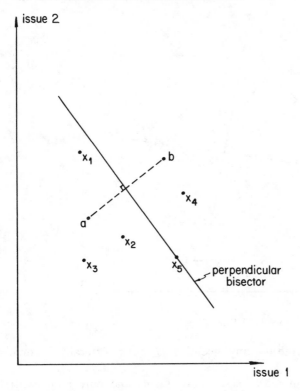

4.10 Identifying the preferences of voters for candidates.

This discussion fills in the equations for a candidate's payoff function, expression (4.7). To secure a positive payoff, a candidate must have a majority of voter ideal points on "his" side of the bisecting line. The question, then, is whether there is an equilibrium to the game that expression (4.7) describes, where strategies lie in some multidimensional issue space and where each voter votes for the candidate whose strategy is closest to his ideal. Before answering this question, however, we can make use of the assumption that the election is symmetric – that neither candidate, as in our one-dimensional model, enjoys an advantage owing to partisan identification, spatial mobility, and the like. If no advantages exist and if an equilibrium exists for the zero-sum or constant-sum election game, then the equilibrium outcome is a tie. And if a position is an equilibrium strategy for one candidate, then it is also an equilibrium strategy for his opponent. Hence, if an equilibrium exists, there must exist a point in the issue space such that if both candidates adopt it, then

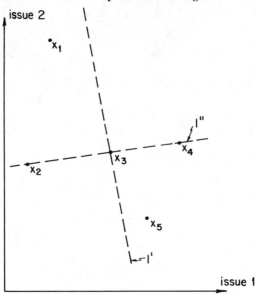

4.11 A median in all directions.

neither candidate has any incentive to shift unilaterally to some other position.

What we intend to show now is that the only point with these properties is the *median in all directions*, by which we mean

> *Median in all directions*: x^* is a median in all directions if *every* line through x^* divides ideal points thus: At least half are on or to one side of the line and half are on or to the other side of the line. Such a line is called a *median line*.

For the five ideal points in Figure 4.11, the ideal of voter 3 is a median in all directions. For example, if voters 1 and 2 are to one side of a line through x_3, say, l', then at least one-half of the ideal points are on or to one side of l' (x_1, x_2, and x_3), as well as on or to the other side (x_3, x_4, and x_5). Similarly, for the line l'', voter 1 is to one side of the line and 5 is to the other; thus, since 2, 3, and 4 are on l'', at least one-half are on or the either side of l''. Thus, both l' and l'' are median lines. Indeed, any line that we draw here such that three ideal points are *not* all to either side of the line (as against one or two being on one side and a third being on the line) must pass through x_3. Hence, x_3 is the median in all directions.

We can now state the generalization of the median voter theorem for

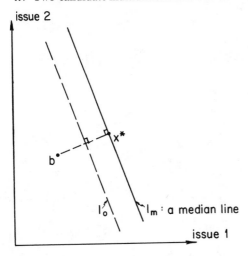

issue 2

l_o

l_m : a median line

issue 1

4.12 The perpendicular bisector.

multi-issue elections, which was first proved by Otto Davis, Morris P. DeGroot, and Melvin J. Hinich:

4.4. (Davis, DeGroot, and Hinich): *For symmetric majority-rule elections, if voters' preferences are given by simple Euclidean distance, a necessary and sufficient condition for a two-candidate pure-strategy equilibrium to exist is that there exists a median in all directions. This median is the equilibrium.*

To see the logic of this result, consider Figure 4.12 and suppose that x^* is a median in all directions. Suppose further that candidate 1 adopts x^* as his strategy while his opponent is at b. Then all voters to the left of the perpendicular bisector between the candidates, l_o, vote for 2, and all voters to the right of l_o vote for 1. But since x^* is a median in all directions, there exists another line, l_m, parallel to l_o, that divides the electorate in half. Hence, candidate 1 at x^* captures at least half of the electorate, those with ideal points on and to the right of l_m, and therefore 1 ties and possibly wins the election. Since this argument holds for any point $b \neq x^*$, the security value of x^* must be zero, and the security value of any other point is less than zero. Candidate 2's best response to x^* is to adopt x^* also, and to accept a tie. Hence, the median in all directions is an equilibrium-strategy pair for the corresponding election game and a Condorcet winner under simple majority rule.

The importance of Theorem 4.4 is not that it establishes a sufficient condition for an election equilibrium in pure strategies, but that, in the

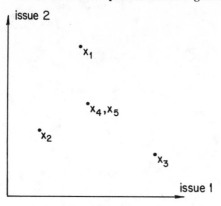

4.13 A median in all directions at x_4, x_5.

context of simple Euclidean preferences, it establishes a necessary condition as well. With it we can judge the extent to which the kind of stability that pure strategy equilibria imply characterize election competition. Theorem 4.4 also establishes a necessary and sufficient condition for the existence of a Condorcet winner. Hence, if preferences correspond to simple Euclidean distance (if indifference contours are circles), then this result establishes the conditions required for avoiding cyclic social preferences and the instability of agenda manipulation that Theorem 2.2 describes.

The critical questions, then, are the following:

1. How restrictive is the requirement that a median in all directions exists?
2. Do all preference distributions have such a point?
3. If the answer to the second question is no, can we supply any geometric intuition for distinguishing between distributions with and without such a point?
4. Are we any more or less likely to have an election equilibrium when preferences are characterized by something more general than simple Euclidean distance?

That we can design preference distributions with a median in all directions is trivial. Such a point exists if we pile up a majority of ideal points at a single location. Fortunately, other sorts of distributions have a median in all directions. Figure 4.13 illustrates a five-voter electorate in which less than a majority, voters 4 and 5, prefer the same point, and in which that point is such a median. But we must answer our first question in the negative: Not all distributions have a median in all

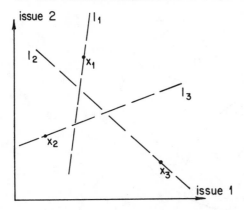

4.14 Three median lines that do not intersect at a common point.

directions. Consider the three-voter electorate in Figure 4.14, which reproduces the example that Chapter 2 uses to illustrate the power of agendas. Briefly, Figure 4.14 graphs three median lines, l_1, l_2, and l_3, which do not have a common intersection. Thus, for this electorate, there is no median in all directions and no pure-strategy election equilibrium.

To illustrate the instability of election competition that this preference configuration creates, consider Figure 4.15 and the point a'. Through a' we have drawn three indifference contours, one for each voter, which reveal that a point such as a'' beats a' two votes to one. The point a''' defeats a'', however, while losing to a'. Thus, we have a cycle, so that the security values of a', a'', and a''' all equal -1. Since the security value of equilibrium strategies must be 0 in the symmetric election game, and since we chose a' arbitrarily (that is, our argument will apply to any point we might have chosen), the game does not have an equilibrium in pure strategies. Hence, it does not have a Condorcet winner.

Since not all preference distributions have a median in all directions, it is important to develop some sense of Theorem 4.4's restrictiveness. If we start with one voter with an ideal at x, then trivially this ideal is the median in all directions. Now let us begin adding voters in pairs, to keep the number of voters odd while at the same time maintaining x as the median in all directions. Referring to Figure 4.16, suppose that the ideal of the first voter added is x_1. To maintain x as the median in all directions, the ideal of this voter's partner, x_2, must fall somewhere on the line running through x and x_1, but to the other side of x. This construction ensures that we cannot pass a line through x and have both

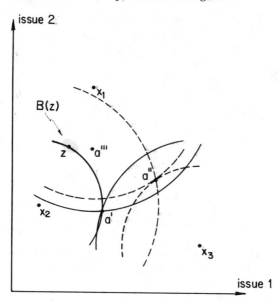

4.15 A cycle with circular indifference contours.

x_1 and x_2 lie on the same side of such a line. And since our electorate at this point only contains three voters, this construction means that all median lines must pass through x. (In Figure 4.16b we locate two points, x_1 and x_2, that violate this construction, so that the line l_o, which is a median line, does not pass through x.) If we add a second pair of voters to Figure 4.16a, with the first partner at x_3, then the second partner's ideal point, x_4, must fall on a position such as the one shown.

If we continue adding pairs of ideal points in the same fashion so that each pair plus x lie on a common line, it is evident that we are describing a distribution of a particular sort, one that is *radially symmetric*, so that for every voter on one side of the equilibrium point there is necessarily another voter on the other side of this point. Thus, the five-voter example in Figure 4.11 satisfies this radially symmetric property, and it yields an election equilibrium at its center, voter 3's ideal point. The three-voter distribution in Figure 4.14, on the other hand, does not satisfy this condition (compare Figures 4.14 and 4.16b), and a pure-strategy equilibrium does not exist.

Radial symmetry is not a necessary condition for a median in all directions, since our construction assumes that at most one voter prefers the presumed Condorcet point and since we have kept the number of voters odd. Nevertheless, the restrictiveness of the assumption that a

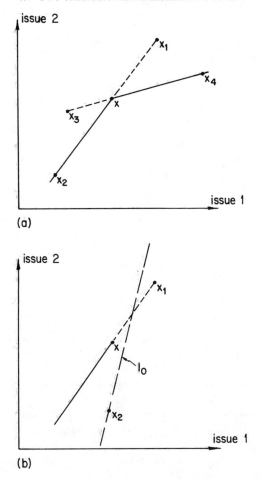

4.16 Building a symmetric distribution of preferences.

median in all directions exists is apparent. Looking at Figure 4.11, suppose that we change one voter's ideal slightly, either right or left. Regardless of how small the change, we violate radial symmetry, and no pure strategy equilibrium exists. Thus, *the existence of an election equilibrium and the corresponding Condorcet point is sensitive to the exact distribution of preferences.*

Answering our fourth question about Theorem 4.4, this sensitivity is not limited to circular, compared with the more general case of convex, preference sets. We can reinterpret the assumption that a median in all directions exists with circular indifference contours to mean that the

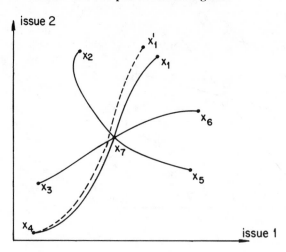

4.17 The Plott condition

Pareto-optimals of all majority collections of voters have a common intersection. Recall that in the case of circular indifference contours, contract curves are straight lines. Hence, if a point is on a median line, then it necessarily is Pareto-optimal for the collection of voters that include all voters to one side of the line plus those voters with ideal points on the line. Since the line is a median, by definition this collection is a majority. Thus, if a point is on every median line, then it is Pareto-optimal for *every* majority. If, on the other hand, indifference contours simply describe convex sets (for example, ellipses), then any Condorcet winner still must be Pareto-optimal for every majority. In this event, contract curves are no longer straight lines necessarily, and thus a median in all directions becomes a moot consideration. But Charles Plott proves that if only one voter most prefers the presumed Condorcet point, then we must be able still to pair voters, so that all contract curves have a common intersection. Figure 4.17 illustrates this pairing and shows that although a Condorcet winner exists if voter 1's ideal point is at x_1, no Condorcet winner exists if we move it even slightly to x_1'.

Here we should ask whether Theorem 4.4's assumptions are so restrictive that it is unimportant to the student of politics. Spatial outcomes and circular indifference contours are, after all, not the only possibilities. Certainly, we should be concerned if a theorem's conditions are restrictive, especially if it presents only a sufficient condition for something to be true, such as Theorem 3.2. In this event, though, we can either regard that theorem as relevant or irrelevant to a particular problem. It is valuable information, provided that we do not misuse it. But Theorem 4.4 is different. Although limited to the case of simple Euclidean preferences, it presents a *necessary*, as well as a sufficient, condition for the existence of a Condorcet winner and an election equilibrium. Other more general assumptions about preference are possible, but experience has taught us that the existence of election equilibria and Condorcet winners is no more likely if we substitute more general asumptions. Thus, Theorem 4.4 leads some scholars to conclude that pure-strategy equilibria are rare in spatial elections. We should keep in mind, of course, that the mathematical and geometric precision of our examples do not match reality. Candidates are uncertain about voters' exact preferences, and voters are only imperfectly informed about candidates' strategies. Such considerations require that we treat theorems such as 4.4 as starting points for a theory, not as ultimate conclusions. Despite this disclaimer, the instability in elections that this analysis reveals is of paramount importance. It signals that identifying the public policies that result from particular configurations of citizens' preferences is not a trivial task.

4.8   Disequilibrium with income redistribution policies

Chapter 1 derives Euclidean preferences in our illustrative discussion from preferences over the specific goods and services that governments provide, defense and social services. Although such preferences are amenable to a broad interpretation, they are not necessarily well suited for modeling another government activity, income redistribution. Hence, we should see how we can model elections that concern such matters. We are curious about whether the issue of redistribution affects the likelihood that equilibria exist in two-candidate elections. Our conclusion is that such issues probably diminish this likelihood.

We begin with an election game concerned solely with income redistribution among three voters. Suppose that the total income of our three-person society equals 1, that x_i is the income of person i, and that $x_1 + x_2 + x_3 = 1$. Finally, suppose that each person's utility for a specific distribution of income $\mathbf{x} = (x_1, x_2, x_3)$ is

$$u_i(\mathbf{x}) = a_{1i} x_1 + a_{2i} x_2 + a_{3i} x_3.$$

Hence, if i is concerned solely with his own income and cares nothing about anyone else's welfare, $a_{ji} = 0$ if $j \neq i$, and $a_{ii} = 1$. Permitting a_{ji} to exceed 0 for $j \neq i$ is one way of assuming that i cares about the income of others. Next, assume that a victorious candidate in a two-candidate

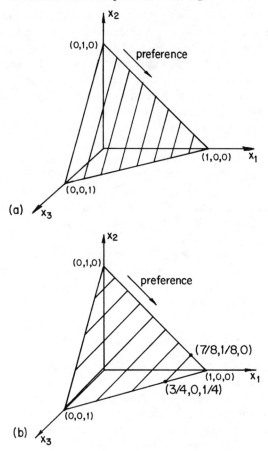

4.18(a) Indifference contours for voter 1 with $a_{11} = 1$, $a_{21} = a_{31} = 0$.
(b) Indifference contours for voter 1 with $a_{11} = 1$, $a_{21} = 0$, $a_{31} = .5$.

election selects the eventual income distribution. Each candidate's strategy is a distribution satisfying the constraint $x_1 + x_2 + x_3 = 1$. We let $\mathbf{x}_A$ and $\mathbf{x}_B$ be the strategies of candidates A and B, respectively, and we assume that each voter votes for his or her most preferred distribution.

Notice that a budget constraint, $x_1 + x_2 + x_3 = 1$, described by a simplex in three dimensions, again determines the strategies available to each candidate. To visualize the nature of election competition over this constraint, consider Figure 4.18, which illustrates some indifference contours for voter 1. In Figure 4.18a, $a_{21} = a_{31} = 0$, so that 1 cares only about his own welfare. Since $u_1(\mathbf{x})$ is a linear and increasing function of x_1, 1's indifference curves are straight lines. Voter 1's ideal point is at $x_1 = 1$ and his utility declines as we move from this point. Since he does not care about the welfare of others, his indifference contours correspond simply to constant values of x_1. In Figure 4.18b, however, $a_{11} = 1$, $a_{21} = 0$, and $a_{31} = .5$. In this instance, voter 1 cares most about himself, somewhat about person 3, and nothing at all about 2. These indifference contours are again straight lines, owing to the linear form of $u_1(\mathbf{x})$, but, unlike those in Figure 4.18a, these contours pitch toward person 3. For example, now voter 1 is indifferent between, say, the distributions $(\frac{3}{4}, 0, \frac{1}{4})$ and $(\frac{7}{8}, \frac{1}{8}, 0)$, both of which yield him a utility of $\frac{7}{8}$.

To appreciate the instability of electoral competition that this structure implies, consider Figure 4.19a, which shows the indifference contours for all three voters over the simplex, assuming that each cares only about his own welfare. Focusing on the point $\mathbf{x'}$, we can see that $\mathbf{x'}$ cannot be an equilibrium strategy. Moving away from $\mathbf{x'}$ in the direction of any of the three arrows moves a candidate to more preferred indifference contours for two voters and, thus, this strategy would defeat a candidate at $\mathbf{x'}$ by two votes to one. Points on the boundary of the simplex such as $\mathbf{x''}$ are also readily defeated. In this instance, no election equilibrium exists because every point can be defeated by some other point. Admitting considerations of equity, moreover, does little to resolve this instability. Figure 4.19b illustrates a situation in which 1 and 3 give some weight to each other's welfare, but none to 2's. Again any arbitrary point, such as $\mathbf{x'}$, can be defeated by strategies in the figure's shaded triangles.

We can establish equilibrium within this model's structure, but it requires an unlikely or trivial set of a_{ji}'s, which exists, for example, when all voters give positive weight only to one specific voter's welfare (in which case an equilibrium strategy is to redistribute all income to that voter) or when all voters weight the welfare of everyone in society equally (in which case any distribution of income is an equilibrium). Another approach, which also illustrates the application of Theorem 3.2, assumes that people *vote probabilistically*. Specifically, assume that the candidates know each voter's preference for redistribution, but that neither candidate is certain that this is the only criterion voters use in the election. Thus, letting p_{iA} denote the probability that citizen i votes for candidate A, and, assuming that there is no abstention, so that $p_{iB} = 1 - p_{iA}$, suppose that

$$\frac{p_{iA}}{p_{iB}} = \frac{u_i(\mathbf{x}_A)}{u_i(\mathbf{x}_B)}.$$

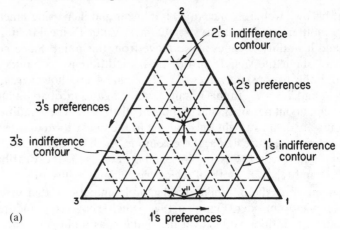

(a)

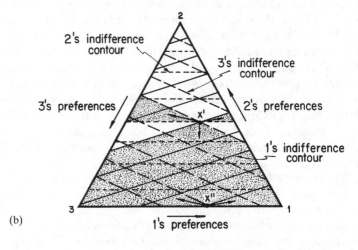

(b)

4.19(a) Disequilibrium for voters with no altruism. (b) Disequilibrium with two altruistic voters.

That is, the ratio of the probabilities equals the ratio of the utilities voter i associates with the candidates' strategies. Finally, to make the analysis tractable, assume that, with respect to public sector activities, voters are wholly self-centered, so that $a_{ji} = 0$ for $j \neq i$. Thus, from the definition of u and from the preceding expression,

$$\frac{p_{iA}}{(1 - p_{iA})} = \frac{x_{iA}}{x_{iB}} \text{ and } \frac{p_{iB}}{(1 - p_{iB})} = \frac{x_{iB}}{x_{iA}},$$

which, after rearranging terms gives,

$$p_{iA} = \frac{x_{iA}}{(x_{iA} + x_{iB})} \text{ and } p_{iB} = \frac{x_{iB}}{(x_{iA} + x_{iB})}.$$

Candidate A's expected plurality, then, is

$$Epl_A(\mathbf{x}_A, \mathbf{x}_B) = \sum_{i=1}^{3} p_{iA} - \sum_{i=1}^{3} p_{iB}$$

$$= \sum_{i=1}^{3} \left[\frac{x_{iA}}{x_{iA} + x_{iB}} - \frac{x_{iB}}{x_{iA} + x_{iB}} \right].$$

With the exception of the electoral college voting weight, e_i, and the fact that this expression concerns 3 voters rather than 50 states, this function is equivalent to expression (4.3). Thus, applying the earlier analysis here, we conclude that our redistribution election game has a unique pure-strategy equilibrium, in which each candidate proposes to divide income equally among all voters. That is, $\mathbf{x}_A^* = \mathbf{x}_B^* = (\frac{1}{3}, \frac{1}{3}, \frac{1}{3})$.

This analysis, in conjunction with the previous section's discussion, shows that the existence of an election equilibrium in our models depends on what assumptions we make about voting – deterministic or probabilistic. Equilibrium is assured in deterministic models only if we impose severe restrictions on the electorate's preference distribution, whereas the equilibrium just described can be shown to exist for far weaker assumptions about this distribution. But we cannot ensure the existence of an equilibrium with probabilistic voting without incurring some cost. That cost is reflected in the fact that not all functions relating voter utilities to probabilities of voting ensure the existence of an equilibrium. Generally, we must assume that utility functions are concave functions of candidate positions and that the probability of voting for candidate k is a concave function of the utility a voter associates with k's strategy. And such assumptions cannot always be justified.

Are probabilistic or deterministic assumptions more reasonable? We cannot provide a definitive answer to this question, but probabilistic assumptions are reasonable if we consider that candidates are rarely certain about voter preferences and who will vote. Information from public opinion surveys is not error-free and is best represented as statistical. Hence, if we want to design models that take cognizance of the kind of data that the candidates are likely to possess, probabilistic models seem more reasonable. On the other hand, the conclusions we deduce using probabilities depend on the specific functional forms that relate those probabilities to candidate strategies. Deterministic assumptions yield cleaner propositions, and are better suited to uncovering fundamental forces that operate in politics. Choosing between deterministic and probabilistic assumptions requires knowing research intent: uncovering general theoretical propositions, or molding those propositions and their accompanying perspective to the limitations of our data.

4.9   Mixed strategies and the uncovered set in elections

Regardless of what assumption about voting we make, it is unsatisfactory to conclude that a pure-strategy equilibria need not exist in two-candidate elections, since this conclusion leaves us without any hypotheses about eventual strategies and outcomes for a wide class of situations. Shortly, we will see how we might use a concept called "the uncovered set" to make predictions. First, though, recall that Chapter 3 introduced the concept of a mixed strategy to provide mathematical closure to the issue of the existence of an equilibrium. Hence, the reader may wonder why we limit the discussion here to pure strategies. Indeed, there are good reasons for believing that mixed strategies are inappropriate in the present context. To grasp this lack of appropriateness, we must first distinguish between mixed and *risky* strategies, which people often confuse, although the two are different. A mixed strategy is a lottery that a player uses in choosing some pure strategy before a game begins. In an election context, if a candidate abides by some mixed strategy, then, ultimately, the electorate sees only the pure strategy he chooses by lottery. If a candidate adopts a risky strategy, though, then the electorate sees the lottery as that candidate's platform. Risky strategies are pure strategies, except that a candidate uses them to augment his alternative platforms to include obfuscation of the policies that he will adopt if he wins the election.

We can envision circumstances in which candidates might prefer risky to riskless strategies. And the analysis of such strategies can be challenging. For example, suppose that an election concerns a single issue, and that we can model a candidate's risky strategies by a normal density function over this issue. Since its mean and variance characterize a univariate normal density, the candidates have two parameters to vary strategically. Similarly, voters must assess candidates by these same two parameters. In effect, admitting risky strategies expands the issue space to two dimensions, a mean and a variance dimension, so that even with one substantive issue, Theorem 4.3 is not relevant. Theorem 4.4 is also not relevant, since the induced preference sets in the mean-variance space, in general, will not be circular or even convex. Hence, the analysis of risky strategies is complex and resistant to simple generalizations. The most straightforward result is that if all voters are risk averse (if utility functions in the original issue space are concave), then all voters prefer zero variance, and a pure-strategy equilibrium exists, as before, at the electorate's median preference.

Although risky strategies make sense, the concept of a mixed strategy seems inappropriate, at least for the theoretical perspective that this chapter adopts. Here, we represent election competition as a normal-form game, in which the sole campaign strategy is the choice of an issue platform. But real elections are dynamic events, with extensive forms that permit candidates to adjust their tactics as the campaign proceeds. Although our models abstract this complexity away, it seems silly to conceptualize candidates spinning spinners or rolling dice to choose policy platforms. Mixed strategies seem even less reasonable because they can result in pure strategies that an opponent can readily defeat. Suppose, for example, that the electorate looks like the one depicted in Figure 4.15 and that, after an appropriate choice and implementation of mixed strategies, candidate 1 chooses a' and his opponent chooses a''. Candidate 1 would lose the election, so it seems unlikely that he will not try to adjust his platform, finding comfort in the fact that he chose his losing strategy in accordance with some game-theoretic concept.

The problem is that our static conceptualization of campaign strategies is an abstraction that we must treat carefully. This abstraction is reasonable if we are concerned primarily with the policy implications of majority-rule systems and if there is some platform that attracts the candidates, even in a dynamic context. Theorem 4.4 establishes a necessary and sufficient condition for a Condorcet winner with spatial preferences. If coupled with Theorem 2.3, Theorem 4.4 provides a necessary and sufficient condition for a transitive social preference order under majority rule. In this context we might reasonably argue that the campaign dynamics that we have ignored are unimportant, because candidates will be led to the equilibrium. But if an equilibrium does not exist, then such dynamics may be critically important. And to argue that candidates are drawn somehow or can otherwise compute a mixed strategy equilibrium supposes a normal form that accommodates dynamic complexity.

To illustrate this argument in a somewhat different context, consider again the problem of presidential-candidate resource allocations discussed in Section 4.3. There, our assumptions about the relationship between resources and the probability that a candidate wins a state result in a continuous concave game that has a pure-strategy equilibrium. If we accept those assumptions, then our conclusion that candidates allocate resources in proportion to each state's electoral weight is reasonable, even with dynamic electoral competition. But suppose that the probability of capturing a state follows a different rule; namely, the candidate who allocates the greater resources to a state wins that state with certainty. In this instance a candidate's payoff function is neither concave nor continuous, and the corresponding election game has an equilibrium only in mixed strategies. To illustrate one such equilibrium, let each candidate's resources total 1 and suppose that the election concerns three identically weighted states, so that a simplex in three dimensions describes the set of all possible allocations. If we inscribe a circle in this triangle that is tangent to its three sides, with a center at the point of equal division of resources among the three states, then a mixed-strategy equilibrium for this game, called a Colonel Blotto game, is given by the density function $f(x) = \frac{9}{2}(1 - 9r^2)^{1/2}$, where r is the distance from the circle's center to x and where $0 \leq r \leq \frac{1}{3}$. (Imagine the complexity of a mixed-strategy equilibrium with 50 unequally weighted states.)

Suppose that, after abiding by this mixed-strategy equilibrium candidate 1 allocates his resources in the proportion $(\frac{1}{2}, \frac{1}{3}, \frac{1}{6})$, and 2 chooses $(0, \frac{1}{2}, \frac{1}{2})$. Thus, 2 wins the election, two states to one. This is possible because a mixed-strategy equilibrium does not promise a tie or a win as the final outcome to this symmetric zero-sum game, but simply that, ex ante, *the expected outcome* is a tie. Certainly, as this strategy unfolds, candidate 1 should adjust his allocations, just as real presidential candidates target their efforts on close states in a campaign's later stages.

The concept of a mixed strategy here makes little sense, since any candidate should defect from it if he is losing. The reason that it makes little sense is that the analysis models a dynamic situation, an election campaign, as static, and if it does not consider a candidate's opportunities for adjustment and recalculation, then the assumed extensive form is not the true extensive form. Earlier, when the game had a pure-strategy equilibrium (a unique spatial strategy that, if adopted, could not be defeated in a majority vote), this abstraction of the extensive form appeared acceptable. It did so because we implicitly (and reasonably) assumed that if such a strategy exists, then candidates would be drawn to it regardless of the game's extensive form. But in the absence of a pure-strategy equilibrium, it no longer seems reasonable to apply the lessons learned from a static model.

One alternative to modeling elections in which pure-strategy equilibria do not exist when we describe these elections in static terms is to model their extensive forms more carefully. Elections are complex

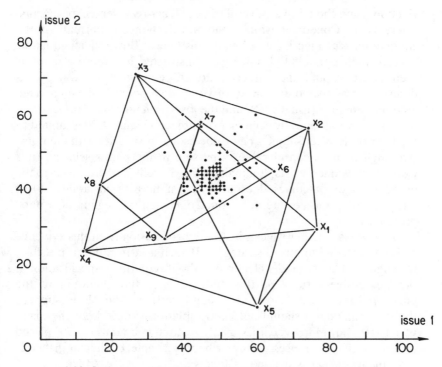

4.20 Outcomes of last five trials of nine-voter election without an equilibrium.

events, however, and their extensive forms are difficult to specify. Yet it seems unreasonable to suppose that we cannot say something about candidates' strategies if pure-strategy equilibria do not exist. Suppose, for example, that the electorate consists of six million and one voters and that one million voters' ideal points are distributed uniformly on each of six concentric circles, with the final odd voter located exactly at the center of these circles. Given this precise symmetry, Theorem 4.4 states that an equilibrium platform to the two-candidate election game exists at this center. But suppose that we move *one* voter's ideal point on the outer ring ever so slightly, to eliminate this symmetry. Since the conditions of Theorem 4.4 are necessary as well as sufficient, a pure-strategy equilibrium no longer exists. But what positions would two candidates choose? It seems apparent that they would not choose a position far from the original center.

There is experimental support for the proposition that points near the center remain attractive, even in the absence of a pure-strategy election equilibrium. Figure 4.20 plots the points that subjects playing a simple election game choose over several trials and across several experiments. There is no Condorcet winner and no election equilibrium for the configuration of nine ideal points in this figure. But subjects in ignorance of their opponent's choice were instructed to choose a point as their strategy, and the subject who chose a winning strategy, as determined by the configuration of voter ideal points, won some fixed amount of money (usually $2), and the loser won nothing. This task was repeated several times by each pair of subjects. Figure 4.20 is limited to the later trials, after subjects had gained some familiarity with the rules. Although there is considerable dispersion in the points selected (if a Condorcet winner exists, then there is virtually no dispersion), the distribution of outcomes does not suggest massive instability. Most outcomes do not deviate far from some central region of voters' preferences.

The issue is whether we can justify the conclusion from this evidence without resorting to mixed strategies. If such a justification is possible, we might be able to model elections without requiring a specification of their extensive forms. To this end we offer the definition of the uncovered set. As we show subsequently, it is reasonable to suppose that two candidates should not choose platforms outside of the uncovered set, and, moreover, reasonably restrictive bounds can be placed on this set if preferences are described by simple Euclidean distance. The uncovered set is defined as follows:

> *Uncovered set*: Let O be the policy space and $W(x)$ the points in O that defeat $x \in O$. Then y *covers* x if and only if y defeats x and the points that defeat y, $W(y)$, form a proper subset of $W(x)$, the set of points that defeat x. If $U(x)$ is the set of points that x does not cover, the uncovered set of O, $U(O)$, is $\{x \in O \mid$ for no y in O does y cover $x\}$, or equivalently
>
> $$U(O) = \bigcap_{x \in O} U(x).$$

Notice that $U(O)$ is nonempty, since, at the very worst, it can include all of O. The relevance of the uncovered set to elections, however, follows from the properties of three sets, $W(x)$, $I(x)$, and $D(x)$. $W(x)$ is, as the preceding definition states, the set of points that defeat x. Similarly, $I(x)$ is the set of points that tie x, and $D(x)$ is the set of points that x defeats. What we want to show is that if y covers x then

$$I(y) \subset W(x) \cup I(x) \tag{4.9}$$
$$D(x) \subset D(y). \tag{4.10}$$

The importance of such results in the context of elections is that if each candidate's payoff is $+1$, 0, or -1, depending on whether he wins, ties, or loses the election, then *if y covers x, y dominates x as an election strategy*: Not only does y defeat x, but any platform that ties y defeats or ties x and any platform that defeats y defeats x also. Thus, candidates should eliminate covered points because they are dominated. Hence, whether or not we believe that a mixed strategy makes sense in an election context, candidates should confine their attention to those points that are uncovered by any other point, the uncovered set.

The proof of (4.9) and (4.10) requires explicit use of the assumption that $I(x)$ is thin, which is equivalent to assuming that indifference curves have no interior and that we can describe them by lines or curves that have no thickness. Referring to Figure 4.15, notice that the points that defeat a', $W(a')$, are bounded by segments of the voter's indifference contours. Thus, points that tie a', $I(a')$, correspond to these segments, so that if indifference contours are thin, then $I(a')$ is thin also. And if $I(a')$ is thin, if z is in $I(a')$, and if we form any arbitrarily small region around z that includes z in its interior, denoted $B(z)$, then this region contains points that defeat a' (points in $W(a')$) and other points that a' defeats (points in $D(a')$).

With these preliminaries, we can prove expressions (4.9) and (4.10). We include their proofs here to illustrate exactly why specific assumptions about preference and utility are important and how we use them to ensure that results follow rigorously.

First, we prove expression (4.10) by showing that if we assume the opposite, we get a contradiction. Suppose, then, that $D(y) \subset D(x)$. This must mean that there is a z that x defeats (that is, z is in $D(x)$), but that y does not defeat (is not in $D(y)$). But then z must be in $I(y)$. Otherwise, if z is in $W(y)$, z must be in $W(x)$ as well since y covers x, and this contradicts the assumption that z is in $D(x)$. The set $D(x)$, though, is an open set, since it does not include its boundary, $I(x)$. From the definition of an open set, we can find an arbitrarily small region around z, $B(z)$, that is wholly in $D(x)$. But because z is in $I(y)$ and $I(y)$ is thin, there exists a $w \in B(z)$ such that w is in $W(y)$. Again, since y covers x (so that $W(y) \subset W(x)$), w must be in $W(x)$, which contradicts that $B(z)$ is wholly in $D(x)$. By contradiction, then, $D(x) \subset D(y)$. Expression (4.9) is established in the same way. Suppose the point z is in $I(y)$ but that it is not in $W(x) \cup I(x)$, so that, contrary to (4.9), $I(y)$ is not a proper subset of $W(x) \cup I(x)$). Then z must be in $D(x)$. But this is impossible, since we began by assuming that z is in $I(y)$ and since, in accordance with expression (4.10), if z is in $D(x)$, then it must be in $D(y)$. Thus, *in a two-candidate, majority-rule election game, if candi-*

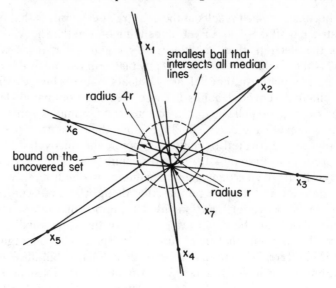

4.21 Example of bounds on uncovered set.

dates limit strategies to undominated points in the issue space, then they choose points in the uncovered set, U(O).

What is interesting about this result is that if indifference contours are concentric circles, then we can place definitive bounds on $U(O)$. Referring to Figure 4.21, suppose that we construct the smallest circle that intersects every median line, and suppose that this circle has radius r and its center is at x_o. Then $U(O)$ is necessarily contained in a circle (a ball in n-dimensions) of radius $4r$, centered at x_o.

The proof of this result is tedious, and we do not present it here. But its importance is evident: It reestablishes the central tendency of majority-rule elections in a policy space, a central tendency that appears to disappear if preferences are not distributed in a radially symmetric fashion or if a median in all directions does not exist. Briefly, the radius r is a measure of the absence of symmetry in the preference distribution so that in the perfectly symmetric case, all median lines intersect at a common point, $r = 0$, and $U(O)$ is the median in all directions. As symmetry declines, r and the bounds on $U(O)$ grow. For our earlier example of six million and one voters, then, slight distortions of ideal points should cause only small increases in r and only small regions in which the candidates would compete.

Both the theoretical and substantive implications of the preceding discussion are important. Often theorists, when modeling the interaction between political and economic forces, make extraordinary assumptions to ensure that political competition concerns a single issue or dimension. They do this because a median in all directions is not likely to exist if there is more than one issue and because of the mistaken belief that if no such median exists, then no prediction about candidate platforms or public policy can be uttered. But the properties of the uncovered set and the bounds we can place on it justify the prediction that the candidates will choose strategies near the median on each issue that is salient in the election. Substantively, these bounds imply that *two-candidate majority-rule elections do not evolve into "chaos" in the likely event that a set of restrictive conditions are not met. Rather, the attractiveness of the center of public opinion is maintained.*

4.10 Rational expectations and voter ignorance

The preceding discussion gives us some confidence that game-theoretic concepts such as a Nash equilibrium and dominated strategies provide predictions about candidates' strategies. But we should be concerned that we have not merely solved some mathematical puzzles that, although challenging, are unrelated to real elections. We cannot review all of spatial theory or the debates over its empirical adequacy, but we do examine an important issue that illustrates the power of formal modeling as well as a new game-theoretic idea.

One obviously inaccurate assumption thus far is that voters have complete information about candidate positions. Indeed, given the costs of becoming informed and the small or zero probability that any vote is decisive in an election, it is probably irrational for people to gather much information about politics and politicians unless they derive the same pleasures from politics as they derive from other spectator sports. Thus, the relevance of complete information models to politics seems suspect. Models that rely on the assumption of complete information may offer a misleading view of democratic processes. If voters are uninformed about the candidates' issue positions, then might not those who are informed or who can afford to incur the cost of information, such as interest-group leaders and bureaucrats, have an excessive voting weight that undermines popular control of public policy and that at least partly restricts the preceding election model's relevance?

One way to treat incomplete information is to assume that voters perceive only risky candidate strategies. Of course, we then have the problem of modeling voters' *beliefs* about the corresponding probability densities. If these densities are not the same for all voters, then how do they differ? Can candidates manipulate uncertainty? Finally, how might the strategic choices of other voters affect a voter's beliefs about candidates' strategies?

These are important research questions; a subtle modification in the definition of a Nash equilibrium illustrates one approach to some answers. First, for any noncooperative game in normal form, we can restate the definition of a Nash equilibrium thus: A strategy n-tuple is an equilibrium if, *given the players' beliefs about the game's structure* (for example, about each player's utility function, what he believes other players believe about him, and so on), no player has an incentive unilaterally to choose some other strategy after the remaining n-1 players reveal their strategies. Thus far, we assume that people know the game they are playing. Although nature may make chance moves and players may choose strategies by lottery, we assume that everyone knows everything there is to know about a game's extensive form, including other players' motivations and the strategies available to everyone. Thus, the italicized part of the preceding definition usually remains unstated.

Suppose, however, that a particular game violates this assumption, and suppose that a player believes something about an opponent that simply is not true. Then, after the game is played, the outcome or the strategic choices of others might reveal the player's belief to be false, in which case he might prefer to revise that belief as well as the strategy that he chose (although it may be too late to do so). This process suggests that we can revise the preceding definition as follows:

> A strategy n-tuple and the beliefs of the players about the game being played are in equilibrium if, after all players reveal their choices, no player has an incentive to change his strategy or to revise his beliefs.

With complete information about a game's structure, the reference to beliefs is redundant: No player should revise his beliefs since they correspond to reality and if everyone acts accordingly, players learn no contradictory information after the game ends. With incomplete information, however, if the same persons play a game a second time, then a different outcome might prevail, since what players learn from the first play could cause some persons to change their beliefs and therefore their strategies.

Suppose that people play a game several times in succession, and that they begin with imperfect or inaccurate information about their opponents. After each game is played, suppose that they revise their beliefs and change their strategies according to what their opponents' previous choices have revealed. Now, one of three things can happen: first, a final equilibrium of beliefs and strategies never occurs; second, an equilibrium of beliefs and strategies prevails, but it is the wrong equilibrium, because some or all people believe something that is incorrect; and third, beliefs converge to reality and players choose the correspondingly correct strategies. We are most interested in this third possibility, since it suggests that people can learn a game's relevant characteristics simply by playing it.

To apply these ideas to elections, notice that two-candidate elections are really $n + 2$ person games, in which n is the number of voters. Thus far in this chapter, however, we have ignored voters as participants and concentrated on candidates instead. We justify this approach theoretically because if voters must choose between the strategies "vote for candidate 1" and "vote for candidate 2," then voting for the preferred candidate is dominant. Thus, our approach has assumed that voters choose dominant strategies, which they can readily identify, given candidates' positions and voter issue preferences. It is legitimate, then, to assert that the election equilibria that previous sections describe are actually $n + 2$-tuples satisfying the following condition:

i. No voter, given his belief about the candidates' positions, and no candidate, given his belief about voter preferences, has an incentive unilaterally to choose some other strategy,

This condition is unexceptional, and it parallels the usual definition of a Nash equilibrium, except that it explicitly defines equilibrium strategies conditional on beliefs. Now we can assert an additional condition pertaining to these beliefs. Instead of assuming that beliefs do not change and either mirror reality or are described by some fixed probability distribution, we assume that they depend on the strategic choices of others. To state this condition formally, and to revise the corresponding definition of an election equilibrium, notice that the relevant beliefs for candidates ordinarily concern the location of the median voter. Limiting discussion for the present to unidimensional elections, let c_k denote candidate k's belief about the location of the median, let a and b denote candidate 1 and 2's strategies on the issue, respectively, let s_j denote voter j's strategy, and let s denote the n-tuple of all voters' strategies. Suppose that candidate 1, believing that he is at the median, loses the election. Thus, he (or his party in its next nominating convention) should revise his belief about the median and perhaps adjust his position in the direction of his opponent. This situation illustrates a belief that does not satisfy the following equilibrium condition:

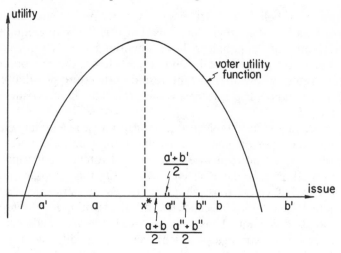

4.22 A single-peaked, symmetric utility function.

ii.   Given a, b, and s, no candidate has any incentive to revise his belief, c_k, about the location of the median preference.

A strategy n-tuple is a Nash equilibrium if no person has an incentive to revise his strategy after all other persons reveal their strategies. Condition ii adds the requirement that no candidate would revise his beliefs about the median preference after the remaining participants (voters and the election opponent) reveal their strategies.

Turning to voters, suppose that each voter's utility function is both single peaked and symmetric about its ideal point. The voter's ability to identify a preferred candidate, then, does not depend fully on his knowing the candidates' exact locations. Instead, if a voter knows which candidate is to the left on the issue and the location of the *midpoint* between the candidates, then he can identify the candidate closest to his ideal. Figure 4.22 illustrates three pairs of candidates' positions and a symmetric, single-peaked utility function. For all three pairs, candidate 1 (who is associated with the points a, a', and a'') is to the left of candidate 2, whereas the midpoints between the candidates [$(a + b)/2$, $(a' + b')/2$, and $(a'' + b'')/2$, respectively] are to the right of the voter's ideal, x^*. Hence, in all three cases, the voter should prefer candidate 1 to 1's opponent. This example, contradicts the supposition that citizens can "vote on the issues" only if they know the candidates' issue positions. Casting a correct vote does not require such specific knowledge about candidates' issue positions; instead, the voter needs to know only that candidate 1 is to the left of 2 and that the midpoint between the candidates is to the right of the voter's ideal. Thus, if we view citizens as investors in information who are concerned solely about casting a correct vote, then it is irrational for them to invest in learning specific positions instead of identifying the location of the midpoint between candidates and their left–right issue orientation.

Suppose, then, that h_j is voter j's belief about the location of the midpoint between the candidates and, to simplify notation, assume that h_j also summarizes j's beliefs about the candidates' left–right placement. Then the equilibrium condition on voter beliefs is:

iii.   Given a, b and s, no voter has an incentive to revise his belief, h_j, about the midpoint between the candidates.

Conditions i, ii, and iii define a system of simultaneous equations that an equilibrium must satisfy. Since i assumes that a voter's strategy is a function of his belief about which candidate he prefers, we write $s_j = s_j(h_j)$. Similarly, to represent the part of condition i that implies that the candidates' strategies are a function of their beliefs about the location of the median, let $a = a(c_1)$ and $b = b(c_2)$. Conditions ii and iii, now, suppose that beliefs are a function of the strategies that voters and candidates choose. Thus, we write $h_j = h_j(s, a, b)$ and $c_k = c_k(s, a, b)$. If s^*, a^*, and b^* are equilibrium strategies, and if h_j^* and c_k^* are equilibrium beliefs, then, in equilibrium, beliefs and strategies must reproduce themselves. That is:

$$a^* = a(c_1^*), \quad b^* = b(c_2^*), \quad \text{and } s_j^* = s_j(h_j^*), \quad \text{from i}$$
$$c_k^* = c_k(s^*, a^*, b^*), \quad \text{from ii}$$
$$h_j^* = h_j(s^*, a^*, b^*), \quad \text{from iii}$$

Conditions ii and iii are unstated in models that assume complete information, because those models assume that beliefs mirror reality, and that beliefs do not change, regardless of the election's outcome. That is, although ii and iii express h and c as functions of s, with complete information they are invariant with s. With incomplete information, however, ii and iii are no longer superfluous, and to describe equilibria, we must specify how beliefs change as voters and candidates reveal their strategies.

What especially interests us are equilibria in which all voters and candidates choose the strategies that complete information dictates. That is, under what conditions is it the case that a unique equilibrium of strategies and beliefs exists in which this equilibrium yields the same strategic choices that prevail if everyone enjoys complete information?

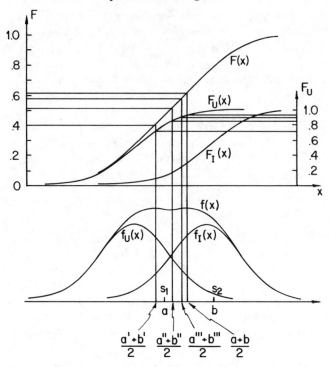

4.23 Inferring midpoints from polls.

To identify such conditions we must first model the way in which voters and candidates form beliefs about the relevant parameters.

Focusing on voters, consider this scenario: In place of a costly study of candidates' platforms, past performance, and the like, voters instead monitor public opinion polls. Suppose that a voter prefers policies to the right of center. But this voter is unsure of how extreme candidate 1 might be even though he knows that 1 is more conservative than candidate 2. The voter's specific concern is that 1 fits the extremist portrait that his liberal detractors have painted. Initial polls, though, might reveal that among those voters responding to questions such as "If the election were held today, for whom would you vote?" 1 and his opponent split about evenly. Several hypotheses are consistent with such a poll. First, respondents might simply be uniformly badly informed, in which case the poll indicates nothing. Second, a reasonable proportion of those responding know the candidates' policy positions, and either (1) 1 and his opponent are on opposite sides of but near the median or (2) 1 is an extremist on the right, but his opponent is an

believe that the midpoint is not at $(a + b)/2$, then some will change their strategies (votes). To see that true beliefs and optimal votes in accordance with those beliefs are an equilibrium, suppose that everyone believes that the midpoint is $(a + b)/2$. Figure 4.23 shows the resulting poll under poll n. The inference that a voter draws from this poll is that $(a + b)/2$ is actually the midpoint. That is, a belief that the midpoint is $(a + b)/2$ simply reproduces itself through the poll. So voters' strategies and beliefs are in equilibrium. Finally, if voters can ultimately infer the true midpoint between the candidates, and if candidates can also follow the polls, but if candidates have not converged to the median, then (ignoring some complications that arise if the midpoint corresponds to the median without convergence), at least one candidate (the one who is losing) must revise his belief about the location of the median. Thus, all beliefs and strategies are in equilibrium only if candidates converge to the electorate's median preference and if voters' beliefs correspond to this reality.

This example does not mean that voters and candidates act as if they enjoy complete information under all conditions, or that an equilibrium requires the assumptions underlying the preceding discussion. This discussion is a simple illustration of the use of an equilibrium concept that looks at beliefs as well as strategies. Although models that look at the historical data that voters possess as well as other sources of information have been explored, the development and exploration of such equilibria are new to political science. Also we must be careful when applying this perspective. If a game is sequential, then people might prefer to misrepresent their preferences, because they prefer that others believe something about them that is false. Strategies in this context become exceedingly complex. It is probably not unrealistic to suppose that citizens vote sincerely in two-candidate elections and that they do not think much about how their expressions of preference affect other voters, but things can be different in, say, a legislature. Thus, Chapter 10 explores one such situation in regard to the development of reputations. The preceding example simply illustrates how we can apply a different equilibrium concept to an important issue, the electorate's ability to make correct choices even if information does not correspond to what pundits usually assume is necessary for "voting on the issues."

This example is also substantively significant. Many students of elections assume that democratic institutions work only with an informed citizenry, arguing that if citizens fail to keep abreast of public debate and fail to understand the positions of election candidates on those issues, then special interests will dominate government action.

Player 3

Player 2:	c_1		c_2	
	b_1	b_2	b_1	b_2

Player 1:					
a_1	3, 1, -4	2, -2, 0		3, 0, -3	0, -3, 3
a_2	1, -2, 1	2, -1, -1		2, -2, 0	1, 0, -1

4.25 Three-person zero-sum game.

The preceding model cannot address such an argument fully, of course, but it does demonstrate that incomplete information is not a sufficient condition for the failure of democratic institutions.

4.11   N-person zero-sum games

Section 4.10 departs from the theoretical confines of two-person games to consider equilibria if the active players include voters. Thus, the analysis there concerns $n + 2$-person games that are not necessarily zero sum. In that analysis, however, we encounter little difficulty in moving to a larger game, since at any stage in the voting, each voter has a dominant strategy depending on his beliefs: Either vote for candidate 1 or vote for candidate 2. In general, however, the theorems of the preceding sections about interchangeability, equivalence, and uniqueness do not apply to n-person games even if we limit the discussion to zero-sum games.

Consider the three-person finite, normal-form game in Figure 4.25 and observe, first, that (a_1, b_1, c_2), (a_2, b_2, c_2), and (a_2, b_2, c_1) are pure-strategy equilibria. These strategy 3-tuples, however, are neither equivalent nor interchangeable. For example, (a_1, b_1, c_1) is not an equilibrium n-tuple. How, then, can we justify the hypothesis that players choose equilibrium strategies? Suppose, for example, that each player chooses his respective minmax strategy. Then, 1 should choose a_2, 2 should choose b_1, and 3 should choose c_2. But (a_2, b_1, c_2) is not an equilibrium. Thus, if one player, say, 1, thinks that his opponents will choose in accordance with the minmax criterion, then he should choose some strategy other than his minmax strategy, namely, a_1. But this reasoning begins the "he thinks" regress anew.

The problems that this simple example occasions also illustrate the problems associated with extending our analyses of two-candidate competition to multiple-candidate contests. For example, if voters' preferences are distributed uniformly over a circle, then a messy paper and pencil exercise shows that arranging three candidates in an equilateral triangle (with sides whose length depend on specific parameters) centered in the circle constitutes an equilibrium. Unfortunately, any rotation of this triangle is an equilibrium, but no other 3-tuples are equilibria. Thus, although all equilibria yield an equal three-way division of the vote and are equivalent, they are not interchangeable.

If we limit discussion to unidimensional contests, then we can better see the problems associated with analyzing multicandidate, plurality-rule contests. First, we must specify the candidates' payoffs. (As before, the strategy spaces consist of a segment of the real line that we interpret as an issue.) Even if we limit ourselves to objectives that depend solely on the votes that candidates win or lose, in a multicandidate, plurality-rule election there are several possibilities. One is that candidates maximize their respective proportion of the vote, which if everyone votes, yields a constant sum game. Another possibility is that candidates maximize plurality, so that candidate *j*'s objective function is

$$\text{Pl}_j(\mathbf{s}) = v_j(\mathbf{s}) - \max_{i \neq j} \{v_i(\mathbf{s})\}, \tag{4.11}$$

where Pl_j denotes the candidate's expected plurality, $\mathbf{s}$ is the strategy *n*-tuple of the *n* candidates, $v_j(\mathbf{s})$ is candidate *j*'s vote, and $\max \{v_i(\mathbf{s})\}$ denotes the strongest opponent's vote. To ensure that Pl_j is defined for all strategy *n*-tuples, we could add that if two strategies give a candidate the same plurality with respect to the strongest opponent, then the candidate prefers the strategy that maximizes his plurality with respect to the second strongest opponent, with respect to the third strongest, and so on. If citizens' preferences are single peaked over the issue, and if citizens vote for the candidate closest to their ideal, then it is straightforward to see that Pl_j is neither a continuous nor a concave function of *j*'s strategy. Hence, Theorem 3.2 is not relevant, and we must ascertain the properties of equilibria by a careful analysis of the particular game.

With one issue and three candidates, no pure-strategy equilibrium exists. For any arrangement of candidates' positions, the candidates at either extreme prefer to shift toward the center candidate, since this shift detracts from the center candidate's vote without causing any loss of support from the extremes of preference (assuming constant turnout). This shift squeezes the center candidate until his vote is reduced to zero (or, at best, to $\frac{1}{3}$ if, whenever all three candidates adopt the same position, they divide the vote equally). Before this happens, though, the center candidate should jump over one of his opponents and place that opponent in the squeezed position. No pure-strategy equilibrium to this process can exist, since either the candidate in the center has an incentive to jump or one or both of his opponents have incentives to shift toward the center. If candidates cannot jump, then the process ultimately eliminates the center candidate, yielding a two-candidate equilibrium at the median preference.

This discussion does not mean that if three or more candidates compete, an equilibrium in a plurality-rule system cannot exist. It means that the existence of an equilibrium requires that special conditions be satisfied. For example, with expression (4.11) Gary Cox shows that if preferences are uniformly distributed over an interval such as the [0, 1] interval in Figure 4.26, and if the number of candidates is *even*, then an equilibrium exists. Letting n be the number of candidates, if candidates that adopt the same position split the vote of citizens that prefer them, if citizens vote for the candidate closest to their ideal, and if voters' preferences are symmetric and single peaked, then a Nash equilibrium exists in which the candidates pair off and choose positions at specific positions across the issue.

Cox's analysis proceeds formally, but a simple four-candidate example illustrates his argument. Referring to Figure 4.26, suppose that two candidates are at $\frac{1}{4}$ and two are at $\frac{3}{4}$. Since voters' preferences are symmetric and single peaked, all voters to the left of $\frac{1}{2}$ prefer the candidates at $\frac{1}{4}$, and all voters to the right of $\frac{1}{2}$ prefer the candidates at $\frac{3}{4}$. Thus, each candidate receives one-quarter of the vote and none has a positive plurality. To check whether this configuration of candidate positions is an equilibrium, we must establish that if any candidate defects unilaterally, then his plurality necessarily drops below zero. Suppose that one of the candidates at $\frac{1}{4}$ shifts to position x. Then all voters with ideal points between $(\frac{1}{4} + x)/2$ and $(\frac{1}{2} + x)/2$ prefer this candidate, and all voters between 0 and $(\frac{1}{4} + x)/2$ prefer the solitary candidate remaining at $\frac{1}{4}$. The vote that the candidate who has shifted receives, then, is the area of the shaded rectangle, which is simply $(\frac{1}{2} + x)/2 - (\frac{1}{4} + x)/2 = \frac{1}{4}$, since the height of this rectangle is 1. This candidate gains nothing in terms of vote share; but his actions increase the share of one opponent – the candidate whose position remains at $\frac{1}{4}$. That candidate now receives the proportion $(\frac{1}{2} + x)/2 - 0 = \frac{1}{4} + x/2$. Thus, in accordance with expression (4.11), and as asserted, the shifting candidate's plurality becomes negative. As distinct from a three-candidate situation, if n is even, then the opportunity to pair pins candidates and keeps them from converging to the center. The final step

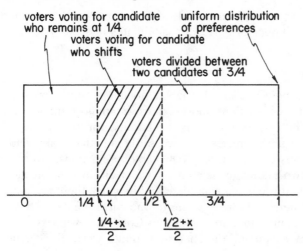

4.26 An equilibrium with four candidates.

in the analysis is to check whether there are other configurations of strategies that are in equilibrium. The usual way to proceed is to conduct an exhaustive examination of possibilities. First, we might suppose that the candidates are not all paired, in which case it is evident that one unpaired candidate must be the furthest to the left or right, and that he has an incentive to shift unilaterally toward the center. Second, we can consider candidates that are paired at other positions. For example, if the candidates are paired at $\frac{1}{3}$ and $\frac{2}{3}$, then the candidates at $\frac{1}{3}$ have a unilateral incentive to shift to the left, and the candidates at $\frac{2}{3}$ have the same incentive to shift to the right. Proceeding in this way verifies that the equilibrium $(\frac{1}{4}, \frac{1}{4}, \frac{3}{4}, \frac{3}{4})$ is unique.

This model provides some justification for supposing that, although two candidate contests yield convergence to the median preference (Theorem 4.3), differentiation will occur among candidates in multi-candidate, single-member constituencies. The assumption that preferences are distributed uniformly, however, is critical for establishing an equilibrium. Other assumptions, of course, should be considered. Hinich and Ordeshook, for example, show how different representations between the candidates' strategies and probabilities of voting can yield equilibria in multicandidate elections. Nevertheless, the general implication of this research is that, unless special assumptions are satisfied, it is difficult to ensure a pure-strategy equilibrium in a plurality-rule system. This accords with the traditional intuition that the equilibrium number of parties or candidates in such systems is two.

4.12   Summary

This chapter looks at a specific class of noncooperative games, zero-sum games, and reviews their most extensive application, two-candidate elections. This class is important also because for two-person games, at least, equilibrium strategies are necessarily interchangeable and equivalent, and thus are especially believable predictions. Moreover, for symmetric two-candidate elections, in which all citizens vote deterministically for their preferred candidate and in which there are no informational asymmetries, there is a one-to-one correspondence between the candidates' equilibrium strategies and Condorcet winners.

Hence, by examining the conditions under which the election modeled as a two-person zero- or constant-sum game between two candidates has a pure-strategy equilibrium, we are in effect examining the conditions under which the spectre of McKelvey's theorem about cyclic social preferences can be avoided. Our general conclusion, however, is that although unidimensionality or symmetric preference distributions ensure the existence of an equilibrium, these are strong conditions, and in general, election games have no pure-strategy equilibria unless we augment the models with additional assumptions (for example, probabilistic voting). And because these conditions are necessary and sufficient for the existence of pure-strategy equilibria and Condorcet winners in the context of spatial preferences, they tell us that the instabilities identified by Theorem 2.2 are likely to be a factor in many political processes.

Chapter 3 offers the concept of a mixed strategy as a possible way to "solve" games without pure-strategy equilibria. Applying this concept to our election models seems inappropriate, however, because those models abstract away too much of the dynamic complexities of real campaigns. But even if we resist applying the concept of a mixed strategy in this context, spatial models also show that the existence of a pure-strategy Nash equilibrium is not a necessary condition for rendering predictions about events. The concept of dominated strategies and the related definition of the uncovered set allow us to say something about candidates' strategies if no Condorcet winner exists. The uncovered set, for example, provides a justification for supposing that candidates will not choose strategies far from the center of voters' preferences, even if the distribution of those preferences does not yield a median in all directions. Thus, the existence of a pure strategy equilibrium is not required to predict a centralizing tendency in two-candidate plurality-rule elections.

Aside from using zero-sum games to identify the conditions under which electoral equilibria and Condorcet winners exist, the discussion of candidate objectives shows how we can use game theory to establish important equivalencies in objectives. Although the discussion focuses on candidates, it provides a general perspective for learning whether our models or speculations are sensitive to the specific way that we measure and represent the goals of key actors in a political game. Similarly, the discussion of incomplete information, although restricted to a specific election mechanism, is a counterexample to the proposition that our democratic institutions work only if information is perfect. Some might argue that if voters are poorly informed about candidates' issue positions, then the electoral imperatives that abstract models of elections describe are at best weak and at worst irrelevant. By expanding the definition of a Nash equilibrium to include beliefs as well as strategies, however, our analysis demonstrates that people can learn in indirect ways, and that as people adjust their strategies and beliefs in response to the decisions of others, incomplete information need not lead them from the decisions they would have made if their information had been complete.

The discussion of multicandidate elections shows the problems of prediction if we cannot satisfy the conditions of interchangeability and equivalence. But again, Cox's analysis reveals that the multicandidate problem does not preclude analysis. Minimally, we find new reasons for believing that plurality-rule systems favor two-party competition; if three or more candidates compete, then, except in special circumstances (an even number of candidates and a uniform distribution of preferences), centrist candidates will find their support eroded as the remaining candidates converge toward them.

And this chapter only scratches the surface of the research that exists under the rubric "spatial models of elections." In addition to the topics that we review here, that research considers alternative formulations of probabilistic voting, multimember district and proportional representation systems, models that accommodate abstentions, elections in which candidates must first secure the nominations of their parties, elections in which interest groups control valued resources other than votes, candidates with policy preferences, sequential elections with complete and with incomplete information, elections with constraints on candidates' strategies, and elections that concern the growth of the public sector, as well as a myriad of empirical applications to elections both in the United States and Europe. The review this chapter offers is simply a starting point. Before readers can approach that literature, they must be familiar with the following terms and concepts:

zero-sum game	preference distribution	interchangeability
equivalence	constant-sum game	minmax strategy
symmetric game	security level	unidimensional
single peaked	maxmin strategy	incomplete information
complete information	median ideal preference	median line
median voter theorem	perpendicular bisector	cover
median in all directions	radially symmetric	uncovered set
probabilistic voting	issue space	beliefs
candidate midpoint	risky strategies	saddle point

CHAPTER 5

Nonzero-sum games: political economy, public goods, and the prisoners' dilemma

We can model many political situations as zero sum, and the tools that Chapter 4 develops are thus an important part of political theory. Nevertheless, the zero-sum assumption is a restriction that precludes modeling many other situations that are central to politics. If politics concerns the collective allocation of resources, and if those resources are not fixed, as they need not be in an expanding dynamic society, then the principal political processes are necessarily nonzero sum in character. This chapter begins the exploration of the more general class of nonzero-sum, noncooperative games by focusing on a particular game, the prisoners' dilemma. We find this focus useful for several reasons. First, prisoners' dilemmas involving two or more persons illustrate many of the problems that one is likely to encounter in applying noncooperative game theory to politics. Second, the prisoners' dilemma itself models a great many key political situations, including the causes of market and governmental failure, the incentives for political participation, paradoxes of vote trading, and the disincentives of people to reveal truthfully their demand for publicly provided goods and services. And with the prisoners' dilemma as our starting point, we can introduce an additional modification of the Nash equilibrium concept, Bayes's equilibria, as well as tax schemes (called demand-revelation mechanisms) that are designed to induce truthful revelation of preferences for the goods and services that governments produce or regulate.

5.1 Nonzero-sum games

In one sense, moving from zero-sum to nonzero-sum games is not a radical theoretical departure. The representation of a situation's extensive form and the analysis of that form depend not at all on whether the game is zero sum or nonzero sum. Further, Theorem 3.2, which showed that every finite zero-sum game has an equilibrium in pure or mixed strategies applies to nonzero-sum games as well. Thus, Theorem 5.1 is simply a corollary to that theorem:

> 5.1. *Every finite nonzero-sum, noncooperative game in normal form has at least one equilibrium n-tuple in pure or mixed strategies.*

Person 2

	b_1	b_2
a_1	2, 1	0, 0
a_2	0, 0	1, 2

Person 1:

5.1 Nonequivalent and noninterchangeable equilibria.

We find the reason for distinguishing between zero-sum and nonzero-sum games in Theorem 4.1, which guarantees that equilibrium strategies for zero-sum games are necessarily equivalent and interchangeable. No such theorem exists for nonzero-sum games, because we can make no such extension. The 2×2 normal-form game in Figure 5.1 illustrates this limitation. Although both (a_1, b_1) and (a_2, b_2) are equilibrium pairs, they are not equivalent; nor are the equilibrium strategies interchangeable.

The failure of Theorem 4.1 to apply to nonzero-sum games means that we must approach more cautiously the concept of a Nash equilibrium as a prediction about outcomes. Referring again to Figure 5.1, suppose that person 1's first instinct is to seek the equilibrium that he prefers, (a_1, b_1). If, however, 1 supposes that person 2 will approach the game as he does, that 2's motivations are the same as his, and that 2 will also seek his most preferred equilibrium by choosing b_2, then 1 should instead choose a_2 and avoid the mutually distasteful outcome $(0, 0)$. But if 1 applies this same reasoning to 2 by conjecturing that 2 will also think that 1 will opt for his best outcome, then 1 would conclude that 2 will choose b_1, since he also wishes to avoid a $(0, 0)$ cell, and that 1 should respond with a_1. But again, for the same reasons that 1 infers that he should choose a_1, he can also conclude that the same reasoning, if applied to 2, would lead 2 to choose b_2. Clearly, this reasoning can continue indefinitely, so if 1 and 2 cannot coordinate their decisions, then the mere existence of Nash equilibria is not sufficient to terminate the infinite "he-thinks-that-I-think" regress.

Solving such games in the sense of arriving at definitive choice predictions may require an appeal to considerations that lie outside of a game's representation. For example, if two persons must each choose a number between 1 and 10, and if their payoffs are $25 each if they choose the same number and zero otherwise, then we suspect that most people will choose 5 because it is central. There is no conflict of interest in this instance, and both persons should attempt implicitly to coordinate their choices: Numbers that are central are a logical focus for such

Person 2

		b_1	b_2
	a_1	.75, −.25	.75, 0
Person 1			
	a_2	0, .75	1, 0

5.2 A game with only mixed-strategy equilibria.

coordination. Thus, (5, 5) is a reasonable prediction. But every pair (x, x), $1 \leq x \leq 10$ is an equilibrium, and if there is no reason to suppose that both persons are attracted to a particular pair, our predictions may be indeterminate. Such indeterminacies are disquieting (although we shall encounter others as we proceed), and thus we shall be concerned with the uniqueness of equilibria (thereby ensuring equivalence and interchangeability) and with concepts such as perfect equilibria, which reduce the number of predicted outcomes.

The analysis of mixed strategies also differs somewhat for nonzero-sum as compared with zero-sum games. Consider the game in Figure 5.2, and notice that no pure-strategy equilibrium exists. If this game were zero sum, then we could find the appropriate mixed strategy for person 1 by graphing 1's expected payoff against his mixed strategies, first when 2 chooses b_1, and then when 2 chooses b_2 (as in Figure 4.3, for example). For the game in Figure 5.2, however, this process yields the conclusion that a_1 maximizes 1's security level, but we already know that a_1 is not an equilibrium strategy. The problem here is that we have not used our knowledge of 2's payoffs. In the zero-sum case, maximizing one's payoff is equivalent to minimizing the opponent's payoff; for nonzero-sum games we must account more explicitly for the opponent's likely responses. We can find mixed-strategy equilibria, nevertheless, by looking at each person's best responses. Suppose that 1 and 2 adopt the mixed strategies $\mathbf{p} = (p, 1 - p)$ and $\mathbf{q} = (q, 1 - q)$, respectively. Hence, if $(\mathbf{p}^*, \mathbf{q}^*)$ is a Nash equilibrium, then $\mathbf{p}^*$ is 1's best response to $\mathbf{q}^*$ and $\mathbf{q}^*$ is 2's best response to $\mathbf{p}^*$. Looking first at person 1's decision and at the term $u_1(\mathbf{p}, \mathbf{q}^*)$, we find that after some algebra the game in Figure 5.2 yields

$$u_1(\mathbf{p}, \mathbf{q}^*) = 1 + p(q^* - .25) - q^*.$$

Thus, if $q^* < .25$, then person 1's best response is $p = 0$ (that is, choose a_2 with certainty), whereas if $q^* > .25$, 1's best response is $p = 1$. But if q^* is exactly .25, then person 1 is indifferent among all mixed strategies and has no incentive to shift from any particular mixed strategy. With respect to player 2, after some algebra we arrive at

$$u_2(\mathbf{p}^*, \mathbf{q}) = q(.75 - p^*).$$

If $p^* < .75$, then 2's best response is $q = 1$, whereas if $p^* > .75$, then 2 should set q equal to 0. But if $p^* = .75$, then 2 has no incentive to shift from any mixed strategy to any other. Hence, for $q^* = .25$ and $p^* = .75$, the players are in equilibrium: Neither has any incentive to shift unilaterally to some other strategy, pure or mixed.

Aside from how we must compute mixed-strategy equilibria and the problems that nonequivalent and noninterchangeable equilibria occasion, the construction and analysis of nonzero-sum games are identical to their zero-sum counterparts. Hence, we can now proceed to the focus of this chapter, *prisoners' dilemma* games. The reasons for this attention are, first, such games have unique equilibria and thus they avoid any problems of equilibria that are neither equivalent nor interchangeable. Second, the equilibrium to which the players seem drawn is not Pareto-optimal, and both players prefer to avoid it. Finally, this game models much that is central to politics. Indeed, versions of the prisoners' dilemma provide the normative economic justification for government itself.

5.2   The two-person prisoners' dilemma

The prisoners' dilemma derives its name from this scenario: Suppose that two prisoners, who are factually guilty of a felony, are locked in separate cells by the district attorney. Contemplating their fates, they each perceive two strategies: (s_1) stonewall the DA and admit nothing, and (s_2) turn state's evidence and confess all. The DA, however, sees an opportunity to extract a confession if he keeps the prisoners from communicating with each other and if he implements the following incentives: Realizing that if neither prisoner confesses, a conviction on the felony is at best doubtful, he nevertheless promises to make life miserable with convictions for several lesser offenses and a 10-year sentence at one of the state's less luxurious incarceration facilities. With a confession, he can ensure a felony conviction, and if only one prisoner turns states evidence, then the DA promises him parole in 8 years, while threatening the less cooperative felon with 20 years and no hope of parole. Of course, if both confess, then both are convicted and the DA regards himself as less bound by any promises made with respect to securing a confession (a confession is now a cheap commodity), and both felons receive 15-year sentences. This game is depicted in Figure 5.3 in normal form, with payoffs in terms of years of incarceration.

Prisoner 2

		s_1(don't confess)	s_2(confess)
	s_1	−10, −10	−20, −8
Prisoner 1:			
	s_2	−8, −20	−15, −15

5.3 The two-person prisoners' dilemma.

The important strategic feature of this game is that both prisoners have a dominant strategy: s_2. Regardless of what the other prisoner does, of what "blood oath" agreements they made upon arrest, and of what each prisoner thinks the other will do, both prisoners should confess. But this calculation yields an outcome that both prisoners prefer to avoid in favor of neither confessing. Yet, the result that they both confess is exactly what the DA intended when he kept them from communicating by locking them in separate cells. That is why we call the game a *prisoners'* dilemma rather than a DA's dilemma.

The distinguishing features of this game can be summarized as follows: (1) each person has a dominant strategy; (2) if each person uses his dominant strategy, then the final outcome is Pareto-inferior, in that both persons can find some other outcome that they jointly and unanimously prefer; and (3) that their strategies are dominant means that even if the players can communicate beforehand and agree to avoid the Pareto-inferior outcome, if they cannot somehow make a binding agreement, then each person ultimately will defect from it.

The prisoners' dilemma would be little more than a curiosity if it did not model important political (and economic) processes. Consider a reinterpretation of this game, in which the strategies now are (s_1) install an automobile pollution-control device in one's car and (s_2) do not install a device. Suppose that a device costs $10, but that every such device contributes to the reduction of automobile air pollution. For purposes of an example, let the value to each person in society from the increment to clean air that one device provides be $7. What we must emphasize about this benefit, though, is that *every* person receives $7 in clean-air benefits if one person installs a device on his car. Clean air, after all, cannot be bottled and consumed only by the persons providing it, just as dirty air affects persons who have no part in polluting. In a two-person society, if both persons install devices in their cars, then they each receive a net benefit of $4: a $7 clean-air benefit from the device each installs on his car; a $7 benefit from the device that the other driver installs on his car; minus the $10 cost of each driver's own device. Figure

Person 2

		s_1(install)	s_2(don't install)
	s_1	+$4, +$4	-$3, +$7
Person 1:			
	s_2	+$7, -$3	0, 0

5.4 The air pollution dilemma.

All persons except i:

		... $j - 1$ choose s_1 $n - j$ choose s_2	j choose s_1 $n - j - 1$ choose s_2	$j + 1$ choose s_1 ... $n - j - 2$ choose s_2
	s_1	$7j - $10	$7(j + 1) - $10	$7(j + 2) - $10
i:				
	s_2	$7(j - 1)	$7j	$7(j + 1)

5.5 N-person prisoners' dilemma.

5.4 describes a simple two-person game that corresponds to this situation. As with the prisoners' dilemma depicted in Figure 5.3, the players here both have a dominant strategy, s_2. Thus, neither installs a device, even though both would be better off with such devices installed on both cars.

Before we draw any substantive conclusions from this example, we must be certain that it works if there are more than two people. After all, with only two persons, some cooperation is likely to emerge, along with an appropriate mechanism to ensure that agreements are binding since the transaction costs of negotiating a resolution are generally lower if fewer people are involved. With more than two persons, the transaction costs of reaching cooperative agreements are increased, in which case noncooperative outcomes are more likely to prevail. Suppose, then, that each of n persons must choose between s_1 and s_2. Focusing on the payoffs to a specific person, Figure 5.5 shows the corresponding game's normal form.

Comparing payoffs in each column reveals that regardless of what strategy anyone else chooses, s_2 is a dominant strategy for i. Since the game is symmetric, this argument holds for everyone: s_2 is every person's dominant strategy. But if everyone chooses s_2 and receives payoffs of 0, then everyone is worse off than if they all install devices on their cars. With everyone installing a device, each person's net payoff is $7n - $10, which is considerably greater than zero for $n > 1$. If n is very large, however, it may be especially difficult for people voluntarily to coordinate their actions to avert the dilemma. Such coordination appears to require leadership and some minimal form of organization, which presently are outside of the model. Furthermore, unless society or its leaders can apply sanctions of some sort, every member initially ascribing to any voluntary agreement shares an incentive to defect unilaterally. Everyone could reason thus: If no one else or if very few defect, I am better off not installing a device, so that I benefit at no cost from whatever clean air anyone else provides; of course, if everyone else thinks as I do and reneges on the agreement to choose s_1, then I would be a fool to be the only one to install a device on my car. Under no circumstances, then, should I install a device.

David Hume, in his *A Treatise of Human Nature* states the problem succinctly,

> It is very difficult, and indeed impossible, that a thousand persons should agree in any such action: it being difficult for them to concert so complicated a design, and still more difficult for them to execute it; while each seeks a pretext to free himself of the trouble and expense, and would lay the whole burden on others.

Similarly, Rousseau, in his *Discourse on the Origin and Basis of Inequality among Men*, comments

> If a group of [men] set out to take a deer, they are fully aware that they would all have to remain faithfully at their posts in order to succeed; but if a hare happens to pass near one of them, there can be no doubt that he pursued it without qualm, and that once he had caught his prey, he cared very little whether or not he had made his companions miss theirs.

Collective action to resolve prisoners' dilemmas, then, requires more than cooperative agreements or good faith. These dilemmas can find resolution only if a means exists to reach and enforce appropriate agreements. One method is for everyone to agree to the invention of an entity (which may be a government) with the authority to coerce by levying fines on anyone caught without a device on his or her car. If the fine is sufficiently great, then s_1 becomes dominant, so that everyone installs a device, the air is cleaner, and no fines are collected.

That solution, too, was apparent to political theorists, as Hobbes's famous passage in *Leviathan* indicates:

> Whatsoever therefore is consequent to a time to Warre, where every man is Enemy to every man; the same is consequent to the time, wherein men live without other security, than what their own strength, and their own invention shall furnish them withall. In such condition,

there is no place for industry; because the fruit thereof is uncertain: and consequently no Culture of the Earth; no Navigation; no use of the commodities that may be imported by Sea; no commodious Building; no instruments of moving and removing such things as require much force; no knowledge of the face of the Earth; no account of Time; no Arts; no Letters; no Society; and worst of all, continuall feare and danger of violent death; and the life of man, solitary, poore, nasty, brutish, and short.

Locke translated this condition directly into a justification for a social order:

To avoid this State of War (wherein there is no appeal but to heaven, and wherein even the least difference is apt to end, where there is no Authority to decide between the Contenders) is one great *reason of men's putting themselves into Society*, and quitting the State of Nature. For where there is an Authority, a Power on Earth, from which relief can be had by *appeal*, then the continuance of the State of War is excluded, and the Controversie is decided by that Power.

What these passages indicate, of course, is that the imperatives of the prisoners' dilemma, if not its mathematical representation, were obvious to political thinkers. This dilemma, although but a specific n-person nonzero-sum, noncooperative game, is central to political science and political theory (and much of economic theory as well). Many of the publicly stated justifications for governmental incursions into our lives and for the formulation of specific public policies are arguments for avoiding real or imagined dilemmas.

5.3   Public goods and externalities

To expand the applications of the dilemma and to evaluate better the justifications for government generally and for specific public-policy proposals, we must first examine in the context of our example what it is about clean air that yields a dilemma. Briefly, the particular feature of clean and dirty air that dictates the form of the preceding example is that they are "public goods," those that exhibit *externalities*. These are not game-theoretic concepts. They are terms that economists used after their rediscovery and formalization of the teachings of eighteenth- and nineteenth-century political philosophers to explain why markets fail and sometimes produce Pareto-inefficient outcomes. Given the connection between the prisoners' dilemma and government activity, however, those terms are sufficiently central to applications of game theory to politics to warrant definition.

With considerable imagination we might think of circumstances in which the activity of one person does not affect the welfare of another, but we would all probably agree that such situations seem contrived and uninteresting. Thus, every individual action implies an externality, an effect on the utility of others. Some of these externalities, such as the impact on our senses of the color of a neighbor's car, are inconsequential and politics largely ignores them. Others, such as the infinitesimal effect I have on the price of bread if I purchase a loaf, are legitimately the concern only of buyer and seller. Other externalities, however, such as those resulting from a decision to heat my home by burning rubber tires, to drive recklessly, to commit murder, or to expropriate the wealth of my neighbor, we regard as the legitimate concern of others. In our example the externality of the strategy s_1 is the benefit of the clean air that a pollution-control device yields, and the externality of s_2 is the resultant cost of the dirty air. Externalities and the decisions about which ones to regulate and which are best left to markets and to resolution by individuals are at the heart of both politics and economics.

Externalities have many sources. Some, such as those that arise when we experience a sense of loss because another person suffers, may derive from cultural traditions and individual moral codes. Others, such as those that arise when governments use their coercive authority to redistribute wealth or when governments declare war, build noisy airports, or construct unsightly expressways, seem obvious in their origins. Still others originate from the specific properties of the goods and services that people trade or produce. This last category of externalities gives rise to a typology of these goods and services, which, although only an abstraction, is nonetheless useful. At one extreme of this typology we find *private goods*, which we usually define as those that, when produced and exchanged among persons, produce only very limited forms of externalities. Examples of such goods are the usual kinds of commodities people exchange in markets, such as groceries, private homes, a ticket to the theater, and the family automobile. The important feature of these goods is that when people produce and exchange them, the principal externality is confined to buyer and seller; only my welfare and that of the grocery store owner are affected when I exchange money for a can of soup. Whatever effects I have on prices nationally, and thus on the welfare of others who must also pay for their soup, are infinitesimal and certainly tend toward zero as the number of persons in the economy grows.

For another class of goods and services, *public goods*, things are quite different. The essential feature of the pure form of public goods is that if one person or firm produces them, then other persons can consume them freely or, minimally, they can be provided to those other persons at zero additional (marginal) cost. Perhaps a classic example of a public good is the lighthouse. If the owner of a shipping firm erects a lighthouse to protect his ships from a treacherous shoal, then the warning that the lighthouse provides is available to all ships passing that shoal. And ignoring congestion of the waterway, as many ships that pass the shoal are warned without any costs beyond those that the builder originally incurred. The lighthouse and its warning are thus a public good.

Various writers also interpret education as another example of a public good. Although certainly the learner benefits directly from his new knowledge, others may also benefit from an educated citizenry. Thus, the benefits from education are not limited to those attending school: There are externalities, such as the added welfare of medical and technological advances that affect us all directly or indirectly. Similarly, inoculation from infectious disease is a public good, since although it protects a specific person, it also protects others to the extent that an inoculated person cannot infect others.

This abbreviated discussion ignores the large variety of public goods that may occur. Some, such as the lighthouse warning may be *nonexcludable*: If someone produces the good, then others can consume it freely. Other public goods, such as public recreation areas and education are *excludable*: The producer (government) can provide the good to as many as possible without incurring additional costs, but it is technically feasible to restrict consumption. Some public goods, like all of our previous examples as well as Hobbes's orderly society, are in fact "goods" in that they are generally desirable. Others, however, such as dirty air, and the noise of low-flying aircraft may be "bads": Consumption may be undesirable but avoidable. Finally, people may be able to choose whether to consume a public good that someone else has provided, such as a recreation area, or they may be unable to choose whether to consume it, as when public goods are the "bads," dirty air and polluted water.

We offer this discussion of public goods and externalities, because, for reasons that the example in Figure 5.2 summarizes, markets often render Pareto-inefficient allocations of some of these goods, and governments may be empowered to correct these market failures. In that example, the public good is clean air, and the correspondence to a market arises because each person can decide independently and without coercion in markets whether to purchase an automobile pollution-control device. And, as with this example, the solution sometimes sought to resolve prisoners' dilemmas in the market is government intervention, either directly by having government enter into the production of various public goods (for example, defense, education, and immunization) or indirectly, by having government regulate the private production of goods with fines, tax incentives, or command-and-control regulation.

The prisoners' dilemma, however, also provides a simple hypothesis as to why governments overexpand and themselves render inefficient outcomes. As Aranson and Ordeshook observe in their formulation of the causes and consequences of government failure, the goods and services that governments provide necessarily have a dual character. Although defense, education, and the like may be public and the costs of providing such goods through a general tax-revenue system may be public as well, there is nonetheless a distinct private character to much of what governments do. An educated citizenry is a public good, but the public provision of education necessarily confers targeted excludable benefits on some. Teachers and their unions benefit directly from increases in education expenditures, and some contractor will benefit from the decision to build a new school, but not all contractors benefit. A highway system may be an approximation to a public good, but again, certain contractors and land speculators benefit to the exclusion of others. Similarly, although certain food regulations ensure the public good of safe foods, they also raise the costs of entry into the market for food production and thereby confer a direct benefit on large firms currently producing in this sector of the market. Finally, governments can confer benefits on a few by resolving prisoners' dilemmas that are best left unresolved, as, for example, when enforcing a nonmarket price through regulation aids a cartel. Much of what we are saying is familiar, of course, to students of interest-group politics.

Suppose, for purposes of a formal example, that society consists of n (odd) disjoint and exhaustive but readily identifiable groups, and that each group perceives the potential benefits to its members if it can secure government support for some specific project. Such a group might consist of the members of an industry that desire import quotas to halt the flow of cheaper and perhaps higher-quality foreign imports. Suppose that the specific benefit to each group from its project is B and its public cost is C. Suppose, further, that each group has two strategies: (s_1) oppose all inefficient governmental activities, and (s_2) lobby for one's own project. Suppose, finally, that majority rule operates. If $(n + 1)/2$ or more groups choose s_1, then the legislature approves no projects and the status quo, evaluated at 0 utility for each group, prevails. But if $(n + 1)/2$ or more groups choose s_2, then those groups lobbying for their projects are successful at securing

Groups 2 through n

	Less than $(n - 1)/2$ choose s_2	Exactly $(n - 1)/2$ choose s_2	$k > (n - 1)/2$ choose s_2
s_1	0	0	$-kC/n$
s_2	0	$B - (n + 1)C/2n$	$B - (k + 1)C/n$

5.6 An interest-group dilemma.

government provision. Assuming no coordination among groups, Figure 5.6 summarizes this game's normal form in terms of payoffs to the first group.

We compute the entries in the cells of this table thus: 0 appears whenever a majority of groups chooses s_1, thereby blocking all projects. For the lower middle cell of the table, if $(n - 1)/2$ other groups join group 1's choice of s_2, then a bare majority of $(n + 1)/2$ groups now support their respective individual proposals, so that these $(n + 1)/2$ groups are successful at securing their programs: Group 1 receives its targeted benefit of B, and each group incurs a cost of $(n + 1)C/2n$ (each approved project costs each group C/n, assuming equal tax shares). For the upper right-hand cell, if a majority of k groups choose s_2, but if group 1 chooses s_1, then the legislature approves k benefits at a per-group cost of k times C/n. Group 1, however, does not receive a targeted benefit, since it does not lobby for one. Finally, for the lower right-hand cell, if k groups choose s_2, and if group 1 chooses s_2, then each group must pay a tax bill of $(k + 1)C/n$, and the $k + 1$ groups choosing s_2, including group 1, each receive their targeted benefit of B.

What is interesting about this game is that if $B > (n + 1)C/2n$, which is approximately equal to $C/2$ for large n, then s_2 is a dominant strategy for group 1 (and, by symmetry, for all other groups as well). That is, lobbying for one's project dominates lobbying against such projects if the targeted benefit of each project exceeds one-half of its cost. But in this event, the outcome can be inefficient. Specifically, if $C = 100$ and $B = 65$, then every group's net benefit in equilibrium is -35. Thus, we have a publicly induced prisoners' dilemma.

The preceding game admits of many interpretations. Instead of conceptualizing the players as interest groups, suppose that they are legislative constituencies, in which we define each strategy thus: Through voting, letters to their representative, and the like, members of a constituency can indicate a preference for one of two kinds of legislators, namely, a legislator who secures benefits for his or her constituents only if

s_1.   The corresponding programs are efficient; otherwise, the legislator should opposes *all* inefficient appropriations.

s_2.   The corresponding program makes the constituency better off given what everyone else does.

Suppose, as before, that constituency programs each have a benefit of B and cost of C associated with them, and that $B < C$. Suppose, finally, and in accordance with much of the lore of Congress, that the legislature operates under a norm of universalism. That is, if a majority of legislators seek benefits for their respective constituents, then all such legislators are successful at securing those benefits. In this instance the resulting game is identical to the one that Figure 5.6 shows. Thus, constituents are in a prisoners' dilemma: Although individually they might prefer efficiency in government, it is irrational for any of them unilaterally to elect representatives who will not work for constituency-interest legislation if most others are working for that kind of legislation. This example suggests, then, that the reelection chances of an incumbent legislator depend less on the general efficiency of legislative action than on a legislator's ability to ensure that his constituents get their fair share of enactments. Further, if constituents express their dissatisfaction with the overall performance of government through their choice of, say, the president, then the electoral imperatives of the president and of Congress differ, and we should anticipate some conflict between these two branches of government, even though the same constituency elects both branches. Thus, this interpretation of a prisoners' dilemma leads to several hypotheses about legislative incentives and executive-legislative conflict.

5.4   An analogous game with a continuum of strategies

The preceding examples do not mean that prisoners' dilemmas require finite strategies, although finite games are the usual examples that people offer of situations in which inefficient outcomes can prevail as equilibria. At this point, however, it is useful to consider a continuous analogue to the interest-group example in Figure 5.6. This analogue demonstrates, first, the application of Theorem 3.2 for nonzero-sum games, and, second, it illustrates the problems that noninterchangeable equilibrium strategies might occasion. And third, we show how continuous models can be analyzed.

Suppose that each of three interest groups, 1, 2, and 3, must allocate its fixed resources between lobbying for some program that benefits it alone at public costs and lobbying against similar programs that would benefit each of the other two groups. Denote group 1's strategy as $s_1 = (x_1, x_2, x_3)$ where x_1 measures the resources that 1 allocates to secure the passage of its own project, and x_2 and x_3 denote the resources that 1 allocates to block group 2 and 3's projects, respectively. Similarly, we use the notation y and z to denote the allocation strategies of groups 2 and 3.

Now we must define the relationship between these allocations and the probability that each group is awarded its project. A great many functional relationships are possible, such as the one in which a group is successful if it allocates more to have its project approved than the other two groups allocate to defeat it. But this relationship is discontinuous for group 1, for example, at the point $x_1 = y_1 + z_1$. For $x_1 < y_1 + z_1$, the group fails to secure its program with certainty whereas if the inequality is reversed, the group is successful with certainty. The corresponding game is much like a Colonel Blotto game, with complex mixed-strategy solutions. Consider, however, the following specification of the probability that group 1 gets what it wants:

$$p_1(\mathbf{s}) = \frac{x_1}{A_1}, \tag{5.1}$$

where $\mathbf{s}$ denotes the strategies of all 3 groups, and A_1 stands for the total resources directed at (both for and against) 1's program. That is, for group i in general,

$$A_i = x_i + y_i + z_i.$$

Thus, although p_1 increases as the resources that group 1 allocates to its own project (x_1) increase, this probability decreases as the resources that groups 2 and 3 allocate against 1's project $(y_1$ and $z_1)$ increase. This functional form for probabilities is identical to the form that we used earlier in formulating an election resource-allocation game. And as before, this probability is continuous and concave in x_1 and continuous and convex in y_1 and z_1. Parallel definitions prevail for payoffs to groups 2 and 3.

Let the benefit to each group if it successfully secures its program be B, and the cost of each program that the government approves be C. The benefit B, however, accrues only to a particular group, whereas all three groups must share any cost. Assuming that each group must share one-third of the cost of each program authorized, group 1's expected net benefit is

$$u_1(\mathbf{s}) = \left(B - \frac{C}{3}\right) p_1(\mathbf{s}) - \frac{C}{3} [p_2(\mathbf{s}) + p_3(\mathbf{s})]. \tag{5.2}$$

Finally, suppose that an identical budget constrains each group's allocation. Hence, for group 1, we require that

$$x_1 + x_2 + x_3 \leqslant 1,$$

where 1 is simply an arbitrary normalizing constant.

This budget constraint is sufficient to restrict a group's strategies to a convex, closed, and bounded set. And if $B > C/3$, then $B - C/3 > 0$, and the first term of expression (5.2) is continuous and concave in x_1, since p_1 is continuous and concave in x_1. Further, since p_k is convex in x_k, for $k = 2$ and 3 (and both are obviously continuous as well), then $-p_k$ is concave in x_k (recall again that the negative of a convex function is concave). Thus, each term of $u_1(\mathbf{s})$ is continuous and concave in the corresponding element of group 1's strategy vector, (x_1, x_2, x_3). Since everything that we have said about group 1 applies to groups 2 and 3 as well, Theorem 3.2 establishes that this game has at least one equilibrium in pure strategies.

Now a problem emerges. The equilibrium strategies of two-person zero-sum games are interchangeable, but n-person games, both zero sum and nonzero sum, do not necessarily enjoy this property. Further, if we could show that each group had a dominant strategy, as we did in the previous section with respect to Figure 5.6, then we could assert that the equilibrium is unique, in which case interchangeability is not an issue. But here all we can assert from Theorem 3.2 is that *at least one* pure-strategy equilibrium exists, and there may be more than one.

Strictly speaking, then, our game is not a prisoners' dilemma. But it nevertheless illustrates a nonzero-sum game in which only Pareto-inferior outcomes correspond to equilibria. To show this result and to derive other properties of equilibria, suppose that $(\mathbf{x}^*, \mathbf{y}^*, \mathbf{z}^*)$ is an equilibrium n-tuple, where $\mathbf{x}^* = (x_1^*, x_2^*, x_3^*)$ denotes the allocations that group 1 makes across the three groups and in which we similarly define $\mathbf{y}^*$ and $\mathbf{z}^*$. Then it must be the case, from the definition of an equilibrium, that $\mathbf{x}^*$ is 1's best response to $(\mathbf{y}^*, \mathbf{z}^*)$. That is, subject to 1's budget constraint, $\mathbf{x}^*$ must maximize $u_1(\mathbf{x}, \mathbf{y}^*, \mathbf{z}^*)$. Maximizing 1's payoff subject to its budget constraint requires that we once again use Lagrange multipliers, differentiate the following function with respect to the three components of 1's strategy vector, and set the result equal to zero,

$$K = u_1(\mathbf{x}, \mathbf{y}^*, \mathbf{z}^*) - H_1(x_1 + x_2 + x_3 - 1). \tag{5.3}$$

Differentiating this expression first with respect to x_1 gives

$$\frac{dK}{dx_1} = \frac{du_1(\mathbf{x}, \mathbf{y}^*, \mathbf{z}^*)}{dx_1} - H_1.$$

From expression (5.2), we learn that

$$\frac{du_1(\mathbf{x}, \mathbf{y}^*, \mathbf{z}^*)}{dx_1} = \left(B - \frac{C}{3}\right)\frac{dp_1}{dx_1}$$

since only the probability p_1 is a function of x_1. And from the definition of p_1,

$$\frac{dp_1}{dx_1} = \frac{A_1 - x_1}{A_1^{\,2}}.$$

Combining these three expressions by substituting the third into the second and the second into the first and setting the result equal to zero so that we can solve for a maximum yields

$$\frac{dK}{dx_1} = \left(B - \frac{C}{3}\right)\frac{A_1 - x_1}{A_1^{\,2}} - H_1 = 0. \tag{5.4a}$$

Differentiating K with respect to x_2 and x_3 (keeping in mind that only the probability p_2 is a function of x_2 and that only p_3 is a function of x_3), and setting the result equal to zero,

$$\frac{dK}{dx_2} = \frac{(C/3)y_2}{A_2^{\,2}} - H_1 = 0 \tag{5.4b}$$

$$\frac{dK}{dx_3} = \frac{(C/3)z_3}{A_3^{\,2}} - H_1 = 0 \tag{5.4c}$$

Solving expression (5.4c) for H_1,

$$H_1 = \frac{(C/3)z_3}{A_3^{\,2}}.$$

Notice that group 2 must solve a problem that is isomorphic to that of group 1. Thus, if we repeat the preceding analysis for group 2, then we will find that H_2 must satisfy

$$H_2 = \frac{(C/3)z_3}{A_3^{\,2}}.$$

Since the right-hand side of the expressions for H_1 and H_2 are identical, it must be the case that if groups 1 and 2 are both choosing their best-response strategies, which they must be choosing in equilibrium, then $H_1 = H_2$. If we now solve for H_1 in expression (5.4a) and in the parallel expression for group 2, then, after equating the results and canceling common terms, we get

$$\frac{A_1 - x_1}{A_1{}^2} = \frac{A_2 - y_2}{A_2{}^2}. \tag{5.5}$$

But group 3 also must be solving an identical problem. That is, there must be a set of expressions for this group that parallel expressions (5.4a) through (5.4c), with appropriate substitutions for x_1, A_1, and so forth. Equating the two expressions that parallel (5.4b) and (5.4c) for group 3 yields

$$\frac{x_1}{A_1{}^2} = \frac{y_2}{A_2{}^2}. \tag{5.6}$$

Substituting this identity into (5.5) and canceling and rearranging terms, we have $A_1 = A_2$. A parallel argument using group 3 instead of 1 or 2 shows further that $A_1 = A_2 = A_3$. This equality states that, in equilibrium, the total resources allocated for and against a group's program must be equal for all groups. From expression (5.5), moreover, the requirement that $A_1 = A_2 = A_3$ also gives us the identity $x_1 = y_2 = z_3$. Hence, in equilibrium all groups allocate the same proportion of their resources to passing their own legislation. Since $p_i = x_i/A_i$ for $i = 1, 2,$ and 3, the identities just established prove that in equilibrium

$$p_1(\mathbf{x}^*, \mathbf{y}^*, \mathbf{z}^*) = p_2(\mathbf{x}^*, \mathbf{y}^*, \mathbf{z}^*) = p_3(\mathbf{x}^*, \mathbf{y}^*, \mathbf{z}^*).$$

If we denote this common probability by p^* and substitute its value into expression (5.2), we get, as the equilibrium payoff to group i,

$$u_i(\mathbf{x}^*, \mathbf{y}^*, \mathbf{z}^*) = (B - C)p^*.$$

Notice, however, that the preceding analysis does not assume anything about the relationship between B and C, other than that $B > C/3$. Thus, if $C/3 < B < C$ and $p^* > 0$, then only Pareto-inferior outcomes correspond to equilibria. Actually, using the various identities just established, we can solve for p^* in terms of B and C. From expressions (5.4a) and (5.4b), we know that

$$\left(B - \frac{C}{3}\right)\frac{A_1 - x_1}{A_1{}^2} = \frac{(C/3)y_2}{A_2{}^2}.$$

Because A_1 must equal A_2 and x_1 must equal y_2, we find, after rearranging terms and suitable cancellations, that $x_1/A_1 = 1 - C/3B$.

But, $p^* = x_1/A_1$. Hence, in equilibrium,

$$p^* = 1 - \frac{C}{3B}.$$

Clearly, $p^* < 0$ if $B < C/3$, which makes no sense for a probability. In this instance, the implicit assumption of the previous analysis, that best-response strategies lie in the interior of each group's strategy set, is incorrect, and it must be the case that $p^* = 0$. But if $B > C/3$, and if, in addition, $C > B$, then the preceding analysis tells us that $p^* > 0$, and only Pareto-inferior equilibria exist. For example, if $C = 1$ and $B = .5$, then in equilibrium, $p^* = 1 - 1/[3(.5)] = \frac{1}{3}$, and each group's expected net payoff becomes approximately $-.167$.

The discussion thus far does not indicate whether equilibria are unique, but it is obvious that they are not unique, and that equilibrium strategies are not interchangeable. The conditions $x_1 = y_2 = z_3$ and $A_1 = A_2 = A_3$ admit of several possibilities. For example, if $C = 1$ and $B = .5$, then $p^* = \frac{1}{3}$, so that from expression (5.1) we can calculate that $x_1/(x_1 + y_1 + z_1) = \frac{1}{3}$, or, equivalently, that $2x_1 = y_1 + z_1$. Similar expressions must hold for groups 2 and 3. One possibility that satisfies this condition is that each group chooses the strategy $(\frac{1}{3}, \frac{1}{3}, \frac{1}{3})$. That is, each allocates as much money for passing its own project as it allocates for defeating the project of each of the other groups. Another admissible set of allocations that are also in equilibrium, however, is

$$s_1 = (\tfrac{1}{3}, \tfrac{2}{3}, 0)$$
$$s_2 = (0, \tfrac{1}{3}, \tfrac{2}{3})$$
$$s_3 = (\tfrac{2}{3}, 0, \tfrac{1}{3})$$

That is, group 1 allocates one-third of its resources to secure the passage of its project, two-thirds toward defeating 2's project, and nothing to defeat 3's project. Groups 2 and 3 make similar biased allocations. The net result is that an equilibrium prevails. But these strategies are not interchangeable: If groups 2 and 3 abide by this equilibrium, and if group 1 chooses $(\frac{1}{3}, \frac{1}{3}, \frac{1}{3})$, then the result is not an equilibrium. Hence, the issue of whether groups arrive at an equilibrium with uncoordinated action is problematical. This analysis nevertheless illustrates the use of Theorem 3.2 for asserting the existence of an equilibrium and the possibility of describing the corresponding strategies and outcomes.

5.5   Other prisoners' dilemmas

The prisoners' dilemma is a powerful model of politics, and people have used it to gain further insight into political processes, including wars, the

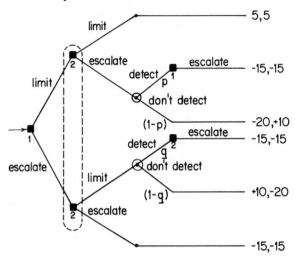

5.7 Arms race with detection.

formation and maintenance of interest groups as well as their failure to form or be maintained, and political participation as well as the absence of participation. We review some of this research to illustrate further the application of noncooperative games.

Arms races

Several scholars suggest that wars are the result of the loose and fragile fabric of international organizations and that when played noncooperatively, international-relations games, especially arms races, resemble prisoners' dilemmas. In the prototypical arms race between two nations, both nations presumably prefer a minimal defense with the bulk of their resources being devoted to consumer goods and domestic public goods. Each nation, however, also prefers some small advantage over the other "just in case." And both nations fear most the other nation's potential expansion of military capabilities without a corresponding response. The result is that both nations escalate their arms stockpiles at the expense of domestic expenditures.

In a slight variation of this theme, R. Harrison Wagner suggests that arms races are more a function of each nation's ability to detect and respond to violations of agreements to limit arms development. Elaborating on his discussion, suppose that after choosing whether to limit or escalate arms, each nation has some probability of detecting the other country's violation of the agreement, so that the game's extensive form

Nation 2

		Limit arms	Escalate
Nation 1:	Limit arms	5, 5	$5p - 20, -25p + 10$
	Escalate	$-25q + 10, 5q - 20$	$-15, -15$

5.8 Arms race.

looks like the form in Figure 5.7. Figure 5.8 depicts this game's corresponding normal form.

This game can have two equilibria: If $-25q + 10 < 5$ and if $-25p + 10 < 5$, or equivalently, if the probabilities of detection are each greater than $\frac{1}{5}$, then the agreement to limit arms is an equilibrium. But if $-15 > 5p - 20$ and $-15 > 5q - 20$, or equivalently, if both p and q are less than one, then joint escalation is an equilibrium. Thus, for all values of p short of certain detection, escalation is an equilibrium, and if the probabilities of detection are both less than $\frac{1}{5}$, then it is the unique equilibrium and the game is a prisoners' dilemma. But if the probability of detection is at least $\frac{1}{5}$, then an arms-control agreement reached cooperatively but enforced noncooperatively is also stable. This is not a surprising result, but the example, which we can complicate further to include multiple moves, discounted payoffs over time, and the like, illustrates the insight that a relatively simple game affords.

Interest groups and collective action

The previous section illustrates the prisoners' dilemma with a three-person model of government failure, in which the players are interest groups. Mancur Olson, in his *Logic of Collective Action*, applies lessons from this dilemma at a more basic level to ask how and why people form and maintain interest groups in the first place. Olson observes that if the purpose of a group, such as a manufacturing association or a labor union, is to provide some collective (public) benefit to its members, then such groups themselves may be trapped in a dilemma. To the extent that the good or service provided is truly public, potential members of the group will be tempted to *free-ride* on the contributions of others, in which case the group will never form or will soon dissolve. For example, if the firms in an industry try to form a pressure group to lobby the legislature for the limitation of competitive foreign imports, and if such limits are secured, then all firms in that industry necessarily benefit (we ignore here some compelling economic arguments that these gains are illusionary or can be maintained only in the very short run). But if this is the case, then might not each firm's management reason that its contribution to this effort ought to be minimal, so as to avoid bad publicity, the actual costs of contributing to the pressure group's maintenance, and the expenditure of political capital that it might spend lobbying for things that benefit that firm specifically.

To illustrate this problem further, suppose that in a society of n persons, each person must choose a level of contribution, x_i, to the public good. Let each person's net benefit be given by

$$u_i(x_1, \ldots x_n) = r(x_1 + x_2 + \cdots + x_i + \cdots + x_n) - x_i,$$

where the variable r is some multiplier. That is, if contributions total X, then the gross benefit to *each* person is rX, and the cost is the magnitude of i's contribution, x_i. And once again, if we arbitrarily constrain each x_i to a closed, bounded, and convex set such as the interval $[0, 1]$, then the conditions of Theorem 3.2 are satisfied. Since utility is linear in x_i, it is also concave in x_i. Thus, the corresponding noncooperative voluntary-contribution game has at least one pure-strategy equilibrium. Rewriting the expression for u_i and letting z_{-i} denote the total contributions of all persons except i,

$$u_i(x_i, z_{-i}) = x_i(r - 1) + rz_{-i}.$$

Clearly, if $r < 1$, i's utility decreases as x_i increases. Hence, the unique equilibrium is one in which all x_i's equal zero. Situations in which r is less than one, however, correspond to situations in which, although it does not pay for any individual to purchase the collective good on his own, it may nevertheless be advantageous for everyone to contribute, since each person benefits from everyone else's contribution. Specifically, if $r > 1/n$, then everyone is better off at the n-tuple $(1, 1, \ldots, 1)$ than at the n-tuple $(0, 0, \ldots, 0)$. Hence, for $1 > r > 1/n$, the preceding game is an n-person prisoners' dilemma.

Olson argues further that small groups are advantaged compared to larger groups. In large groups, everyone might reason that since his contribution is so small relative to the whole, a defection is likely to go unnoticed and is unlikely to have an adverse effect on the group's performance, in which case everyone defects and the group fails in its objectives or does not form at all. Smaller groups have an easier time policing their membership, and with everyone also believing that their contribution is more nearly crucial to success, membership is more ·easily maintained. We can discern some rationale for Olson's hypothesis in our earlier continuous example of government inefficiency growing out of interest-group pressures. A generalization of that model to n

groups shows that a group's probability of passing its program, in equilibrium, is $1-C/nB$. Thus, as n increases, ceteris paribus, p^* increases as well. If $B < C$, the dilemma of uncoordinated action becomes more severe. Further, only if B is less than C/n is $p^* = 0$, thereby avoiding the dilemma. Hence, as n grows, the incentives for escaping the dilemma weaken.

The complete version of Olson's argument, however, requires a different model, one that takes into account the ability and willingness of persons to sanction those who defect from any cooperative agreements. That is, Olson's hypothesized relationship between group size and its inability to supply collective or public goods presupposes some relationship between group size and various transaction costs related to effective cooperative action within groups. But the ability of groups to communicate and coordinate effectively, ceteris paribus, declines with group size, in which case it should be possible to formulate an appropriate model in which the hypothesized relationship between size and inefficiency follows.

How, then, do groups resolve the centrifugal forces of the prisoners' dilemma? Olson's answer is that groups must provide private benefits and sanctions to individual members, in much the same way that a government levies injunctive proscriptions, prescriptions, or fines and subsidies. Labor organizations, for example, offer group-insurance programs that benefit workers who maintain their membership in the union as well as the proverbial brick through the front window of those who cross picket lines. Other scholars suggest that a different method is often used to form and maintain groups, especially in politics. First, many large groups exist to lobby for legislation. If a group is sufficiently large, then it is in the interest of elected officials, serving as *political entrepreneurs*, to assist in the statutory maintenance of these groups and to provide their members with the benefits that they seek. Group nonmembership, then, becomes relatively more costly.

Similarly, in the context of regulation, we might observe that industries in which there are only a few firms (those in which market forces may be weak owing to cartelization and collusion) are often only loosely regulated, but industries containing a great many firms (those industries that are highly competitive in principle) are heavily regulated. The explanation that we can offer parallels Olson's. If a few firms dominate an industry, then a cartel is relatively easy to maintain, but if an industry is highly competitive, as in trucking, food processing, and milk production, then any cartel is likely to fail. Consequently, producers must seek other means of maintaining the cartel and excluding entry. The solution is regulation, whereby the government itself maintains the cartel, either by regulating rates and routes and effectively forestalling competition (as in the trucking industries), imposing high health standards and raising the costs of entry (as in the food processing industries), or simply subsidizing profits (as in milk production).

Political participation

A recurring question in the application of decision theory to politics is: Why do people vote? Or more generally: Why do they participate in any way? The source of such questions is the observation that political outcomes are like public goods. Everyone must live with the policies of a president, whether one voted for him or not, or even whether one voted at all. Thus, the benefits and costs are public. But voting is not costless. Although these costs, which include the time it takes to go to the polls as well as the costs of becoming sufficiently informed to vote "correctly," are small, the benefits of voting instead of abstaining appear minuscule. In recalling the analysis of Figure 1.1 in Chapter 1 and the derivation of expression (1.9), we see that if a voter prefers candidate 1 to 2, then his expected utility of voting less that of abstaining equals

$$u(a_1) - u(a_o) = [u(x, x_1) - u(x, x_2)](p_2 + p_3)/2 - C, \qquad (5.7)$$

where u is the citizen's utility function; x is his ideal point; x_j is candidate j's position on one or more issues; p_2 is the probability that the citizen's vote creates a tie between 1 and 2; p_3 is the probability that his vote breaks a tie in favor of candidate 1; and C is his cost of voting. Hence, the citizen who prefers candidate 1 to candidate 2 votes if and only if $u(a_1) - u(a_o) > 0$, which is equivalent to requiring that

$$2C < (p_2 + p_3)[u(x, x_1) - u(x, x_2)]. \qquad (5.8a)$$

A citizen who prefers candidate 2 votes if and only if

$$2C < (p_3 + p_4)[u(x, x_2) - u(x, x_1)]. \qquad (5.8b)$$

The problem with either expression with respect to explaining turnout, however, is that zero is a good approximation to the objective value of $(p_2 + p_3)$ or $(p_3 + p_4)$ in national elections, in which case voting is simply an unrewarded costly act.

Attempts at squaring this equation with the observation that so many people vote include the argument that, given the constant reminders in the media of elections that are decided by a few votes (the "closeness" of some contests), subjective probabilities overestimate objective reality. There is also the hypothesis that a private benefit to voting

Person 2 in $T2$

		Abstain	Vote
Person 1 in $T1$:	Abstain	$\frac{1}{2}, \frac{1}{2}$	$0, 1 - C$
	Vote	$1 - C, 0$	$\frac{1}{2} - C, \frac{1}{2} - C$

5.9 The dilemma of voting.

accompanies and often outweighs the private cost. Referred to as the "sense of citizen duty," such a term recognizes that much of our political socialization leads people to value the symbolic nature of political acts at least as much as their instrumental impact. Palfrey and Rosenthal, as well as Ledyard offer two alternative approaches that provide interesting examples of the application of game theory to politics.

Beginning with Palfrey and Rosenthal's analysis, suppose that the electorate consists of two teams, $T1$ and $T2$. Next, assume that the cost of voting is identical for all voters, that simple plurality rule operates in which a coin toss breaks a tie, that each voter must choose between the strategy of voting for one's preferred candidate and abstaining, and that, between the teams, payoffs are symmetric. For a voter in $T1$, let the rewards from candidate 1's victory equal 1, and if 2 wins let them equal 0; for a member of $T2$, the payoff if 1 wins is 0, but if 2 wins it is 1. The game-theoretic nature of voting, then, is that if everyone votes, then one's contribution is so small that it pays to abstain; but if everyone else abstains, then it pays to vote, since any person is decisive. Hence, an equilibrium in terms of who votes and who abstains is neither everyone voting nor everyone abstaining.

But the situation is more complicated. Voters, in effect, are playing two games simultaneously. First there is the game between opposing teams (those who prefer candidate 1 and those who prefer 2), which is illustrated in Figure 5.9. Only two voters are depicted there, one from $T1$ and the other from $T2$, and C is the cost of voting. This game is a prisoners' dilemma if $C < \frac{1}{2}$. Thus, although both persons prefer to abstain, avoid the cost of voting, and have a coin toss decide the election, the fear that abstention leaves the other voter decisive leads both persons to vote. A coin toss still decides the election, but both persons incur the cost of voting.

Although no one wants members of the opposition to be decisive, people prefer that other members of their team vote, so that they can abstain and costlessly enjoy their preferred candidate's victory. Figure

Person 2

		Abstain	Vote
Person 1:	Abstain	$\frac{1}{2}, \frac{1}{2}$	$1, 1 - C$
	Vote	$1 - C, 1$	$1 - C, 1 - C$

5.10 Common teams.

5.10 illustrates this game within teams for two people, assuming that if either person votes, then that person is decisive for the team's candidate. If $C < \frac{1}{2}$, there are two pure-strategy equilibria, such that one voter abstains and the other votes. Thus, attainment of one of these nonequivalent, noninterchangeable equilibria requires some form of explicit or implicit coordination.

Assuming more generally that the size of $T1$ or $T2$ exceeds 2, Palfrey and Rosenthal find a vast array of equilibria, most of which require mixed strategies, which binomial densities of considerable complexity describe. For the reasons that Figure 5.10 illustrates, however, the plethora of equilibrium strategies that they identify need not be equivalent or interchangeable, and thus although turnout need not be zero, team members may never actually attain an equilibrium.

This discussion establishes the interdependent nature of the voting decision, and Ledyard's analysis draws on this insight to offer a slightly revised analysis that yields unique equilibria. Notice that two parameters can characterize each potential voter: an ideal issue preference, x, and a cost of voting, C. Each citizen knows his preference and cost, but he may not know the preference and cost of any other vote. Thus, for fixed candidate position a citizen can decide whether he should vote or abstain, conditional on what others do [that is, conditional on the values of p_2 and p_3 in expression (5.8a)], but he does not know what others will do. The usual assumption is that people estimate $p_2 + p_3$ from news commentaries and polls, and on the basis of some estimate, they decide whether or not to vote. Ledyard, however, explores the implications for the voting decision of citizens who "think things out to the fullest."

First, looking at the problem from a single citizen's perspective, let $P_1 = p_2 + p_3$; that is, P_1 is the probability that candidate 1 ties the election or loses by one vote if our citizen abstains. Similarly let $P_2 = p_3 + p_4$. Ignoring for the present how people form such estimates, let Q_1 and Q_2 be the citizen's estimate that any randomly selected person votes for candidate 1 and 2, respectively (the probability of abstention, then is $1 - Q_1 - Q_2$). Clearly, P_1 and P_2 are functions of Q_1 and Q_2. This relationship, however, involves a complicated binomial expression, so we summarize it simply as

$$P_1 = f(Q_1, Q_2) \text{ and } P_2 = f(Q_2, Q_1). \tag{5.9}$$

Thus, from expressions (5.8a) and (5.8b), the citizen should vote for candidate 1 if

$$C < (P_1/2)[u(x, x_1) - u(x, x_2)], \tag{5.10a}$$

for candidate 2 if

$$C < (P_2/2)[u(x, x_2) - u(x, x_1)], \tag{5.10b}$$

and abstain otherwise. The particular difficulty here now is that the P's in these expressions depend on the likelihood that others vote, which depends on their preferences and costs of voting and which thereby varies from citizen to citizen. But suppose that citizens know how costs and preferences are distributed among the entire electorate. Specifically, let the densities $h(C)$ and $g(x)$ describe the distributions of C and x, and suppose that all citizens know h and g. This information permits voters to compute Q_1 and Q_2. For example, from (5.10a),

$$Q_1 = \Pr\{C < (P_1/2)[u(x, x_1) - u(x, x_2)]\}.$$

For a fixed value of x, this probability is the area under the function $h(C)$ from the lower bound on C, 0, to $(P_1/2)[u(x, x_1) - u(x, x_2)]$. Denoting this result by the function $H(P_1, x)$, then Q_1 is the area under H for all values of x for which citizens prefer candidate 1 to 2. This conceptualization establishes Q_1 as a function of P_1, and by a similar argument, Q_2 as a function of P_2. That is, in addition to expression (5.9), we have,

$$Q_1 = r(P_1) \quad \text{and} \quad Q_2 = r(P_2). \tag{5.11}$$

The citizen's decision task represented here is that each is trying to estimate his probability of affecting the outcome, P_1 or P_2, which is a function of what he believes all others will do (represented by Q_1 and Q_2). A citizen knows, however, that the Q's are functions of his P's; just as what he does depends on what he believes others will do, what they do depends in part on what he decides. Hence, (5.9) and (5.11) are simultaneous equations that the citizen must solve in deciding whether to vote.

Proving that there is a solution requires a fixed-point theorem such as the one that we use in the proof of Theorem 3.2. Rather than introduce

Citizen 2

		Vote for 2	Abstain
Citizen 1:	Vote for 1	$1 - C_1, -C_2$	$1 - C_1, 0$
	Abstain	$0, 1 - C_2$	$1, 0$

5.11 Voting when ties favor the incumbent.

even more complicated mathematics, we simply assert that certain assumptions guarantee a solution, and we proceed to a specific numerical example. Suppose that the first of two citizens prefers candidate 1 and that the second prefers candidate 2. All uncertainty about preferences vanishes, but each citizen knows the cost of voting for the other citizen only up to some probability distribution. As a slight modification of earlier assumptions, suppose that if the election is a tie, then the incumbent, candidate 1, wins. Letting C_1 and C_2 be citizen 1 and 2's voting costs, respectively, the noncooperative, complete-information game appears in Figure 5.11.

In this instance an equilibrium requires a mixed strategy (assuming that both C_1 and C_2 are between zero and one). Hence, if q_1 and q_2 are the respective probabilities that citizens 1 and 2 choose to vote, and if both citizens know C_1 and C_2, then the equilibrium is the mixed-strategy pair $(q_1^*, q_2^*) = (1 - C_2, C_1)$. [To check this result, let $C_1 = C_2 = .25$, so that $(q_1^*, q_2^*) = (.75, .25)$. For these values of C, however, the game in Figure 5.11 is identical to the game in Figure 5.2, and this is the mixed-strategy equilibrium we derived earlier in Section 5.1.] The problem is that citizen 1 does not know C_2 and 2 does not know C_1, in which case neither voter can compute his mixed strategy. Thus, we have a game with incomplete information.

Although this game does not satisfy the usual assumptions that all persons are fully informed about everyone's payoffs and characteristics, we can offer a solution. First, suppose that citizen 1 knows C_1 and 2 knows C_2 (that is, each knows his own cost of voting). Suppose further that 1 knows that C_2 is distributed according to the density $g_2(C)$ and 2 knows that C_1 is distributed according to the density $g_1(C)$. On the basis of these densities, let G_1 and G_2 be the cumulative densities of g_1 and g_2, respectively; that is, $G_1(y)$ is the probability that C_1 is less than y. Next, suppose that i votes if C_i is less than some critical number, say C_i^*; otherwise, he abstains.

The critical number C_i^*, of course, depends on what the other person does. Referring to Figure 5.11, notice that person 1 affects the outcome only if 2 votes. If 2 abstains, then candidate 1 wins and citizen 1 prefers to abstain as well and avoid the unnecessary cost of voting. Similarly, citizen 2's vote has an impact on the outcome only if 1 abstains. Summarizing:

> Citizen 1 is decisive only if 2 votes;
> Citizen 2 is decisive only if 1 abstains.

From the usual expression for the calculus of voting, expression (5.7), a person votes for his preferred candidate only if $pB - C > 0$, where p is the probability that he is decisive, B is the differential utility between the two candidates, and C is the cost of voting. In our example, $B = 1$, so a person votes if his cost of voting, C, is less than p. That is, for citizen 1 the critical value below which he votes is

$$
\begin{aligned}
C_1^* &= \text{prob[person 1 is decisive]} \\
&= \text{prob[person 2 votes]} \\
&= \text{prob}[C_2 < C_2^*] = G_2(C_2^*).
\end{aligned}
\tag{5.12a}
$$

Hence, if C_1, which citizen 1 knows, exceeds $G_2(C_2^*)$, then 1 abstains; otherwise, 1 votes. Similarly, 2 is decisive only if 1 abstains. Person 2's critical value, then, is

$$
\begin{aligned}
C_2^* &= \text{prob[person 1 abstains]} \\
&= \text{prob}[C_1 > C_1^*] = 1 - G_1(C_1^*).
\end{aligned}
\tag{5.12b}
$$

Any equilibrium must satisfy expressions (5.12a) and (5.12b) simultaneously. That is, both persons' cutoff levels must be consistent; in other words, each person knows that the other's decision depends on what he does, and vice versa.

For a specific numerical example, suppose that C_1 is distributed uniformly in the interval $[\frac{1}{2}, 1]$ and that C_2 is distributed uniformly in the interval $[0, \frac{1}{2}]$, so that citizen 1 faces higher voting costs than does citizen 2. That is, let the distributions of C_1 and C_2 be

$$
\begin{aligned}
G_1(C_1) &= 0 & &\text{for } C_1 < \tfrac{1}{2} \\
&= 2(C_1 - \tfrac{1}{2}) & &\text{for } \tfrac{1}{2} \leqslant C_1 \leqslant 1, \\
G_2(C_2) &= 2C_2 & &\text{for } 0 \leqslant C_2 \leqslant \tfrac{1}{2} \\
&= 1 & &\text{for } \tfrac{1}{2} < C_2.
\end{aligned}
$$

Equation (5.12a) requires that $C_1^* = G_2(C_2^*)$, or that

$$
\begin{aligned}
C_1^* &= 2C_2^* & &\text{if } 0 \leqslant C_2^* \leqslant \tfrac{1}{2} \\
&= 1 & &\text{if } C_2^* > \tfrac{1}{2}
\end{aligned}
\tag{5.13a}
$$

and equation (5.12b) requires that $C_2^* = 1 - G_1(C_1^*)$, or that

$$C_2^* = 1 \qquad\qquad \text{if } C_1^* < \tfrac{1}{2}$$
$$\quad = 1 - 2(C_1^* - \tfrac{1}{2}) \qquad \text{if } 1 \geqslant C_1^* \geqslant \tfrac{1}{2}. \qquad\qquad (5.13\text{b})$$

It may not be obvious that these equations have a solution. For example, if we set $C_2^* = \tfrac{1}{3}$, then (5.13a) implies that $C_1^* = 2C_2^* = \tfrac{2}{3}$, in which case (5.13b) implies that $C_2^* = 1 - 2(C_1^* - \tfrac{1}{2}) = \tfrac{2}{3}$, in which case (5.13a) implies that $C_1^* = 1$, in which case (5.13b) implies that $C_2^* = 0$, and so on. This situation, then, is equivalent to the "he-thinks-that-I-think" regress in noncooperative games. That is, "I think that his value of C_2^* is ... in which case he thinks that my value of C_1^* is ..., in which case my C_1^* should be ..., but then if he knows this," and so forth. Hence, the proof that a solution to (5.13a) and (5.13b) exists requires a fixed-point theorem. Notice that the two-valued function (referred to as a vector-valued function) $\mathbf{f} = (G(C_2^*), 1 - G(C_1^*))$ takes points in a two-dimensional coordinate system (where the dimensions correspond to C_1^* and C_2^*, and where the coordinates of this system are bounded by 0 and 1, since C_1^* and C_2^* are probabilities) and assigns new values of C_1^* and C_2^* in the same coordinate system. As in our discussion of Figure 3.16, it is possible to show here that at least one point is assigned to itself and that this point solves (5.13a) and (5.13b) simultaneously.

In our numerical example it is easy to check whether $C_1^* = \tfrac{4}{5}$ and $C_2^* = \tfrac{2}{5}$ is an equilibrium, and whether person 1's probability of voting is $G_1(\tfrac{4}{5}) = \tfrac{3}{5}$, and 2's probability of voting is $G_2(\tfrac{2}{5}) = \tfrac{4}{5}$. Suppose, first, that $C_1^* = \tfrac{4}{5}$, in which case expression (5.13b) requires that $C_2^* = 1 - 2(\tfrac{4}{5} - \tfrac{1}{2}) = \tfrac{2}{5}$. Now suppose that $C_2^* = \tfrac{2}{5}$, in which case (5.13a) requires that $C_1^* = 2C_2^* = \tfrac{4}{5}$. Thus, (5.13a) and (5.13b) have a simultaneous solution.

Notice that these probabilities differ from those that we obtain if we simply plug the expected values of C_1 and C_2 into the game in Figure 5.11 and solve for mixed-equilibrium strategies. Given our assumptions about their distributions, the expected values of C_1 and C_2 are $\tfrac{3}{4}$ and $\tfrac{1}{4}$, respectively, in which case Figure 5.11 becomes the game in Figure 5.12. The equilibrium probabilities for Figure 5.12 are $\tfrac{3}{4}$ and $\tfrac{3}{4}$, which differ from the preceding argument that these probabilities are $\tfrac{3}{5}$ and $\tfrac{4}{5}$. Thus, we must ask: Which probabilities are correct?

Before answering this question, consider why different answers are possible. In the face of uncertainty, the preceding analysis requires internal consistency in people's decisions and beliefs. "Beliefs" in this instance refer to estimates of the critical costs, C_i^*. By plugging in expected values and solving for equilibrium strategies to the game in Figure 5.12, citizen 1 is decisive with probability $\tfrac{3}{4}$ (the probability that 2

Citizen 2

		Vote for 2	Abstain
Citizen 1:	Vote for 1	$\frac{1}{4}, \frac{1}{4}$	$\frac{1}{4}, 0$
	Abstain	$0, \frac{3}{4}$	$1, 0$

5.12 Expected values substituted.

votes) and 2 is decisive with probability $\frac{1}{4}$ (the probability that 1 abstains). If 1 and 2 think no further, then each has a strategy that is a best response to the other's strategy. But suppose that they do think further. From (5.12a), we find that if 2 votes with probability $\frac{3}{4}$, then $C_1^* = \frac{3}{4}$. But 1 can reason that if 2 believes that C_1^* is $\frac{3}{4}$, then from (5.13b), we find that 2's critical cost, C_2^*, should be $\frac{1}{2}$, in which case citizen 2's probability of voting should be the probability that $C_2 < \frac{1}{2}$, or from the definition of $G_2(C_2)$, it should be 1 rather than $\frac{3}{4}$. The incomplete-information equilibrium, then, presupposes that each person anticipates the full consequences of his actions, that beliefs and strategies must be completely consistent, and that each person expects that everyone else is doing the same thing.

Looking again at the question "Which answer is correct?" our answer depends on what we hypothesize is true about people. If we hypothesize that people abide by simplifying heuristics, then the probabilities $\frac{3}{4}$ and $\frac{3}{4}$ are more appropriate. But if we hypothesize that people take full advantage of all available information, including what they believe others believe about them, then the probabilities $\frac{3}{5}$ and $\frac{4}{5}$ have a more logical base. Admittedly, this analysis seems to stretch the credulity of game-theoretic reasoning applied to mass electorates. Indeed, neither hypothesis – that they calculate mixed strategies or that they think things through to the fullest – is probably reasonable. In small voting bodies such as legislative committees, however, for us to require internal consistency of beliefs only compels us to adopt the reasonable hypothesis that legislators will decide whether to vote on controversial bills after they have assessed the likelihood that their votes are decisive, knowing that opposing legislators are probably thinking the same.

As a caveat to this analysis, it is interesting to explore whether this approach explains or justifies turnout levels in elections. Ledyard shows that if candidates do not adopt identical policies, then the equilibrium turnout generally is positive. But if candidates are free to adjust their strategies, if preferences are distributed symmetrically so that a multi-dimensional median exists, if voting costs are distributed independently

Payoff if bill i passes

Legislator	Bill 1	Bill 2	Bill 3	Bill 4	Bill 5	Bill 6
1	3	3	2	−4	−4	2
2	2	−4	−4	2	3	3
3	−4	2	3	3	2	−4

5.13 The paradox of vote trading.

of preference, and if utility functions are Euclidean, then candidates converge to this median and turnout is zero. Hence, these game-theoretic approaches to voting do not explain why people vote. The preceding treatment of incomplete information, however, is only now emerging as a fruitful area of inquiry. The conceptual structure of this approach and attendant mathematics are challenging and represent an advanced area of inquiry. Our examples, though, serve as a warning that we should be careful, especially for games involving a few players, about simply plugging in estimated parameter values. Those values (such as the critical costs below which a person votes) may depend on what people believe others will do, and therefore they are not truly exogenous variables.

Vote trading

Chapter 2 shows that if a legislature is considering a series of bills, if preferences over the bills are separable, and if a Condorcet winner exists, then sincere voting yields that winner (Theorem 2.4). But suppose that a Condorcet winner does not exist. What outcomes are likely to prevail in that event?

We cannot now answer this question fully, since such an answer requires an analysis of coordination and coalition formation within legislatures. Riker and Brams, however, suggest that if complete coordination and cooperation are not possible, then a form of the prisoners' dilemma can arise, so that vote trading yields Pareto-inefficient outcomes. Their argument rests on the six-bill situation that Figure 5.13 describes in part.

As in Chapter 2, we assume here that the utility associated with a bill's defeat is zero, so that if all bills fail, then the utility 3-tuple $(0, 0, 0)$ prevails. If all bills pass (the sincere vote outcome), then the utility 3-tuple $(2, 2, 2)$ prevails (for example, legislator 1's payoff is $3 + 3 + 2 - 4 - 4 + 2 = 2$). Hence, the sincere-vote outcome of passing all bills Pareto-dominates that of failing all bills. Riker and Brams argue,

Legislators 2 & 3

		Trade on bills 1 & 2	Do not trade
Legislator 1:	Trade	All bills fail (0, 0, 0)	Bills 1 & 2 pass (6, −2, −2)
	Don't trade	Bills 3, 4, 5, 6 pass (−4, 4, 4)	All bills pass (2, 2, 2)

5.14 The vote-trading dilemma.

however, that failing all bills is exactly the result that prevails with unrestricted and uncoordinated vote trading.

Reviewing their argument, notice first that the sincere outcome is not a Condorcet winner. For example, passing only bills 1 and 6 yields the utility 3-tuple (5, 5, −8), which a majority (legislators 1 and 2) prefers to (2, 2, 2). From Theorem 2.4, then, there is no Condorcet winner. But suppose that legislator 1, after reviewing the situation (we assume that he knows 2 and 3's preferences) chooses to forgo the passage of bills 3 and 6 if he can ensure the defeat of bills 4 and 5. After all, the losses from the passage of bills 4 and 5 (−4 in both cases) exceed the gains from the passage of bills 3 and 6 (+2 in both cases). Figure 5.13 reveals some trading possibilities to bring about this result. Legislator 2 prefers to defeat bill 3 even if that entails defeating bill 4, and legislator 3 prefers to defeat bill 6, even if he must give up bill 5. Legislator 1, then, could approach 2 and propose a vote trade on bills 3 and 4, and he could also approach 3 and propose a trade on bills 5 and 6. If these trades occur, then all four bills (3, 4, 5, and 6) fail. The situation is wholly symmetric, however, and legislators 2 and 3 enjoy similar incentives to negotiate a trade that results in the defeat of bills 1 and 2. Thus, if all trades are made, the Pareto-inferior outcome (0, 0, 0) prevails.

Should legislator 1, then, refrain from trading? The answer is that unless he can control the trade between legislators 2 and 3 on bills 1 and 2, he should not do so because the situation is much like a prisoners' dilemma. To see this similarity, consider Figure 5.14, which simplifies the situation for legislator 1 by giving him two strategies, trade and not trade, and by giving legislators 2 and 3 two strategies, to trade between themselves or not to do so. Clearly, for 1, trading dominates the strategy of refraining from trading.

We suspect that in such a situation (especially if it repeats itself sufficiently often), a legislature will protect itself against such outcomes. Either legislative leadership will police possibilities, or some mechanism for fuller cooperative action that might avoid Pareto-inefficient out-comes will emerge. This example, although not necessarily modeling voting decisions in legislatures, reveals possibilities that might explain certain rules, norms, and procedures that evolve in legislatures to avoid such possibilities. Legislative public works committees are hardly known for economic efficiency, but centralization of authority in the committee permits policing these inefficiencies to avert bankrupting the treasury. This lesson warrants emphasis. *Often we might not find prisoners' dilemmas in a specific situation or institution because certain rules or traditions evolve to avoid them. But this absence of dilemmas does not make an understanding of their logic less relevant, because the only way to understand why such rules and traditions persist is to discern the dilemmas that arise without them.*

5.6 Demand-revealing mechanisms

The preceding examples, especially those pertaining to Olson's analysis of collective choice, highlight a problem for the formulation of public policy that Chapter 2 mentions, namely, the advantages of misstating one's preferences for public goods. In the pollution example that Figure 5.2 depicts, for instance, the dominant strategy essentially corresponds to an understatement of one's true preferences for clean air. Hence, if public policy analysts or practitioners of cost–benefit analysis study people's choices in similar situations without incorporating the strategic context of choice, then they might conclude mistakenly that people care little for clean air, or about national defense, for that matter.

Recall Theorem 2.3's implication that nondictatorial mechanisms are susceptible to manipulation by misrepresented preferences. The question we now address is: Can we violate any of the axioms that we used to prove that result in an acceptable way and find a mechanism that leads people to a truthful revealing of their preferences? The answer is yes, and the axiom we violate is the same one, universal admissibility of preferences, which we violate to ensure pure-strategy equilibria in elections when we assume single-peaked preferences or Euclidean preferences and the existence of a median in all directions. Here we violate this axiom in a different way. Suppose that we define people's utility functions over some m-dimensional space, and that we write u_i' as a function of the values of the m-dimensions thus: $u_i' = u_i'(x_1, x_2, \ldots, x_m)$. In this instance the x_j's might correspond to commodities such as those purchased in the market, in which people prefer more of each commodity to less, or the x_i's might measure levels at which an agency of government produces various public goods. We embody the specific

Voter	Project A			Project B		
	Benefit	Tax cost	Net value	Benefit	Tax cost	Net value
1	$60	$40	$20	$65	$50	$15
2	40	40	0	30	50	−20
3	25	40	−15	25	50	−25
4	75	40	35	140	50	90
5	90	40	50	90	50	40
Sum	290	200	90	350	250	100

5.15 Valuations and costs of projects.

violation of universal domain in the assumption that preferences over the mth commodity are separable from the rest, so that we can express each utility function as

$$u'_i = u_i(x_1, x_2, \ldots, x_{m-1}) + g_i(x_m).$$

Thus, the utility that person i derives from the mth commodity is independent of the values of the first m-1 commodities. Further, suppose that the function g is linear and monotonic in x_m. That is, $g_i(x_m) = ax_m + b$, where a and b may be idiosyncratic to person i. By a suitable linear transformation (subtraction of b, division by a, and corresponding redefinition of u_i), i's utility function can be reexpressed generally as

$$u'_i = u_i(x_1, x_2, \ldots, x_{m-1}) + x_m.$$

Although this assumption about utility is restrictive, it remains theoretically powerful and anchors much of economic theory, in which the mth commodity is money and the assumption of linearity, like the zero-sum assumption elsewhere, permits considerable simplification of analysis.

What we now seek is a mechanism, called a *demand revealing mechanism*, that for each person yields the truthful revelation of the utility function u_i, as an equilibrium strategy. That is, if we suppose that reported utility functions themselves are strategies, then we want to design an n-person, noncooperative game, such that if $(u_1^*, u_2^*, \ldots, u_n^*)$ is an equilibrium n-tuple, then $u_i^* = u_i$ for all i. The specific approach is to use the known preferences on the mth commodity as rewards and punishments that induce the appropriate equilibrium choices.

To illustrate this procedure, consider a situation in which the government must choose between two alternative programs for achieving some objective and suppose that Figure 5.15 shows five voters' individual

Excluded voter	Summed net valuations		Incremental tax
	A	B	
1	$70	$85	$0
2	90	120	0
3	105	125	0
4	55	10	45
5	40	60	0

5.16 Sums without individual valuations.

dollar valuations of these programs and those programs' tax costs. Thus, project B costs more, but on average it is valued more highly. The government must decide whether B's greater average benefit offsets its greater cost. Under simple majority rule, project A defeats B: On balance, voters 1, 2, 3, and 5 prefer A to B. But if the government's objective is to choose the program with the greatest net benefit summed over all voters, and if it is privy to the data in Figure 5.15, then it chooses B, largely because of 4's considerable benefits from B. It is possible that 4 might be willing to compensate 1, 2, 3, and 5 for any losses incurred by the coercive selection of B over A. The problem is that these individual valuations are unknown, and our task is to find a mechanism that elicits their honest revelation.

One method that will not work is to ask people directly for their valuations. Knowing that the government will choose on the basis of maximum net benefits, 4 should overstate his preference for B, and 1, 2, 3, and 5 will overstate their valuations of A. But consider another possibility. As before, suppose that we ask each person to state his net evaluation of each project, and that after we announce these amounts, we choose the project with the greatest net valuation. However, in addition to the usual tax charges required to fund a project, suppose that the government imposes an additional levy as follows: For each project we compute the sum of the net valuations excluding voter i. If the government selects the same project as before, then i pays no tax. But if the exclusion of i alters the outcome, then i's tax is the difference between the valuation of the first-ranked project and the project chosen if we had included i's reported valuation. Figure 5.16 summarizes this calculation for our example and shows that only if we exclude voter 4 does the outcome change (from B to A). Since the difference in summed valuations between B and A without 4 is $45, 4 pays a $45 tax (in addition to the $50 that he pays for B).

The surprising result is that with this tax, truthful revelation is a dominant strategy for all voters, and therefore it is an equilibrium for the noncooperative game. To see first that truthful revelation is an equilibrium, consider voter 4, holding constant everyone else's actions. If truthful revelation is an equilibrium, then it must be a best response if everyone else acts truthfully (that is, when everyone else chooses their presumed equilibrium strategies). Notice that 4's incremental tax is a function of the winning alternative and the valuations that others report. Since he cannot change the evaluations of others, 4 can affect his incremental tax only by changing the outcome selected if all valuations are summed. Thus, voter 4 can lower the valuation of B that he reports or raise his reported evaluation of A so that A wins, in which case he pays no incremental tax. Although this choice would eliminate his $45 tax, 4's overall net benefit declines from $90 - $45 = $45 from B to $35 from A. Hence, he is worse off with A without the incremental tax than he is with B, even if he must pay the incremental tax to get B.

Consider voter 5, who prefers A to B. He pays no special tax and associates a net value of 40 with B. Suppose, instead, that 5 raises his reported net valuation of A, his first choice, to $65, so that A wins when the valuations are summed. Figure 5.16 shows, however, that if the outcome shifts from A to B without 5's evaluation, then 5 pays an incremental tax of $20. Thus, his true net payoff from lying is $50 - $20 = $30, whereas his net payoff from B originally is $40. By overstating his preference for A to make it the winner, then, 5 reduces his payoff from $40 to $30.

This argument, extended to all five voters, shows that truthful revelation is an equilibrium n-tuple. But we can show in addition to being an equilibrium, that truthful revelation is also a dominant strategy, and thus the unique equilibrium. Establishing dominance is important since it eliminates the problems that nonequivalence and noninterchangeability engender, and it means that coordination is not required to attain the appropriate equilibrium n-tuple. To show this result, we leave our example, and shift to a more abstract and general argument.

Assume that people must choose some amount of a commodity such as education, where we denote by x the amount supplied. To simplify matters, suppose that the per unit cost of this commodity is $1, so that the cost of x is simply x. Now assume that people report dollar valuations of all alternative values of x; that is, they report the utility function $u_i^o(x)$, which may or may not correspond to their true utility, u_i. Finally, suppose that the government maximizes the total net dollar valuation of the project selected; that is, it maximizes

$$\sum_{i=1}^{n} u_i^o(x) - x.$$

Let x^o be the alternative that maximizes this function. Our objective is to establish a tax structure that induces people to reveal their true utility functions. Later, we relate our analysis to the preceding example, but consider the following tax scheme for person i:

$$t_i = x^o - \sum_{j \neq 1} u_j^o(x^o) + \max_x \left[\sum_{j \neq i} \left(u_j^o(x) - \frac{x}{n} \right) \right]. \tag{5.14}$$

Notice that nothing that i does affects the third term on the right side of this expression, so that when examining the dominance properties of i's strategies, we can ignore this term. That is, we suppose that i tries to influence x^o by strategically revealing a utility function, u_i^o in a way that maximizes his after-tax welfare. Stated differently, he tries to maximize $u_i(x^o) - t_i$. But, since anything that i does only affects the first two terms in expression (5.14), we can ignore the third term and suppose that i chooses u_i^o to maximize:

$$u_i(x^o) - x^o + \sum_{j \neq i} u_j^o(x^o), \tag{5.15}$$

where the government chooses x^o to maximize

$$\sum_{j=1}^{n} u_j^o(x) - x = u_i^o(x) + \sum_{j \neq i} u_j^o(x) - x. \tag{5.16}$$

But if we substitute i's true utility function, u_i, into (5.16) for u_i^o, then the function that the government maximizes is exactly the function that i seeks to maximize, expression (5.15). That is, i can get the government to do his job for him if he reveals the truth. On the other hand, if i reports a utility function that is not the truth, and as a result the government chooses x', then, although x' maximizes (5.16), it does not maximize (5.15). Since this argument holds regardless of whether the other u_j^o's are true or false, telling the truth remains a dominant strategy.

The third term in expression (5.14), although it does not affect i's incentives, ensures that the government collects enough money to fund whatever project it chooses. To see that this is so, we first rewrite (5.14) as follows:

$$t_i = x^o/n - \sum_{j \neq i} \left(u_j^o(x^o) - \frac{x^o}{n} \right) + \max_x \left[\sum_{j \neq i} \left(u_j^o(x) - \frac{x}{n} \right) \right]. \tag{5.17}$$

We compute the term "max[...]" by searching over all possible values of x, including x^o, for the value that maximizes $[\sum_{j \neq i}(u_j^o(x) - x/n)]$.

By definition then, no other value of x can yield a larger value for the term in brackets. Hence,

$$\max_x \left[\sum_{j \neq i} \left(u_j^o(x) - \frac{x}{n} \right) \right] \geq \sum_{j \neq i} \left(u_j^o(x^o) - \frac{x^o}{n} \right).$$

But this inequality states that the third term in expression (5.17) is greater than the second term. Thus, the third term, less the second term, is positive, which, from (5.17), means that $t_i = x^o/n + K$, where K is positive. Hence, $t_i \geq x^o/n$, so that if we sum this tax across all persons, we find that the total revenues collected are at least x^o, which is the cost of the level of supply that the government chooses.

To interpret the preceding analysis with the numerical example from Figure 5.15, consider expression (5.17) again and notice that the second term, $\sum_{j \neq 1}(u_j^o(x^o) - x^o/n)$, is the summed valuations, excluding voter i, of the project we would select by maximizing the sum of everyone's net valuations. In the example, that project is B, and this second term corresponds to the second column of numbers in Figure 5.16. The last (third) term of the right side (5.17) gives the net summed valuation of the chosen project if we ignore i. Since we would still choose project B if we ignore voters 1, 2, 3, and 5, this term equals the second in (5.17) and these two terms cancel for these voters. But if we ignore voter 4, then we would choose A. From Figure 5.16 we know that the value of this third term is $55. Hence, the third term, $55, less the second, $10, yields 4's incremental tax, $45. Truthful revelation thus remains not only a dominant strategy, but it also provides sufficient taxes to fund the project chosen.

This result does not mean that the proposed tax scheme is without problems. First, there is the problem of what to do with the incremental tax. To redistribute it among voters or to spend it on other projects destroys the incentives that it seeks to establish, so this tax may be Pareto-inefficient. Those who advocate its use in formulating public policy argue, however, that for large populations this problem is minimal, and as a practical matter, we can ignore it. To the extent that this surplus is small, this argument is probably correct, because such small inefficiencies pale in comparison with the inefficiencies that result if we base public policy decisions on the crude technology of cost–benefit analysis.

But there is another threat to any large-scale application of such mechanisms: coalitions. Although people acting noncooperatively have incentives to reveal their preferences correctly, the preceding analysis finds no way to constrain coalitions of voters. Several other variants of this scheme have been devised with different properties. In an alternative mechanism that Groves and Ledyard formulate, for example, people propose increments to the supply of some public good, contingent on what they believe others will propose. Their scheme balances receipts with costs, but truthful revelation is simply an equilibrium and not a dominant strategy. Their scheme also is not immune to coalition strategies, so we must take care before we can apply it to formulating public policy.

There are other reasons besides considerations of coalitions and surplus tax revenues for objecting to the full implementation of such tax mechanisms. First, the assumption that each person's utility for money is linear may be a reasonable approximation for small sums, such as household budgets. But if large sums are involved, as they must be for public works projects, this assumption is probably inaccurate. Second, persons with limited budgets may find that they value projects far more than the taxes they fear paying if they reveal preferences honestly. Persons with lower incomes, then, may report that no project is worth anything to them, and thus they never have to pay a special tax and need not fear the improbable threat of bankruptcy. These different attitudes toward risk (called income effects) distort the results obtained from a demand-revealing mechanism. Nevertheless, each of these schemes reveals that we have not yet ended the search for institutions that escape the prisoners' dilemma and minimize strategic manipulation.

5.7   Summary

Nonzero-sum games are much like their zero-sum counterparts, except for one critical difference: Equilibria need not be interchangeable or equivalent. As a consequence, games with unique Nash equilibria receive special attention, and of these games the two-person prisoners' dilemma is perhaps the most important and illuminating. Because we can represent markets with public goods as prisoners' dilemmas, and because people often call upon governments to correct for market failures, the prisoners' dilemma stands at the interface between economics and political science. But the prisoners' dilemma illustrates far more than market failures. It also augments our understanding of why governments may be inefficient, why interest groups may have difficulty forming or maintaining themselves, why industries sometimes support the regulatory agencies that regulate them, why people might not vote, why nations war, and why people strategically misrepresent their evaluations of the goods and services that governments provide.

The prisoners' dilemma, however, is only one of a great variety of nonzero-sum games, and this chapter also considers in some detail two other applications. The first, concerning the decision to vote in a game of incomplete information, illustrates an extension of the Nash equilibrium concept, in which people make full use of the information available to them. The second, which is not concerned specifically with modeling any existing institution to predict action, illustrates how game-theoretic reasoning can serve as an analytical tool for institutional design. A sizable literature already exists on the theory of games with incomplete information and on demand-revelation mechanisms, and our discussion hardly scratches the surface in these areas. But it does illustrate the great variety of research being done in noncooperative game theory.

The reader should be familiar with the following concepts and definitions after reading this chapter:

prisoners' dilemma	nonzero-sum game	externalities
public goods	private goods	excludable goods
nonexcludable goods	demand-revealing mechanism	political entrepreneur
free rider		

CHAPTER 6

Institutions, strategic voting, and agendas

What differentiates political science from other social sciences is its concern with the actions and decisions that people take within specific institutions: elections, legislatures, the courts, regulatory agencies, and so forth. Because these institutions are human creations, they serve a purpose that reflects human design or evolutionary durability. Those purposes may involve inducing people to choose specific actions, or to lead people not to choose actions that might occur under alternative arrangements. Institutions tend to survive because they satisfy some prevailing objectives and they may evolve or disappear if their damage to human welfare becomes patent. Hence, to understand why one institution survives whereas another disappears, we must try to understand the outcomes that one institution produces but another does not.

Viewing institutions as explicit or implicit products of individual choice seems only logical. We learn why the Constitution establishes a bicameral legislature and an electoral college, for example, by knowing the outcomes that the Framers anticipated under different arrangements, the outcomes of conflicts between large and small states, and of election by an informed elite instead of a mass of uninformed voters. People sometimes forget this purposive view of institutions when they propose utopian schemes for some international order without full consideration of the imperatives of national interests, or when they espouse the establishment of federal regulatory agencies to correct some market failure, without an analysis of the interests that those agencies might serve in the future.

To learn why people choose certain political institutions instead of others, and to understand how these institutions might affect people's actions, we must analyze how institutions affect choice. Chapters 1, 2, and 3 present the theoretical foundations that we need to study any institution. Chapters 4 and 5 illustrate how we can analyze two institutions – two-candidate, majority-rule elections and demand-revelation mechanisms – using that foundation. This chapter focuses on formal voting mechanisms in committees to highlight the importance of institutional procedural details and how an analysis of them might proceed.

We already know some important things about individual choice in

243

formal voting bodies. Theorem 2.3 reveals that in any institution, including the various voting procedures that a legislature might use, the opportunities for strategic maneuver exist given an appropriate configuration of individual preferences. And with respect to legislative voting procedures in particular, Theorem 2.2 states that under majority rule, the stakes of such maneuvering can encompass the entire range of feasible outcomes and that the way that people pose and dispose of political issues can have as significant an impact as the preferences of the participants themselves.

Theorems 2.2 and 2.3 simply assert abstract possibilities, however. In this chapter we consider some specific rules and procedures that a legislature follows: issue-by-issue voting and finite agendas. Our analysis of issue-by-issue voting assumes a spatial context, in which a committee can only consider and vote on one issue at a time. In Section 6.1 we show that for the variety of procedures that this restriction admits, a stable outcome prevails even if no overall Condorcet winner exists, provided that all voters reveal their preferences sincerely and that preferences are separable (Theorem 6.1). And if a Condorcet winner does exist, it will be chosen (Corollary 6.1'). Theorem 6.2 generalizes Theorem 6.1 to show that separability is not a necessary condition for stability, but now the final outcome that the committee chooses will depend on finer procedural details. Furthermore, without separability, that outcome will again correspond to an overall Condorcet winner if such a winner exists. We then turn our attention in Section 6.2 to the possibility that committee members might prefer to vote strategically by misrepresenting their preferences, and we provide a sufficient condition for a stable outcome to prevail in this circumstance (Theorem 6.3). But that condition is restrictive, and in Section 6.3 we show that if it is violated, stability generally disappears. Section 6.4 leaves issue-by-issue voting procedures and turns to the restrictions on committees occasioned by agendas. But unlike the discussion of Theorem 2.2 in Chapter 2, the analyis in this chapter focuses on what occurs if all voters are sophisticated, if all committee members contemplate the potential advantages of preference misrepresentation. Section 6.4 formally introduces the notion of a binary agenda and shows that such agendas yield determinate outcomes (Theorem 6.4), and Condorcet winners if such a winner is included on the agenda (Theorem 6.5). Section 6.5 turns to the important issue of an agenda setter's power when other members are sophisticated. Theorem 6.6 sets the groundwork by establishing a simple procedure for deducing the sophisticated vote outcome in a special but important type of binary agenda, amendment agendas. Theorem 6.7 shows that sophistication reduces the profitable length of an agenda, and Theorem 6.8 establishes the constraints that sophisticated voting places on an agenda setter. Section 6.6 reviews some experimental evidence about procedural effects as well as examples of how people or procedures manipulate the outcomes of specific legislation. Since the research that we review here assumes that everyone knows everyone else's preferences, this chapter concludes with a discussion of how we might analyze agendas if members have incomplete information about other committee members' preferences.

6.1 Sincere voting, issue-by-issue

Theorem 2.2 states that for pairwise majority voting and spatial preferences, if one person controls the agenda, if everyone votes sincerely, and if no Condorcet winner exists, then an agenda can yield any outcome. Theorem 4.3, in turn, shows that for the spatial preferences that Theorem 2.2 assumes, Condorcet winners rarely occur, and the conditions ensuring their existence are fragile. Hence, we can expect that the sensitivity of outcomes to agendas described by Theorem 2.2 is a general possibility, and we might offer the conjecture that simple voting bodies are at the mercy of committee chairmen. Further, since agendas can lead anywhere depending on the particular agenda chosen, the status quo can be relatively unimportant in determining outcomes (if it is the initial motion on the agenda).

We must keep in mind, however, that Theorem 2.2 is not a prediction about outcomes. Instead, it describes the majority-rule, social preference relation for a particular choice rule (pairwise majority voting), and a particular, albeit important, class of preferences. It is legitimate to make predictions about outcomes *only* after we have modeled the specific rules and procedures that people use to aggregate preferences, after we have specified the strategies available to the relevant participants that those rules and procedures admit, and after we have identified equilibrium strategies for the participants. For example, if citizens' spatial preferences do not yield a Condorcet winner in a two-candidate election (if a median in all directions does not exist), then we do not predict that anything can happen. Although Theorem 2.2 establishes that the social preference order is wholly intransitive, for example, we can still predict final outcomes in the uncovered set (see Section 4.8). Hence, the particular institution of two-candidate, majority-rule elections, the assumption that candidates seek to win in those elections, and the conclusion that a voter should vote for the candidate who is closest to his ideal preference establish that two-candidate elections mediate between preferences and choice to yield outcomes in a particular subset of the policy space. This chapter shows that other institutions can mediate peoples' choices in committees so as to modify both an agenda setter's effect on outcomes and the relevance of the status quo.

In an important sense, however, our analysis in this chapter is necessarily incomplete. A thorough study of institutions and procedures, and especially committee procedures, must await developments that permit the study of individual decisions if cooperation and coordination are possible. Only then can we study the phenomena of coalitions. Presently, though, we examine a relatively simple committee, the members of which must act without coordination. One of our purposes is to suggest that the relevance of the status quo to the final outcome may be no less great than that of an agenda.

What gives the status quo its authority in many legislative and parliamentary institutions is that rules often require that members first divide legislation into subparts corresponding to the substantive domains of standing committees. A proposal in the U.S. Congress to build an aircraft of radically new design, for example, would most likely pass through defense committees and subcommittees to assess its military applications, through appropriations committees to determine funding levels, and through aeronautics and space committees to evaluate its commercial value. Other legislative bodies, even informal committees, often structure decisions by imposing rules that have much the same intent as the committee structure in Congress. If the members consider complex matters, they simplify discussion by breaking those matters into substantive subparts, debating each part separately from the others. Thereafter, a germaneness rule focuses discussion on specific issues and precludes the introduction of new ones that the members might otherwise consider simultaneously.

To assess the impact of issue-by-issue voting, suppose that we represent the set of possible outcomes before a committee with an m-dimensional space and that we model individual preferences in this space by simple Euclidean distance, so that indifference contours once again are concentric circles. Instead of proceeding as we did in Chapter 4 by modeling the choice of a policy in a two-candidate election, suppose that the n voters are legislators who must vote for or against certain motions using majority rule, and that if a motion exists that defeats all others, then it is the social choice. Suppose further that there exists a status quo point $\mathbf{y}_o = (y_{1o}, y_{2o}, \ldots, y_{mo})$ in O, where y_{jo} denotes the status quo on issue j. This status quo serves the following function: When voting, the committee can consider motions that affect only one issue, relative to $\mathbf{y}_o$, at any time. Thus, at any stage in the deliberations, the only motions that the committee can consider are of the form $\{(y_{1o}, y_{2o}, \ldots, y_j, \ldots, y_{mo})\} \subset O$. That is, debate and voting are limited to outcomes that differ only on one issue, issue j in this instance.

Many variants of *issue-by-issue voting* are available, including these four:

Procedure 1: After a member makes a motion different from the status quo, another committee member can move to divide the question, so that the committee must vote between the status quo and the motion one issue at a time. Those parts of the original motion that pass become features of the new status quo.

Procedure 2: The committee considers issues in a prespecified order and only motions that differ from the status quo on the issue in question are ruled germane. Amendments to the original motion are permitted, but motions to consider other issues are out of order until the committee takes a final vote to change the status quo on the issue in question. Once the committee considers an issue, it cannot reconsider it.

Procedure 3: No prespecified order exists among the issues, and the committee can reconsider any issue at a later date.

Procedure 4: Members form subcommittees representative of the whole committee, and each subcommittee's jurisdiction is a single unique issue. The committee's final decision is an aggregation of the decisions that each of its subcommittees takes.

There are other variations, but with these possibilities we can indicate a method of analysis and the impact of some parliamentary rules. To begin, consider Figure 6.1, which portrays five voters' ideal preferences, denoted x_1 through x_5, some indifference contours for those voters, and a status quo point, y_o. This configuration of preferences fails to yield a Condorcet winner, because there is no median in all directions. Hence, Theorem 2.2 proves that there exists an agenda that leads from y_o to any other point in the issue space. But suppose that the committee must first vote on issue 1, and then on issue 2. The dashed line l_1, then, corresponds to the admissible proposals that the committee can consider in the first stage of voting.

What we want to deduce now is how a committee member should vote when the set of alternatives confronting him is the line l_1. Here we make a critical assumption, namely, that all committee members

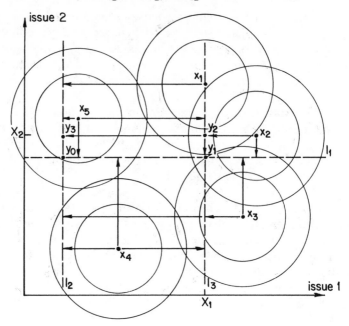

6.1 Voting on one issue at a time.

disregard the fact that after dispensing with issue 1, they must vote on issue 2. That is, no member of the committee anticipates issue 2, and all members vote their preferences on l_1 sincerely. (Later, Section 6.2 reviews the implications of this assumption.) What, then, are committee member preferences on l_1?

Recall from Chapter 1 that with two issues, if preference sets are convex, and if we impose a linear constraint on the admissible proposals, the preferences induced on that constraint are single peaked (see Figure 1.15). Thus, if we restrict voting to l_1, then a Condorcet winner exists at the median ideal point on l_1. We can identify the location of this point exactly. Because indifference contours in this example are circles, we find the point on l_1 closest to voter i's ideal (the point on l_1 that i most prefers) by dropping a perpendicular from x_i to l_1. The projection of ideal points to any horizontal line preserves their left-right order. So if the rules restrict voting to l_1 and if issue 2 is disregarded for the moment, then a Condorcet winner, $\mathbf{y}_1$, exists and on issue 1, and corresponds to voter 1's ideal preference on that issue, X_1.

Now consider voting on the second issue after a vote is taken on the first issue. Two possibilities exist here. In accordance with procedure 1,

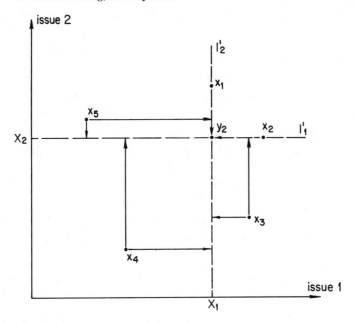

6.2 Induced ideal points.

if voting proceeds with a status quo that the first vote modifies, then all admissible motions lie on the line l_3. But suppose that the committee operates under procedure 4. If the committee takes votes on each issue independently, as when distinct legislative committees consider the subparts of legislation during the same session, then the admissible motions lie on the line l_2. Given circular indifference contours, however, it does not matter which procedure is in force, since ideal points projected perpendicularly to l_2 and l_3 retain their relative positions. In either case, X_2 is the policy on issue 2 that corresponds to the Condorcet winners on l_2 (y_3) and l_3 (y_2).

Hence, even though the social preference order under simple majority rule is intransitive, even though no overall Condorcet winner exists, sincere voting (one issue at a time in sequence), whether by subcommittee or by the committee as a whole, induces a stable outcome at the intersection of X_1 and X_2 (point y_2). But we should find out what happens if the rules permit this committee to revote on the issues with its status quo now at y_2, as in procedure 3. Referring to Figure 6.2, we see as before that the induced ideal points on either l_1' or l_2' are prependicular projections to these lines. Hence, y_2 is the median ideal point on both lines (the *issue-by-issue median*), and it is stable if the committee only considers motions that propose changing one issue at a time.

We can summarize this discussion by a general result that does not depend on circular indifference contours, but instead on the relevant feature of such contours, which is that they illustrate separable preferences. That is, if a person's preferences on one issue do not depend on what occurs on other issues, then the following theorem holds:

> 6.1.   *If preferences in an m-dimensional issue space are separable, if preference sets are convex, and if all committee members always vote sincerely, then the unique issue-by-issue majority-rule stable point is the issue-by-issue median preference. Reconsideration of issues does not affect the stability of this point.*

A corollary to this result follows from the observation that if preferences are separable and if a Condorcet winner exists, then that winner must be a median in *all* directions. Hence,

> 6.1'.   *Under the conditions of Theorem 6.1, if a Condorcet point exists, then issue-by-issue majority-rule voting leads to that point.*

Corollary 6.1' shows that if a median in all direction exists, then issue-by-issue voting yields the same outcome as does simple majority rule or two-candidate competition. But Theorem 6.1 shows that a determinate outcome prevails under sincere issue-by-issue voting even if the social preference order defined over the whole space of alternatives is intransitive. Some persons refer to the issue-by-issue median preference, then, as a *structure-induced equilibrium*, and to Condorcet points and the like as *preference-induced equilibria*. Theorem 6.1 illustrates a structure-induced equilibrium, since no equilibrium exists necessarily if we remove that particular structure and replace it with some other.

Naturally, we should check the sensitivity of these results to the separability assumption. Separability, after all, is a special case and it is unlikely that the major issues that confront committees such as Congress will satisfy it. Consider Figure 6.3, then, which depicts some rotated (and hence, nonseparable) elliptical indifference contours for a three-member committee, and suppose that voting proceeds again on issue 1 from the status quo point y_o. As before, preference sets are convex, so that the preferences induced on the linear constraint l_1 are single peaked. But these preferences do not correspond to the perpendicular projection of ideal points onto l_1. They depend instead on the specific form of the indifference contours and the precise location of l_1. To illustrate this dependency, notice that, as with circular contours, a person's ideal point on a line such as l_1 corresponds to the tangency of

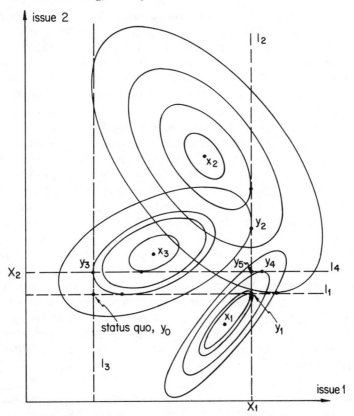

6.3 Issue-by-issue voting with nonseparable preferences.

this line to his indifference contour. But owing to the rotation of the indifference contours, the location of this tangency, relative to the issue under consideration, may change. Looking at voter 3 in Figure 6.3, for example, notice that the tangency of l_2 with the appropriate indifference contour, the point $\mathbf{y}_2$, is above the tangency of l_3, $\mathbf{y}_3$. Similarly, the tangency of l_1 is different (to the left of) the tangency of the line l_4. And unlike what occurs with circles or other forms of separable preferences, none of these tangency points can be deduced from any perpendicular projection of ideal points.

To check what kinds of outcomes prevail now under issue-by-issue voting without separable preferences, notice that the tangency for voter 1 in Figure 6.3 on the line l_1 falls between the tangencies for voters 2 and 3. Thus, 1 is the median voter on l_1 at $\mathbf{y}_1$, and this point is the Condorcet winner on the first ballot if everyone votes sincerely. Yet suppose that $\mathbf{y}_1$

is the new status quo and that voting proceeds with the second issue constrained to the line l_2. The median preference on this line is y_2. Thus, if voting proceeds first on issue 1 and then on issue 2, and if the committee cannot reconsider earlier votes, then the final outcome is y_2.

What happens if we reverse the order of voting, so that the committee votes on issue 2 first and then issue 1? In the first vote, constrained to the line l_3, y_3 is the median preference. On the second vote, with y_3 the new status quo, the median ideal point on the line l_4 is y_4. Hence, since y_2 (the outcome if the committee votes on issue 1 first) is not equal to y_4 (the outcome if the committee votes on issue 2 first), *without separable preferences on the issues, the order in which the committee votes on the issues can affect the final outcome.* Generally, if a committee uses a procedure such as 2 and considers issues only once in a specified order, then there may be as many possible outcomes as there are ways to order the issues.

Each of these outcomes, moreover, differs from what prevails if voting on the issues occurs independently, as is more likely in congressional committees that consider several issues simultaneously. If a committee votes on l_1 and independently on l_3, then the final outcome is y_5, since this point corresponds to what the committee decides on issue 1 (X_1) and on issue 2 (X_2). That is, *without separability, voting on issues independently (procedure 4) can produce an outcome that differs from what a committee arrives at by voting on the issues sequentially (procedure 2).*

Inspection of Figure 6.3 reveals another difference between this example with nonseparable preferences and one that assumes circular indifference contours; neither y_5, nor y_2 nor y_4 is stable if, as with procedure 3, any one of them becomes the new status quo and if the committee can reconsider issues one at a time. Refer to Figure 6.4 and notice that if y_2 is the status quo, then motions to move to the left on issue 1 gain the majority support of voters 2 and 3; if y_4 is the status quo, then motions to move up on issue 2 gain the majority support of voters 2 and 3; if y_5 is the status quo, then motions to move to the right on issue 1 gain the majority support of voters 1 and 2, and motions to move up gain the support of 2 and 3. One question, then, is: If preference sets are simply convex but not necessarily separable, is there an outcome that, after successive iterations of voting, is ultimately stable, or can the process cycle indefinitely? Theorem 6.2 answers this question

6.2. *If preference sets in an m-dimensional issue space are convex, and if every member of the committee always votes sincerely, then there is at least one point that is stable under issue-by-issue, majority-rule voting.*

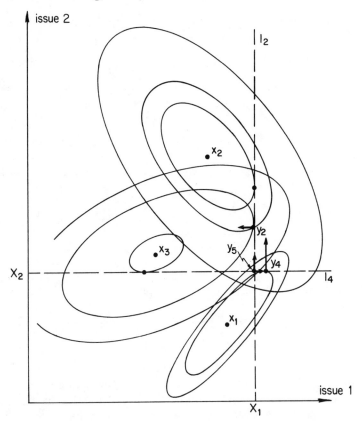

6.4 Instability of various issue-by-issue outcomes.

Before presenting a formal proof of this result, we illustrate its geometry. Imagine a series of parallel vertical lines in two dimensions. On each line we record the induced median ideal preference, which must exist because indifference contours are convex, and which is a unique point if the contours are strictly convex. Let the curve S_2 in Figure 6.5 denote the locus of all median preferences that these vertical lines induce. Further, since a median preference is induced on every vertical line that we draw, we know also that S_2 traverses the entire issue space from left to right. Now, repeat the same process with a set of horizontal lines, denoting the set of induced median preferences by the curve S_1, which for identical reasons traverses the entire issue space from top to bottom. Clearly, S_1 and S_2 must cross, and their intersection is at $\mathbf{y}^*$. For example, if we begin in Figure 6.5 at the point $\mathbf{y}_o$ and vote first on issue 1, then the Condorcet winner is $\mathbf{y}_1$. Voting next on issue 2,

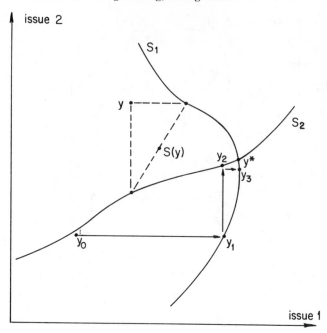

6.5 Locating the stable point for issue-by-issue voting.

y_2 wins a majority vote. Turning again to issue 1, y_3 is the Condorcet winner, and so on. This process leads to y^*. No further movement from y^* is possible, since by the definition of S_1 and S_2, y^* is an issue-by-issue Condorcet winner, regardless of whether we vote on issue 1 or issue 2.

Although we can use this argument to establish Theorem 6.2 if there are two issues, we must generalize it carefully if the committee is concerned with more than two issues. A curve that traverses a square from right to left necessarily intersects a curve that traverses the same box from top to bottom. But three curves, each going from one side of a cube (in the case of three issues) to the opposite side need not intersect. A formal proof proceeds, then, by letting $S_j(y)$ denote the Condorcet winner if the committee votes on only issue j, and if the status quo is the point y. Hence, $S_1(y)$ is the curve S_1 in Figure 6.5. The convexity of preference sets ensures that such points exist, and to simplify discussion, we suppose that it is unique for each y and j (i.e., preference sets are strictly convex). Also, without proof, we assert that $S_j(y)$ is a continuous function of y.

Now define the function

$$S(\mathbf{y}) = \sum_{j=1}^{m} \frac{S_j(\mathbf{y})}{m},$$

which takes each point, **y**, in the issue space and relates it to some other point in that space [because each $S_j(\mathbf{y})$ is a point in the issue space, because $S(\mathbf{y})$ is a convex combination of these points, and because we can assume that the issue space is convex and closed without much loss of generality, therefore $S(\mathbf{y})$ is in the issue space as well from the definition of convex sets]. Figure 6.5 illustrates a point **y** and $S(\mathbf{y})$. Hence, we can apply a fixed-point theorem to establish the existence of a point $\mathbf{y}^* = S(\mathbf{y}^*)$. What remains, then, is to show that $\mathbf{y}^*$ is stable. Specifically, we must show that $\mathbf{y}^* = S_j(\mathbf{y}^*)$ for all j; that is to say, if the status quo is $\mathbf{y}^*$, then no movement along any issue is possible. From the definition of $S(\mathbf{y})$, we can write that $\mathbf{y}^* = \mathbf{y}_1/m + \mathbf{y}_2/m + \cdots + \mathbf{y}_m/m$, where $\mathbf{y}_j = S_j(\mathbf{y}^*)$. The point $\mathbf{y}_j$, then, is the policy outcome that wins in a majority vote if the status quo is $\mathbf{y}^*$ and if the committee is only considering issue j (that is, $\mathbf{y}_j$ is the Condorcet point on the line through $\mathbf{y}^*$, parallel to issue j). Thus,

$$\mathbf{y}_j = \mathbf{y}^* + L_j\,\mathbf{e}_j, \tag{6.1}$$

where L_j is a positive or negative number, and $\mathbf{e}_j$ is an m-element vector with all zeros as entries except for the jth entry, which is 1. Hence,

$$\mathbf{y}^* = (1/m) \sum_{j=1}^{m} \mathbf{y}_j = (1/m) \sum_{j=1}^{m} \mathbf{y}^* + (1/m) \sum_{j=1}^{m} L_j\,\mathbf{e}_j$$
$$= \mathbf{y}^* + (1/m) \sum_{j=1}^{m} L_j\,\mathbf{e}_j.$$

But this equation requires that

$$\sum_{j=1}^{m} L_j\,\mathbf{e}_j = \mathbf{0}.$$

Some simple geometry of vectors, though, shows that we can satisfy this last equality only if $L_j = 0$ for all j (since the $\mathbf{e}_j$ are what mathematicians call *linearly independent* vectors: that is, they all point at right angles to each other). To see this in two dimensions, consider Figure 6.6, which shows the vectors $L_1\mathbf{e}_1 = (L_1, 0)$ and $L_2\mathbf{e}_2 = (0, L_2)$. The arrow $\mathbf{E} = (L_1, L_2)$ denotes the sum of $L_1\mathbf{e}_1$ and $L_2\mathbf{e}_2$. But the length of this arrow is zero only if $L_1 = L_2 = 0$ (just as the length of the hypotenuse of a right triangle is zero only if the length of the triangle's other two sides are zero). Hence, since the numbers L_j must all be zero, equation (6.1) implies that $\mathbf{y}_j = \mathbf{y}^*$, which completes the proof of 6.2.

There is a corollary to Theorem 6.2 that parallels Theorem 6.1'. Briefly, suppose that preferences yield a Condorcet winner at $\mathbf{y}^*$ without issue-by-issue voting and that at most one person has an ideal point located at this point. Then, from our discussion of Figure 4.17, we know that it must be possible to pair voters such that the contract curves of all

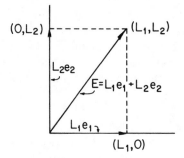

6.6 Summing vectors.

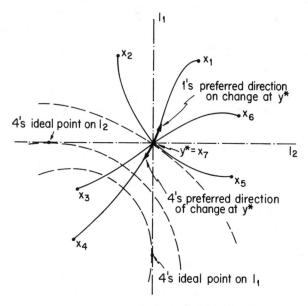

6.7 A Condorcet winner with issue-by-issue voting.

pairs pass through y^*. Similarly, at y^* each voter's preferred direction of movement (the direction in which his utility increases most rapidly) must be along his contract curve toward his ideal point. Thus, the dark arrow in Figure 6.7, which reproduces the preferences and contract curves in 4.17, illustrates the preferred direction for voter 4. Since contract curves represent points of tangency, this arrow must be perpendicular to 4's indifference curve at y^*, in which case 4's indifference curve through y^* must look like the dashed curve in Figure 6.7. It is obvious now that if preference sets are convex, then 4's most preferred point on the line l_1 is below $\mathbf{y}^*$, and that his ideal point on l_2 is to the left of $\mathbf{y}^*$. Similarly, voter 3 and 5's ideal points on l_1 are below $\mathbf{y}^*$, and those of voters 1, 2, and 6 are above $\mathbf{y}^*$. Thus, $\mathbf{y}^*$ is a median along l_1. By a similar argument, $\mathbf{y}^*$ is a median along l_2, and hence it is the equilibrium with issue-by-issue voting. The corollary that this argument establishes, then, is

6.2'.   *Under the conditions of Theorem 6.2, if a Condorcet point exists, then issue-by-issue voting selects that point.*

Although Theorem 6.2 establishes a voting equilibrium if a committee follows procedure 3, it does not imply that the committee can reach this outcome from the original status quo in a straightforward way. The discussion of Figure 6.5 illustrates the problem, and in general – unless indifference contours are of the sort depicted in Figures 6.1 and 6.2 so that the loci S_1 and S_2 are straight horizontal and vertical lines – the committee must reconsider issues sequentially to reach $\mathbf{y}^*$.

In sum, it is important to understand that, except in the case of separable preferences, the specific outcome that a committee achieves with sincere issue-by-issue voting depends on procedural details. With procedure 2, for example, this outcome depends on the specific order in which the committee considers the issues, and there are as many different possible outcomes as there are ways to order the issues. But none of these outcomes necessarily corresponds to those that prevail if the committee uses procedure 4, and the final equilibrium that prevails under procedure 3 need not correspond to any outcome that the other procedures render.

This discussion suggests the importance of institutional details in politics. Procedures can induce stability even if social preferences are intransitive, whereas the outcome that prevails depends on the particular procedure used. Although these findings are important, they are tentative, however, and we should regard them merely as starting points. One reason for this attitude is that we have not considered what happens if committee members coordinate their actions by forming coalitions. We postpone this feature of committees until later chapters and instead consider another possibility – that, in accordance with the implications of Theorem 2.3, members attempt to manipulate outcomes to their own advantage by insincere voting.

6.2   Sophisticated voting, issue-by-issue

In one sense, the preceding results strengthen Theorem 2.2's apparent implication, that an agenda setter has inordinate power to control outcomes. Although the imposition of one of the procedures that Section 6.1 describes is not an agenda in the sense of requiring the committee to vote on specific motions in a specific order, such an imposed procedure does restrict the outcomes that can be paired in a vote. Hence, anyone who has the authority to select a procedure can have an important effect on outcomes, at least if preferences are nonseparable and if everyone always votes sincerely.

Although the restrictiveness of the separability assumption is an empirical issue, the assumption that all members of a committee vote sincerely is theoretically tenuous. Sincere voting means that committee members do not look ahead and anticipate the future consequences of the issue currently under consideration. Perhaps this is a reasonable assumption in committees that meet infrequently (so that members have not learned how to act other than sincerely) or in committees in which the members know little about each other's preferences (and hence cannot fully anticipate the consequences of their actions). Indeed, we might interpret the preceding analysis more properly as modeling amateur committees, such as those of a parent–teacher organization. But in professional committees such as those found in a legislature, it is an unreasonable assumption, since there the members are likely to respond to specific procedures by voting strategically, misrepresenting their preferences if it is in their interest to do so.

The assumption of sincere voting is, of course, the same assumption that we imposed in analyzing two-candidate elections. But there sincere voting is a dominant strategy. This need not be true for other institutions, and Theorem 2.3 states that, with an appropriate preference configuration, every nondictatorial voting procedure with more than two alternatives is susceptible to strategic manipulation. Thus, we have no reason to suppose that the procedures that we discussed in Section 6.1 render sincere voting a dominant strategy. From another perspective, votes are resources, and rational choice requires that voters make the best use of these resources. Sometimes this means revealing preferences sincerely, although at other times this requires misrepresentation. Instead of viewing a particular act of voting as a simple registering of preference, then, we should view it as the choice of one strategy over another.

Robin Farquharson offers a definition of sophisticated voting from this perspective, one based on the observation that for many procedures, elimination of (weakly) dominated strategies (a strategy here is a rule for voting at each stage of the voting procedure) greatly simplifies the normal form of the voting game. Farquharson's definition relies on the observation that by using the domination concept, participants in an

n-person noncooperative game can successively reduce a game without injuring their prospects. First, no voter should choose a strategy that, regardless of how others vote, is never better and is sometimes worse than some other strategy. Since everyone knows this, and knows that everyone else know this, then everyone should simply eliminate all dominated strategies and focus on the resulting reduced game. After this first round of elimination, there is no reason to suppose only unique choices remain. But after we eliminate dominated strategies, we might be able to reduce the voting game further. Each voter should now eliminate weakly dominated strategies under the assumption that all other voters limit their choices to undominated strategies. Farquharson calls the completed application of this algorithm *sophisticated voting*, and he calls the possible choices that remain after no further reductions are possible *sophisticated voting strategies*.

Generally, however, implementation of Farquharson's algorithm is cumbersome. For example, consider the simple three-voter, three-alternative voting situation described by Figure 3.12. This procedure requires two votes: first on A versus B, then the winner against C. A strategy in this instance might read: Vote for A on the first ballot, and if A wins vote for C; but if B wins, vote for B. For each voter there are eight such strategies, so that with three voters, the normal-form game consists of $8 \times 8 \times 8 = 512$ cells. Rather than construct and analyze such a game, however, we can more simply apply the definition of a subgame-perfect equilibrium that we introduced in Chapter 3.

To illustrate this and to see that things can change with sophisticated voting, consider the sincere equilibrium, $\mathbf{y}^*$, in Figure 6.5, which we reproduce again in Figure 6.8 after adding some indifference curves. Next, consider the voting sequence "issue 1 then issue 2" and a procedure that first puts $\mathbf{y}^*$ against $\mathbf{z}$ and then proposes to move down to either $\mathbf{q}$ or $\mathbf{w}$, depending on whether $\mathbf{y}^*$ or $\mathbf{z}$ wins in the first vote. Each committee member has several strategies, such as: Vote for $\mathbf{y}^*$, and if $\mathbf{y}^*$ wins, vote for $\mathbf{w}$; but if $\mathbf{z}$ defeats $\mathbf{y}^*$, vote for $\mathbf{q}$. To avoid complex normal-forms, we can represent the situation in its extensive-form and deduce the corresponding subgame perfect equilibrium strategies. (Recall that a subgame perfect equilibrium must be a subset of the equilibria that Farquharson's method identifies, so subgame perfectness provides a more refined definition of sophistication.)

Figure 6.9 depicts the voting sequence as an extensive-form game that we abbreviate by not describing the simultaneous choices that confront each committee member. That is, at every node of such a tree, every member must make the choice that is indicated, and these choices are made simultaneously (or by secret ballot) by all members. Nothing is

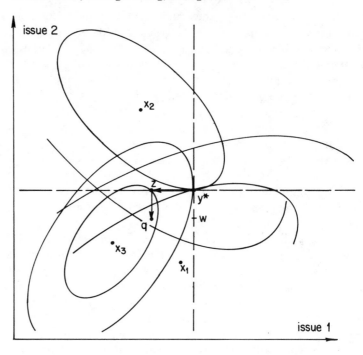

6.8 Instability of issue-by-issue voting with sophistication.

lost by abbreviating matters thus, since, with simultaneous choice, no one can condition their decision on the decisions of others at that node. Figure 6.9, then, indicates that each committee member must first vote between y^* and z; then, depending on the outcome of this vote, each must decide whether to vote for y^* or w or to vote for z or q.

Although we suppress the choices of individual committee members in Figure 6.9, we can still solve for the subgame perfect equilibrium. Working backward up this tree, notice that since a majority of the committee (1 and 3) prefers q to z, and since no one has any incentive to vote other than sincerely on the last ballot, voting in the first ballot for z over y^* is equivalent to voting for q. That is, alternative q is the *sophisticated voting equivalent outcome* of voting for z. Similarly, since y^* defeats w (only member 1 prefers w to y^*), the sophisticated voting equivalent of y^* is y^* itself. Since everyone is assumed to know everyone else's preferences and, therefore, they know what an initial choice of y^* or of z implies ultimately, then everyone should know that the first ballot is a vote between y^* and q. And since voters 1 and 3 both prefer q to y^*, q and not y^* is the sophisticated outcome. Put differently,

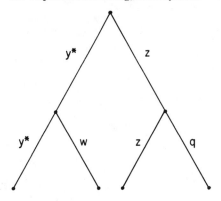

6.9 Voting tree.

on the first ballot between the motions y^* and z, voter 1 should vote strategically and reveal a preference for z. Hence, the procedural equilibrium that prevails under sincere voting is not an equilibrium under sophisticated voting. Furthermore, notice that q is not an equilibrium with sophisticated voting, either: The committee can reach points to the right even under sincere voting. The question, then, is whether a sophisticated voting equilibrium ever exists.

To answer this question we should reconsider Figure 6.8, to discern the specific property of preferences that leads outcomes away from y^* under sophistication. The critical observation is not that a majority prefers q to y^*. After all, we are considering situations that have no Condorcet winner and in which we induce stability, if we can, using a restrictive voting procedure. Thus, Theorems 6.1 and 6.2 apply to situations in which a great many things defeat y^*. Instead, the reason why y^* is not a sophisticated equilibrium is that in the second (last) ballot, a majority sincerely prefers the new proposal, q, to z. But consider the preferences in Figure 6.10, which shows circular indifference contours. There the issue-by-issue median, y^*, is a sophisticated equilibrium. Although q defeats y^* in a direct vote, the committee cannot reach q by any voting order. For example, if the committee first votes on y^* against z, then, unlike the situation in Figure 6.8, the one here has z winning if z is paired against q. Thus, on the first ballot, voter 1 has no incentive strategically to indicate a preference for z.

The property that the preferences in Figure 6.10 satisfy, but that those in Figure 6.8 do not, is separability, the assumption that there is no interaction between the issues in each person's evaluations of alternative positions. *With* separability Gerald Kramer establishes the following result:

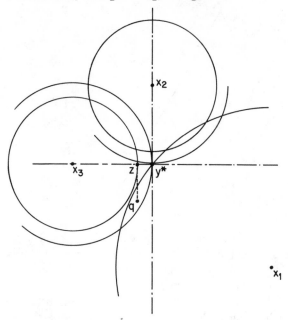

6.10 Sophisticated equilibrium.

6.3.   (Kramer): *Under the conditions of Theorem 6.1, there exists a sophisticated voting equilibrium.*

The proof of this result consists of formalizing the preceding geometric argument and showing that if preferences are separable, then there cannot exist two points such as q and z with a majority preferring q to z. Suppose, to the contrary, that there exists a q and z like the points illustrated in Figure 6.10 such that under majority rule qPz. Then, for each person in this majority, $u(q) > u(z)$. Now consider any point that lies between z and q, which we can represent by the convex combination, $tq + (1 - t)z$. Since preference sets are strictly convex, it must be that

$$u(tq + (1 - t)z) > u(z). \tag{6.2}$$

That is, if a person prefers q to z, then he prefers all points between z and q to z. Suppose, now, that z is identical to y^* on all issues except issue k (issue 1 in Figure 6.10), in which case the committee can vote directly on z against y^*. And let z be identical to q on all issues except j (issue 2 in Figure 6.10), in which case the committee can pair z and q

directly in a vote as well. Let $\mathbf{q} = \mathbf{z} + \mathbf{d}_j$ and $\mathbf{z} = \mathbf{y}^* + \mathbf{d}_k$, where $\mathbf{d}_j$ and $\mathbf{d}_k$ are m-element vectors that contain d_j and d_k in their jth and kth elements, and 0 otherwise. Hence, from expression (6.2), we arrive at

$$u(t\mathbf{q} + (1 - t)\mathbf{z}) = u(t(\mathbf{z} + \mathbf{d}_j) + (1 - t)\mathbf{z})$$
$$= u(\mathbf{z} + t\mathbf{d}_j) > u(\mathbf{z}).$$

But separability permits us to take the utility of any outcome, say $u(\mathbf{x})$, and write it in the form $\sum_{i=1}^{m} u_i(x_i)$, where u_i is the utility from $\mathbf{x}$ on the ith issue. Thus, we can rewrite the previous inequality as,

$$u_j(z_j + td_j) + \sum_{i \neq j} u_i(z_i) > \sum_{i=1}^{m} u_i(z_i),$$

which, after canceling all appropriate terms, reduces to

$$u_j(z_j + td_j) > u_j(z_j).$$

But since $\mathbf{z}$ and $\mathbf{y}^*$ are identical on issue j (they differ only on k), we can substitute y_j^* for z_j in this inequality and get

$$u_j(y_j^* + td_j) > u_j(y_j^*).$$

It then follows that

$$u(\mathbf{y}^* + t\mathbf{d}_j) = u_j(y_j^* + td_j) + \sum_{i \neq j} u_i(y_i^*)$$
$$> \sum_{i=1}^{m} u_i(y_i^*) = u(\mathbf{y}^*).$$

But this inequality violates the assumption that $\mathbf{y}^*$ is a sincere equilibrium on all issues. That is, if this inequality holds, then the voter prefers the point $\mathbf{y}^* + t\mathbf{d}_j$ to $\mathbf{y}^*$, and since this point differs from $\mathbf{y}^*$ only on the jth issue, we can pair them in a direct vote. Hence, the original supposition that $u(\mathbf{q})$ exceeds $u(\mathbf{z})$ is incorrect, so this representative voter of the majority, even if his vote is sophisticated, will not vote for $\mathbf{z}$ over $\mathbf{y}^*$ in the hope of eventually getting $\mathbf{q}$.

6.3   The disappearance of stability without separability

Theorem 6.3 is important because it establishes a condition, separability, under which sincere and sophisticated voting yield identical outcomes. But this theorem is important not only because of what it proves, but also because of what it does not prove. It does not show that a sophisticated-vote equilibrium exists if preferences are not separable. Indeed, the discussion of Figure 6.8 demonstrates that such a conclusion would be false. The implications of this fact are important to the analysis of procedural effects. Consider again a legislature that must decide both

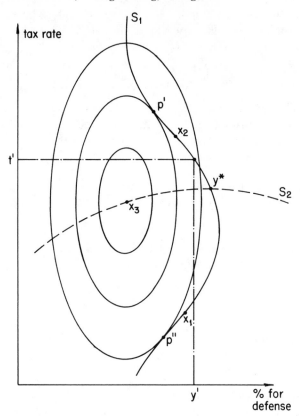

6.11 Inducing nonsingle-peaked preferences with foresight.

the size of the public sector (a tax rate) and the relative weights to give to defense and social-welfare spending. Although the Office of Management and Budget might try to coordinate federal decisions about these matters, these decisions are seldom coordinated in Congress. Suppose that the committee takes decisions sequentially on each issue, but that legislators anticipate what will happen on the second vote, given a specific outcome on the first. For example, each legislator, before voting on a revision of tax rates, tries to anticipate what weights defense and social-welfare spending would receive if rates rise or fall.

To grasp why Theorem 6.3 applies in this context only if preferences are separable, consider Figure 6.11, which, except for voter 3, reproduces the preferences from Figure 6.3, and in which we label the vertical issue "tax rate" and the horizontal issue "percentage of the budget to defense." Suppose that the legislature first votes on a tax rate, t.

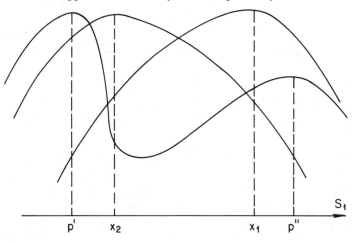

6.12 Induced preferences on S_1.

Anticipating a subsequent vote on defense versus social-welfare spending with a predetermined tax rate, each legislator knows that a specific
rate, such as t', induces a horizontal constraint and a median ideal point
on that constraint that yields the outcome y' as the eventual proportion
of the budget allocated to defense. Thus, each legislator can assume that
by voting for t', he is choosing the outcome (t', y'). The curve S_1, then,
which is the locus of all medians given fixed tax rates, describes the
specific outcomes that prevail if the legislature votes on taxes first and
specific appropriations second, with legislators fully anticipating the
effects of early votes on subsequent preferences.

From its construction and from the orientation of individual indifference contours, we find that S_1 goes through the ideal points of voters 1
and 2, so that, from the convexity of preference sets, we know that 1 and
2's preferences on S_1 are single peaked. More important, however, we
have drawn 3's indifference contours so that 3 has two tangency points
on S_1, one at p' and the other at p'', with p' standing slightly higher on
his preference order than p''. Thus, 3's preferences along S_1 are not
single peaked. This is the problem that separability avoids, since with
separability, S_1 is a straight line and the preferences on it are single
peaked.

Suppose that we now take the curve S_1 and lay it out as a straight line,
as in Figure 6.12, simultaneously depicting the legislators' ordinal
preferences. Thus, by anticipating all subsequent votes, legislators
convert a two-dimensional situation into a unidimensional one. But
preferences are not single peaked on S_1, so there is no guarantee that

the social ordering is transitive or that a Condorcet winner exists. Indeed, with the preferences in Figure 6.12, there is no Condorcet winner. For example, x_1 defeats x_2, x_2 defeats p', but p' defeats x_1. So even though institutional structure separates the consideration of fiscal and appropriations policies into two issues, that separation does not establish an equilibrium.

This example does not mean that procedures have no effect on outcomes. Although Theorem 2.2 shows that, for the conditions of this example, the social ordering is wholly intransitive throughout the policy space, nevertheless separation limits the intransitivity to the line S_1. Furthermore, suppose that the committee takes votes in reverse order, first on defense and then on welfare spending. The dashed curve S_2 in Figure 6.11 depicts the locus of medians that a sequence of vertical constraints generates. Owing to the shape of the indifference contours, all preferences on S_2 are single peaked, with the median preference falling at y^*. Thus, if the legislature follows procedure 2, which does not permit the reconsideration of issues, then a stable outcome prevails if the legislature votes on defense allocations first rather than second. With procedure 3, on the other hand, no equilibrium exists, since in any vote legislators should anticipate that some majority will prefer to revote on the tax rate. And if the legislature takes votes on the two issues simultaneously (or independently), as with procedure 4, then Theorem 6.2 once again establishes the existence of a unique equilibrium outcome, since legislators cannot anticipate that the vote on one issue affects the vote on the other issue when the votes are simultaneous. (We suspect, however, that the rules force few important legislative decisions to be independent in this fashion.)

Instead of showing that procedures have no effect, the preceding example shows exactly the opposite, that the details of procedure can be critically important in determining whether outcomes are stable or unstable. And, in addition, it shows that specific assumptions such as separability can affect profoundly the legitimacy of various claims about the stabilizing or destabilizing effects of procedures.

6.4   Sophisticated voting and agendas

The procedural constraints that Sections 6.1–6.3 discuss concern a form of control in which tradition, or a prescribed rule, or a committee chairman limits the order or the way in which the committee can consider different issues. But there is no limit to the number of motions that the legislature or committee can consider: Subject only to an issue-by-issue constraint, any point on an issue can be voted on. Hence, the procedural constraints the previous sections discuss might model decision making in several legislative committees, in which each committee evaluates all alternative forms of legislation in its domain. For the legislature as a whole, however, voting is typically limited to a few motions that are ordered by an agenda: A bill reported to the floor by some committee, several amendments and substitutes, and the status quo. Perhaps the procedures that we have just discussed model debate over the design of amendments and substitutes. But here let us suppose that debate has ended and that the legislature is confronted with a fixed set of motions that it must vote on in accordance with an agenda.

There are numerous questions that we should ask with respect to the use of agendas in a legislature. First, if we model voting as an n-person game, under what conditions are we assured that a determine equilibrium prevails? Chapter 3 provides a simple three-alternative, three-voter situation in which the final equilibrium under sophisticated voting is unique. Can we generalize this example? Second, Theorem 2.2 establishes an agenda setter's potential for manipulating outcomes if there is no Condorcet winner and if voting is sincere. But if the motions on an agenda include a Condorcet winner, what types of agendas are guaranteed to pick that winner, and what is sophisticated voting's role in that selection? Finally, if there is no Condorcet winner, does sophisticated voting limit an agenda setter's ability to manipulate outcomes, or is the ability to manipulate no less great if everyone is sophisticated than if, as Theorem 2.2 assumes, they are sincere?

Binary agendas

We postpone answering this last question until later; however, we can answer the first two here. We begin by observing that there are many different types of agendas, two of which are illustrated in Figure 6.13. In Figure 6.13a the committee first votes on alternative A against B, and if A wins, then voting ends and A is the outcome. But if B wins, then the committee pairs B against C. If B wins again, then it is the outcome. But if C wins, then the committee pairs it against D, and the winner of this third vote is the final outcome. We call agendas of this sort *elimination agendas*. The agenda in Figure 6.13b, on the other hand, which also concerns the four alternatives A, B, C, and D is much like those that we considered earlier. It is unlike the agenda in Figure 6.13a, however, in that if A is to prevail in this agenda, it must defeat B, C, and D (in Figure 6.13a, A need only defeat B to prevail). Notice that this second agenda introduces the same new alternative at every level. Thus, regardless of whether A or B wins, the committee must consider

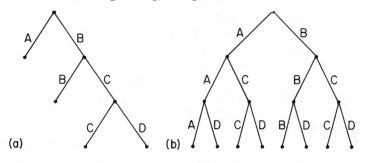

6.13(a) Elimination agenda. (b) Amendment agenda.

C next, and regardless of whether A, B, or C makes it to the final round, the committee must then consider D. We call agendas of this sort *amendment agendas*.

Although both of these agendas are substantively distinct – and as we shall see both can yield different outcomes under otherwise identical circumstances – both illustrate a general class of voting procedure. Suppose that voters face a set of outcomes or motions $O = \{o_1, o_2, \ldots, o_n\}$, and that if a particular motion is before a committee, each member must vote for or against it. This property yields the following definition:

> *Binary agenda*: An agenda is binary if at every stage of the voting a person can either vote in support of one set of outcomes in O or in favor of some other set of outcomes in O.

Binary agendas first divide the outcomes into two exhaustive (but not necessarily disjoint) subsets, O' and O'', and voters choose between these sets. Suppose O' wins: The agenda now divides this set into two exhaustive subsets, and voters choose between them. This process continues until only two sets remain, each consisting of a single outcome, say, $\{o'\}$ and $\{o''\}$. In the elimination agenda in Figure 6.13a, for example, voters must first vote on A versus B; but this is equivalent to a choice between the sets $\{A\}$ and $\{B, C, D\}$; if $\{A\}$ loses, then they must choose between $\{B\}$ and $\{C, D\}$; if $\{B\}$ loses, then they must choose between $\{C\}$ and $\{D\}$. Thus, this agenda is binary. Similarly, in the amendment agenda in Figure 6.13b, the first ballot pairs $\{A, C, D\}$ against $\{B, C, D\}$; if $\{A, C, D\}$ wins (that is, if A defeats B so that B is eliminated in the first round), then the second ballot pairs $\{A, D\}$ against $\{C, D\}$; and so forth. Notice that the *voting tree* that represents a binary agenda must have exactly two branches emanating from every decision node. So plurality voting among three alternatives is not

Voter	1st	2d	3d	4th	5th	last
1	A	B	D	C	E	F
2	B	D	C	A	E	F
3	C	B	A	D	E	F

6.14 A Condorcet winner at B.

binary, since in the first and only ballot, voters must choose between three branches, labeled $\{A\}$, $\{B\}$, and $\{C\}$.

Now recall our discussion of sophisticated voting, which supposes that members of the committee can work backward up a voting tree to deduce their subgame-perfect equilibrium strategies. It is evident from this construction that if everyone's preferences are strict or if there is a determine mechanism for breaking ties (such as a chairman's vote), then the outcome of any final vote in a binary agenda is determinate: One of the two surviving outcomes must win. But this means that the outcome of the next-to-last vote is determinate as well since it effectively compares two winners in the final ballot. With respect to Figure 6.13b, if A defeats D and if D defeats C in a simple majority vote, then the next to last vote between A and C is effectively a vote between the sophisticated agenda equivalent outcome of the left and right branch, A and D, respectively. And again the outcome is determinate: A defeats D. Since we can reiterate this argument as many times as we choose, the following theorem is established:

6.4. *If tie votes are impossible or if they are broken in a determinate way, then under sophisticated voting, every binary agenda is determinate.*

Thus, it isn't a fortuitous selection of preferences that produced a unique outcome in our analysis of the amendment procedure over three alternatives in Chapter 3 (see Figure 3.12). Indeed, our analysis in that chapter followed the proof that we have just reviewed.

For another example consider the preferences in Figure 6.14. With these preferences and with simple majority rule: B is a Condorcet winner; A, C, and D cycle; and all four of these alternatives defeat E, which in turn defeats F. Suppose that this three-voter committee uses the agenda in Figure 6.15, in which case the members first vote between D and B (or equivalently, between the sets $\{A, D\}$ and $\{A, B, C, E, F\}$). If B wins, then they must choose between B and A (or equivalently, between the sets $\{B\}$ and $\{A, C, E, F\}$), and so on; but if D wins in the first ballot, then D is paired against A. Each node of this representation contains exactly two branches, so it is a binary agenda procedure.

Although the definition of sincere voting is unclear here, suppose that

1st ballot ⟶

2nd ballot ⟶

3rd ballot ⟶

4th ballot ⟶

5th ballot ⟶

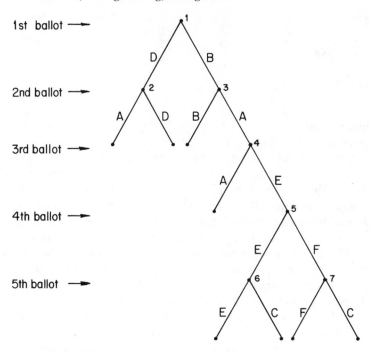

6.15 Binary agenda.

each voter chooses the branch of the voting tree that contains his most preferred alternative. Thus, on the first ballot all voters choose to go right on the tree (voter 1 has to consult his second most preferred alternative to break a tie, since both branches contain A). On the second ballot at node 3, voter 2 chooses left, whereas voters 1 and 3 choose right. At node 4, 1 chooses left, but 2 and 3 choose right. And at node 5, all voters unanimously choose left, and C is the decision on the final ballot.

This agenda thwarts the selection of the Condorcet winner, alternative B, so it serves as a counterexample to the proposition that Condorcet winners necessarily prevail in any binary voting procedure. But this conclusion presumes that all voters cast sincere ballots. Suppose instead that voter 1 recognizes that if 2 and 3 continue to vote sincerely, he should choose the left-hand branch at node 3 by voting for B rather than the right-hand branch. Working backward up the tree, we see that the choice between E and F is effectively a choice between C and C (that is, the sophisticated agenda equivalent of a vote for E and a vote for F is the same, motion C). Thus, at node 4 choosing between A and E is equivalent to choosing between A and C. But since C defeats A, at node 3 the voters' choices effectively become B and C. Since B defeats C, the choice at the first node reduces to a vote between B and a runoff between A and D. Finally, since A defeats D, the first ballot is equivalent to a vote between B and A, in which case B, the Condorcet winner, prevails. Again, this conclusion is no accident, as Farquharson's theorem states:

6.5. (Farquharson): *If a Condorcet winner exists among the alternatives that the committee considers, then sophisticated voting in a binary agenda procedure yields this outcome as the eventual winner.*

By definition, if we pair a Condorcet winner against any other alternative, then it defeats that alternative. Thus, if a Condorcet winner enters the voting agenda at any stage, then it defeats all subsequent alternatives. Since we assume that voters have perfect foresight as well as complete information about each other's preferences, we should view any initial branch of the tree containing the Condorcet winner as containing only that alternative. Hence, in the first stage a majority will vote for that branch. This argument proves the theorem.

Amendment agendas

Binary agendas represent a general class of procedures that order and constrain committee decision making, but if we turn to a special type of such an agenda, the amendment agenda, we can prove some additional theorems. Amendment agendas correspond closely to the procedures that many committees use. Our study of them actually began with Theorem 2.2, which shows the power of an agenda setter if everyone always votes sincerely. But there are a number of reasons for believing that although this theorem is important to our understanding of the social preference relation under majority rule, it provides a poor model of legislative or parliamentary committees. First, it is unlikely that the members of a committee such as the Congress might long tolerate the concentration of power in an agenda setter that Theorem 2.2 implies. If agendas are so important and if our theories have any empirical content, then committee members should recognize this importance. Second, the selection of an agenda or an agenda setter may only push the paradox of voting back to some earlier more fundamental decision, and this shift merely resurrects cyclic social preferences and the instability that they imply. Since, with spatial preferences, sincere voting, and majority rule, Theorem 2.2 shows that if a Condorcet winner does not exist, agendas can be designed to lead to any outcome, this theorem seems to predict a never-ending struggle for agenda control within any legislative committee, as members vie for the agenda they most prefer.

Now, suppose that instead of submitting to a nearly dictatorial concentration of power in an agenda setter, debating the selection of an agenda setter endlessly, or abandoning the institution of majority rule entirely, committee members can protect themselves against extreme and undesirable possibilities by sophisticated voting.

To see how this protection might work in a finite alternative case, first consider the three-alternative, three-voter example of agenda control and sophisticated voting from Chapter 3, which illustrates an extensive-form game and backward reduction (see Figure 3.12). There, three voters and the usual preferences over three alternatives yield the Condorcet paradox. The following outcomes prevail for each agenda under the assumption that everyone is either a sincere or sophisticated voter:

Amendment agenda	Sincere outcome	Sophisticated outcome	Voter with misrepresented preference
(x, y, z)	z	y	1
(x, z, y)	y	x	2
(y, z, x)	x	z	3

Virtual dictatorial power still seems to remain with the agenda setter. If the setter prefers z to y to x (person 3), then he can secure z by adopting the agenda (x, y, z) if he believes that everyone votes sincerely, whereas if he believes that people vote in a sophisticated manner, then the agenda (y, z, x) secures his most preferred alternative. But this analysis ignores the constraints that the rules of most committees impose on admissible agendas. Commonly, the committee must vote first on the alternative introduced last. For example, if x must be included in the first vote, then 3 cannot secure his ideal, z, if the other two voters are sophisticated. The best that 3 can do is to secure his second choice. Thus, by taking advantage of an institutional rule that restricts the admissible amendment agendas to a prescribed starting point, and by voting in a sophisticated manner, persons 1 and 2 can protect themselves against z.

In a spatial context, Theorem 2.2 asserts that with an amendment agenda the committee can reach any outcome *from any starting point*, but the theorem does not permit sophisticated voting. Thus, we now want to extend the lessons of the previous example to the context of Theorem 2.2 to see how sophistication might constrain an agenda setter's power. We begin by letting $W(y)$ be the outcomes in O that a majority prefers to y. That is, all outcomes in $W(y)$ defeat y under sincere voting. Next, let $\mathbf{V} = (y_1, y_2, \ldots, y_m)$ be a specific amendment agenda, so that the committee first pairs motion y_1 against y_2, the winner against y_3, and so on. Finally, we define

> *Innocuous*: A motion y_i is innocuous in an amendment agenda if it cannot defeat the noninnocuous motions that follow it in the order of voting.

This definition appears circular, but since no motion follows y_m, it is necessarily noninnocuous. But if y_{m-1} does not defeat y_m, then y_{m-1} is innocuous. We label a motion innocuous so that we can eliminate it from consideration. Recall that under the sophisticated voting hypothesis, everyone should vote sincerely in the last vote (at the last node of a tree such as the one in Figure 6.13b). Thus, as we have done in the past when working up a voting tree, we should eliminate the innocuous proposal, y_{m-1}, from consideration, because it is certain to lose to y_m. Now consider how people think when they contemplate the pairing of y_{m-2} against whatever motion has survived from the earlier stages, which we represent by y'. If people are sophisticated voters, then they know that a vote for y_{m-2} cannot yield y_{m-1}. If y_{m-2} defeats y', and if the committee then pairs y_{m-2} against y_{m-1}, the real vote is between y_{m-2} and y_m, since y_{m-1} is innocuous. And if y_{m-2} also loses to y_m, then it is innocuous also. Hence, when voting between y' and y_{m-2}, if y_{m-2} is innocuous, a vote for y_{m-2} is actually a vote for y_m, and the members can ignore y_{m-2} as a serious proposal. Working backward up a voting tree, we can thus derive the *sophisticated amendment agenda* $\mathbf{Z} = (z_1, z_2, \ldots, z_h)$, which consists only of noninnocuous motions and which is the *sophisticated equivalent to* $\mathbf{V}$.

To illustrate these definitions, suppose that voters' preferences over six motions yield the following dominance relation among the alternatives (where + indicates that, in a simple paired comparison, the row motion defeats the column motion, and where − indicates that the row motion loses to the column motion):

	A	B	C	D	E	F
A		+	+	−	−	+
B	−		+	−	−	−
C	−	−		+	−	+
D	+	+	−		−	+
E	+	+	+	+		−
F	−	+	−	−	+	

With the amendment agenda $V = (A, B, C, D, E, F)$, F defeats E, so E is innocuous: E cannot be the final outcome, and any vote for E earlier in the balloting is a vote for F. It matters little that E defeats everything else in the agenda if it cannot defeat F, the last motion on which the committee votes. Motion D now loses to E and defeats F. That D loses to E, however, is irrelevant, because if everyone votes in a sophisticated manner, then everyone knows that a vote for E is a vote for F. And since D defeats the noninnocuous proposal that follows it in the agenda, F, D is not innocuous. Similarly, C defeats D and F, so it is not innocuous. Motion B, however, although it defeats C, does not defeat the noninnocuous motion D, so B is innocuous. Again, under sophisticated voting, everyone knows that if B survives to be voted on against D, B will lose, and thus it cannot be the final outcome: A vote for B is a vote for D. Finally, we can eliminate motion A, because it also does not defeat D. The sophisticated agenda, then, is $Z = (C, D, F)$.

Our first theorem about amendment agendas and their sophisticated equivalents establishes why we are interested in the definition of innocuous motions and sophisticated-agenda equivalents. Letting $W(z_i)$ denote the outcomes in O that defeat z_i in O,

6.6.   *The first element, z_1, of the sophisticated-agenda equivalent, $Z = (z_1, \ldots, z_h)$, of the amendment agenda V is the sophisticated vote outcome. Furthermore, z_1 is necessarily in the intersection of the sets $W(z_2)$ through $W(z_h)$.*

This theorem follows almost immediately from what we already know about working backward up voting trees. Returning to our example, Figure 6.16 portrays the voting tree that corresponds to the amendment agenda $V = (A, B, C, D, E, F)$, and at each node we indicate the sophisticated-vote outcome. Consider the motion A, which defeats F. To become the outcome under sophisticated voting, A must survive the voting to be paired against F. Working backward up the tree, we find that to survive, A must defeat the noninnocuous proposal that precedes F, D (which it does not defeat in our example, but let us suppress this fact for the moment). If A can survive this vote, then it must survive the noninnocuous proposal that precedes D, motion C. Continuing this argument, we learn that motion A emerges as the sophisticated outcome if and only if it defeats all noninnocuous motions that follow it. But if it does so, then it must be noninnocuous, and it must be the first element of Z.

Theorem 6.6 provides a convenient constructive proof for identifying the outcome that results from a specific amendment agenda under sophisticated voting. But it does not tell us how far members can force

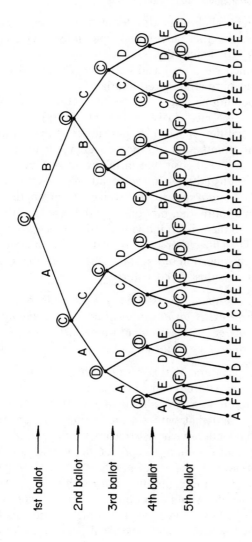

6.16 The agenda $\mathbf{V} = (A, B, C, D, E, F)$.

outcomes to wander if, paralleling the development of Theorem 2.2, the committee rules prescribe that any initial vote must include some specific motion, say y_1. To analyze this issue we return to the concept of the uncovered set, which we introduced earlier in Chapter 4 to treat election games without pure-strategy equilibria.

Recalling earlier definitions, let $C(y)$ be the set of outcomes or motions in O that *cover y*. Thus, if x is an element of $C(y)$, x defeats y, and the set of points that defeat x is a proper subset of the points that defeat y. The set of points that do not cover y, denoted $U(y)$, corresponds to $O - C(y)$. Suppose that in the event of a tie between two motions, the motion listed later in the agenda wins. Alternatively, suppose that voting is a *tournament*, by which we mean that if O corresponds to all and only the motions on the agenda, then all preferences are strict, so there is no individual indifference over O and ties are impossible. Then in either case we can prove the following result, which answers the question of how far outcomes can be forced to wander if everyone is a sophisticated voter:

6.7.   *If preferences are spatial or if voting corresponds to a tourna-*
        ment, then there exists an amendment agenda beginning at y_1
        such that y_m emerges as the eventual winner under sophisticated
        voting if and only if y_1 is not in $C(y_m)$.

To illustrate the implication of this result with a spatial example, consider the three voters with circular indifference contours in Figure 6.17. Given these preferences, we already know that a Condorcet winner does not exist, and that with sincere voting, we can find an amendment agenda that leads anywhere from any starting point, such as y_1. The regions A, B, and C denote the points that defeat y_1 directly. The hatched five-leaf clover is the set of points that y_1 does not cover and exhausts all points that the committee can reach from y_1. Hence, this figure implies something quite different from the notion that amendment agendas can lead anywhere. Outcomes cannot lead anywhere, and indeed they do not include the ideal point of at least one voter, voter 3. Hence, even if 3 is the agenda setter, he cannot manipulate amendment agendas to achieve his ideal.

Proving Theorem 6.7 for spatial preferences is tedious since we must be careful about how ties are broken and how people vote if they are indifferent. Instead, we focus on tournaments, and to shorten the argument further we suppose that y_m enters the agenda in the final stage. We begin by noticing that if alternative y_1 must enter the agenda in the first ballot, then this first stage must look like Figure 6.18a, where the circled outcome indicates that, in accordance with the theorem, y_m

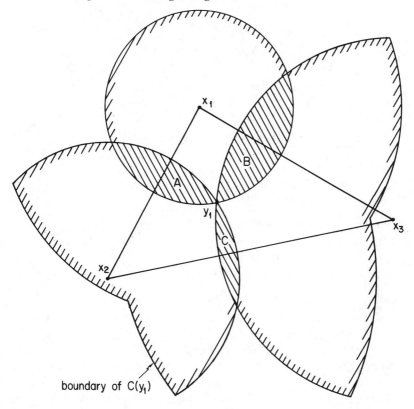

boundary of C(y_1)

6.17 Points that y_1 does not cover and that can be reached by some two-stage agenda.

prevails under sophisticated voting. Since all preferences are strict in a tournament and since social indifference cannot arise, we have two possibilities. First, if y_m defeats y_1, then y_1 does not cover y_m. So, as the only other possibility, suppose that y_1 defeats y_m. In this event, if y_m is to emerge as the eventual winner, then the committee must eliminate y_1 at some stage in the voting, say the kth. If this elimination did not occur – if y_1 is the sophisticated equivalent of the left branch of the voting tree in Figure 6.18a – then, contrary to the assertion of the theorem, y_m could not possibly prevail as the eventual outcome: To prevail, y_m would have to be the sophisticated equivalent of the right branch, but if everyone is sophisticated, then everyone knows that choosing the left branch yields y_1 while choosing the right branch yields y_m, and a majority of voters prefer y_1 to y_m and would thereby choose the left

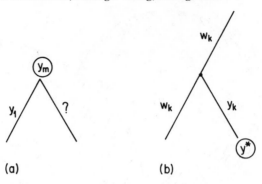

6.18 Proving Theorem 6.7.

branch and y_1. Hence, looking at the level in which y_1 is eliminated, the kth, notice that the agenda must introduce one new alternative here, say, y_k. The important characteristic of amendment agendas now is that y_k is introduced necessarily as a new alternative at every branch at this level. That is, at the kth level, every branch must look like Figure 6.18b, where w_k varies with the particular branch. (Referring to the amendment agenda in Figure 6.16, for example, notice that alternative D is introduced in the third level, so that if $k = 3$, then $y_k = D$ and w_k is either A, C, or B.) Furthermore, regardless of which kth level branch we are looking at, "beneath" each occurrence of y_k here there is a new subtree that is identical for each such branch. (Again using the agenda in Figure 6.16 to illustrate, notice that every subtree beneath those branches labeled D in the third level are all exactly alike.) Let y^* be the sophisticated voting equivalent of this subtree (which we indicate by y^* circled in Figure 6.18b), so that any motion eliminated by the committee at the level portrayed by Figure 6.18b is either y^* or a motion that y^* defeats. But since y_1 is eliminated at this level, y^* must defeat y_1. Correspondingly, everything that survives this level is either y^* itself or something that defeats y^*. Thus, since we assume that y_m is the eventual outcome and, therefore, is the sophisticated equivalent of at least one such kth level branch, either $y_m = y^*$ or y_m defeats y^*. If $y_m = y^*$, then y_m defeats y_1. But we have already assumed that y_1 defeats y_m. So suppose that y_m defeats y^*. Then there is some motion on the agenda, (namely, y^*), that defeats y_1 but that does not defeat y_m. So y_1 does not cover y_m, which proves our result.

Theorem 6.7, then, is a valuable counterpoint to Theorem 2.2, which states that amendment agendas can lead anywhere if preferences are spatial and if all voting is sincere. Theorem 6.7, in contrast, defines specific limits on the motions that a committee can reach if members vote in a sophisticated manner. An amendment agenda cannot lead anywhere from some initial point, but it is limited to the points that the first motion on the agenda does not cover.

Sophisticated voting also changes another feature of Theorem 2.2. Although that theorem does not exclude the possibility that a committee must use lengthy agendas to reach various outcomes, the following result shows that if everyone is sophisticated, then lengthy amendment agendas are unnecessary.

6.8. *If preferences over O are spatial or if the procedure is a tournament over O, and if y_1 and y_m are in O, but y_1 does not cover y_m, then there is a one- or two-stage amendment agenda from y_1 to y_m.*

As with Theorem 6.7, proving this result in a spatial context is cumbersome, so we focus again on tournaments. Notice that if y_1 does not defeat y_m, then in a tournament, y_m defeats y_1, so that the simple agenda "y_1 against y_m" works. But if it is not the case that for all z in O, z defeats y_1 implies that z defeats y_m, then there exists a z in O such that z defeats y_1 but y_m defeats z. Hence, the agenda "y_1 against y_m, the winner against z" yields y_m as the final outcome under sophisticated voting.

Although we might anticipate lengthy agendas in committees in which voting is sincere (myopic), Theorem 6.8 proves that committees with members who vote in a sophisticated manner and that uses only amendment agendas should confront relatively short agendas. Of course, 6.8 assumes that every member knows every other member's preferences. If people's information about each other's preferences is incomplete, then members may design agendas to learn the preferences of others and to "test the waters," in which case lengthy agendas may result.

Lengthy agendas might also result if the committee must consider all amendments and substitutes that members submit. Indeed, we can use this suggestion, that agendas are designed endogenously, to build a model that restricts outcomes that might prevail not only to the elements of O that the first motion entered in the agenda does not cover, but also to the uncovered set, $U(O)$, of the entire feasible set. Let $\mathbf{V} = (y_1, \ldots, y_m)$ be a specific amendment agenda, let $U_v(y)$ be the alternatives in $\mathbf{V}$ that y does not cover, and let $= U_v(\mathbf{V}) = \cap_{y \in V} U_V(y)$. Then, if y^* is the sophisticated-vote outcome from $\mathbf{V}$, y^* must be in $U_v(\mathbf{V})$. Theorem 6.7 already tells us that the agenda's first item cannot cover y^*. Notice that no other motion in $\mathbf{V}$ can cover it as well. To see this, suppose that y_k and y^* are on the amendment agenda, $\mathbf{V}$, and that

y_k covers y^*. If y_k comes after y^*, then y_k renders y^* innocuous, and from Theorem 6.6, we know that y^* cannot be the sophisticated outcome. Suppose, then, that y_k comes before y^* in **V**. If y^* is noninnocuous (which it must be if it has a chance of emerging as the final outcome), then no motion that follows y^* can render y_k innocuous (since our assumption that y_k covers y^* means that every motion that defeats y_k also defeats y^*). And if a noninnocuous proposal that comes between y_k and y^* in **V** renders y_k innocuous, then by Theorem 6.6, y^* cannot be the outcome, since only the first element of the sophisticated-agenda equivalent can be the final outcome.

The implication of this argument is that any person unilaterally can kill a motion, y, by entering some other motion, y', anywhere on the agenda that covers $y \in O$. Indeed, by introducing y', a person can ensure that the committee will not choose any alternative in the set of points in O that y' covers. Suppose that instead of there being a centralized agenda setter, members individually propose motions without knowing the specific agenda that the committee will use, and afterward, an agenda setter orders the motions that the members have introduced. Let y_i be person i's motion, let **V** be one of the agendas that the setter can construct from the n motions entered, and let V be the set of all possible agendas that the setter can construct. Suppose that in evaluating the set of motions introduced, $y = \{y_1, \ldots, y_n\}$, voter i assumes the worst with respect to how the setter will order the motions. That is, he assumes that the setter will establish the worst possible agenda. Thus, i's payoff from y is $u_i(y) = \min[u_i(y_v^*)]$, where the minimum is taken over all feasible agendas and y_v^* is the sophisticated outcome of the agenda $\mathbf{V} \in V$. We can interpret $\min[u_i(y_v^*)]$ as i's payoff function in an n-person, noncooperative game, in which each player i chooses y_i as his strategy. An equilibrium to this game necessarily exists because $u_i(y)$ is really only a function of member i's strategy, the motion he introduces, y_i. If i's evaluation of y reflects only the worst outcome that can arise with y_i, then from the preceding discussion, we know that outcome must be one that y_i does not cover. This payoff function, then, decouples the players, so that an equilibrium is a strategy n-tuple such that each person maximizes the value of the worst outcome that can prevail if y_i is somewhere in the final agenda. It follows that $y_i \in U(O)$ for all i. If not, then there exists a $y_i' \in U(O)$ that yields a higher payoff. Specifically, y_i' defeats y_i, and the points that defeat y_i' are a subset of the points that defeat y_i. So the security value of y_i' must be at least as great as the security value of y_i. Thus, if y_i is not in $U(O)$, then y_i cannot be an equilibrium strategy for i.

One implication of this analysis is especially interesting. Chapter 4 suggests that the outcomes from two-candidate, plurality-rule elections fall in the uncovered set, and the preceding discussion offers a similar conclusion for a different institution, majority-rule committees with endogenous amendment agendas. Of course, this conclusion imposes a particular assumption about how the committee forms agendas after its members introduce motions and about how people evaluate their strategies. Certainly, we can imagine a great many other processes. Nevertheless, this analysis suggests that we can say something quite general about outcomes for broad classes of institutional arrangements. Instead of finding that social outcomes and processes are wholly chaotic, an implication that some draw from Theorem 2.2, we find quite the opposite. The outcomes that we predict in two wholly different institutions – two-candidate elections and committees with endogenous agenda and sophisticated voting – exhibit the same centralizing tendency with respect to final outcomes. Thus, although the absence of a Condorcet winner may preclude the possibility of uttering precise predictions about outcomes, carefully modeling the institutional context of choice permits us to narrow predictions considerably. And insofar as we might want to develop a comparative theory of institutions, we have also learned that two institutions produce quite similar results in the uncovered set.

Congressional agendas

To make the appropriate theoretical contrast with Theorem 2.2 under sophisticated versus sincere voting, the previous discussion assumes, first, that the status quo enters the voting in the first stage of an agenda. Second, it assumes that the agenda must be of the amendment type; regardless of the identity of the previous round's winner, that winner enters the next round against a new alternative. If we look at the actual agendas that the U.S. Congress employs, however, we find that neither of these assumptions is necessarily satisfied.

Typically, the status quo enters the agenda only in the last stage. And if the following alternatives are all simultaneously on the floor, then an amendment agenda is not used: a status quo (q), a bill as reported out of some committee (b), an amendment proposing an amended version of b (a), a perfecting amendment to the amendment (aa), the bill amended by a substitute amendment (s), and a perfecting amendment to the substitute (sa). Rule XIV of the House, for example, requires that the first vote taken be between a and aa. That is, the Congress must first decide whether or not to perfect the first amendment. But this case is unlike an amendment agenda, in that the winner of this vote does not enter into the next stage of the agenda. Instead, a vote must now be taken between s and sa to perfect the substitute amendment. The winner of this vote is then placed against the winner of the first vote (*a* or *aa*), the winner then placed against *b*, and the survivor against the status quo, *q*.

Indeed, congressional agendas can take even more complicated forms, and can be combined with elimination agendas, as when members graft an elimination agenda onto its usual procedures by introducing a succession of *killer amendments*. For example, if there is no restriction on the number of amendments that can be introduced and voted on, then opponents of a bill might attempt a succession of killer amendments. As soon as one such amendment is approved, no further motions are offered since the objective has been meet, which is to render the bill unacceptable to a majority. Furthermore, permissible forms of agendas differ between the House and the Senate, the general implication being that amendment agendas are observed only if the Congress considers a single amendment and perfecting amendment, or of it considers only a single amendment and substitute.

Naturally, we should ask whether admitting these alternative forms affects our conclusion with respect to amendment agendas and sophisticated voting, namely, that sophisticated voting limits an agenda setter's power to dictate outcomes. The most general conclusion in this regard is that *if we require only that the agenda be binary, then even when everyone is a sophisticated voter, outcomes are no longer restricted to the uncovered set and the general implication of Theorem 2.2 in the context of spatial preferences and majority rule reasserts itself: If a median in all directions does not exist, then agendas can lead anywhere.*

To illustrate the first part of this assertion, consider the elimination agenda in Figure 6.13a, which is binary, and suppose that the social preference relation over $O = \{A, B, C, D\}$ is: A preferred to B preferred to D preferred to C preferred to A, C preferred to B, and D preferred to A. In this instance, A is covered by C since the set of outcomes that are preferred to C, $\{D\}$, are a subset of the outcomes that are preferred to A, $\{C, D\}$. Nevertheless, if we work backward up the tree in Figure 6.13a to model sophisticated voting, then we learn that D eliminates C, C eliminates B, and A eliminates B, so A, which is not in the uncovered set of O, prevails.

Proving the full assertion involves arguments that are not unlike those that we presented earlier. But this example is sufficient to show that it is not sophisticated voting, per se, that restricts the power of an agenda setter. Rather, it is sophisticated voting in conjunction with the requirement that the setter use an amendment agenda. If setters are free to select from a wider class of agendas, then their opportunities for manipulation increase. In this way, then, we can begin to understand who might favor the liberalization of procedures (those who are likely to control agendas) and who will prefer the rigid adherence to some specific structure, such as amendment agendas (those who merely vote).

Although this discussion seems to imply that agenda setters in Congress are likely to have more power that an analysis of amendment agendas reveals, there is one important reason to believe that they have less. Specifically, notice that our previous analysis assumes throughout that the committee first votes on the status quo and that it builds agendas forward from some starting point. The usual procedures of Congress, however, require that members vote on the status quo, q, last. Such agendas are especially conservative with respect to giving the status quo an advantage since the set of points that the committee can reach with sophisticated voting is restricted to $W(q)$ or q itself. That is, the final outcome is q or something that defeats q directly.

Congress can make its own rules, of course, including opening the floor to revote on any motion that it considered previously. Thus, ignoring the transaction costs of the complex game that ensues, we find that any agenda is possible. If there is a bias, it need not favor the status quo, per se. Instead, those members can prevail who are intensely concerned with the issues and who are intelligent enough to map carefully the strategies that take advantage of those who vote sincerely, either because of indifference or because their attention is focused elsewhere.

Partially sophisticated committees

Nothing that we have said thus far should be interpreted to mean that a game-theoretic treatment of voting in agendas requires that committee members are either all sincere or all sophisticated. Although there are no general theorems to report, we can analyze any binary agenda by the backward reduction method, assuming that any subset of a committee is sophisticated. To illustrate this, consider the following preference orders and the amendment agenda that Figure 6.13b illustrates:

	Person			
1	2	3	4	5
B	B	D	C	A
C	D	B	A	C
A	A	A	D	D
D	C	C	B	B

Notice that if the committee uses majority rule and if everyone is sincere, then B defeats A and C, but loses to D. So D is the sincere vote outcome. But if everyone is sophisticated, then the sophisticated equivalent agenda of (A, B, C, D) is (A, C, D), so A prevails in this circumstance.

Suppose now that only person 1 is sophisticated while everyone else is sincere. Of course, everyone should vote sincerely on the last ballot, so the sophisticated equivalents of the last vote in the agenda are A, C, D, and C, respectively. Thus, the first vote of the left subtree (the second ballot in the agenda), which ostensibly is a vote between A and C, is indeed a vote between these two alternatives. Hence, regardless of whether a person is a sincere or a sophisticated voter, he should vote his sincere preferences at this node, and since a majority prefers A to C, the sophisticated equivalent outcome of the agenda's left branch is A. Indeed, regardless of who is sincere and who is sophisticated, A is this branch's sophisticated equivalent. Turning now to the agenda's right branch, we find that the vote here is ostensibly between B and C. Among the sincere voters, 2 and 3 choose B, and 4 and 5 choose C. Person 1, however, is sophisticated and recognizes that this vote is between D and C, because choosing B yields D. Thus, person 1 joins 4 and 5 and votes for C (as against a sincere vote by him for B), so that sophisticated equivalent of the right branch is C. Finally, the first ballot is ostensibly between A and B, so persons 2 and 3 choose B, and 4 and 5 choose A. Person 1, however, sees this as a vote between the sophisticated equivalents, A and C, so he votes C. Thus, C prevails as the final outcome if only person 1 is sophisticated.

The discrepancy in this example between the outcome that prevails if everyone is sophisticated and the outcome that prevails if only person 1 is sophisticated need not arise in every example of an agenda. Different agendas and different preferences can yield a variety of outcomes, depending on who is sincere and who is sophisticated. The point of the example, however, is to show how we apply the backward reduction method regardless of who is sincere and who is sophisticated. This opens the door for a variety of research enterprises, including learning when it is in the interest of one sophisticated voter to teach another to be sophisticated, identifying those he ought to teach, and describing games in which the strategies of certain participants determine who is sincere and who is sophisticated.

6.5   Some experimental and empirical evidence

Although we should not interpret this chapter's theorems to state that institutional structure remains entirely decisive, they do imply that institutions affect the form of political argument and strategy. This section reviews some empirical support for this argument. We begin with a series of experiments that this author and Richard McKelvey conducted to test for the equilibrium that, according to Theorem 6.1, issue-by-issue voting induces – the issue-by-issue median preference. Figures 6.19 shows two preference configurations for five voters, each of which differs from the other by a rotation of the ideal points, denoted x_1 through x_5. Owing to this rotation, the issue-by-issue median in Figure 6.19a, y^*, is near the Pareto-optimals of the player set $\{2, 3, 4\}$; in Figure 6.19b, y^* is nearer the Pareto-optimals of the set of voters $\{1, 2, 5\}$. This variability in y^* with the rotation of ideal points provides a means for testing the influence of institutions. Specifically, if we implement appropriate procedures, and if procedures do matter in the way that Theorem 6.1 predicts, we should observe the effects of this rotation experimentally.

To induce preferences consistent with these indifference contours with money, we let each subject's payoff from the experiment's outcome decline monotonically with the distance between that outcome and the subject's ideal. To model the conditions of Theorem 6.1, we imposed procedure 3 and prohibited voters from engaging in any discussions with each other (in accordance with the game's noncooperative features). Beginning with a status quo point at $(0, 0)$, voters could offer motions to change the value of the status quo on issue 1 or issue 2 but not on both issues simultaneously. If a motion received a second, then it was voted either up or down, without discussion. If it was agreed to, then it became the new status quo and the floor was open for consideration of new motions. The experiment ended when one voter offered a motion to terminate that received a second and was approved in a formal majority vote.

The dots in Figures 6.19a and 6.19b show the outcomes of experiments using this procedure. Although there is some dispersion, clustering around the appropriate equilibrium, y^*, is obvious. So for at least this contrived situation, issue-by-issue voting induces a theoretically predicted bias. We should also emphasize that outcomes track y^* as we rotate ideal points. We know little about how people define issues in politics, about how people conceptualize alternatives, and about how political debate affects the formal structuring of a legislative scenario. But these experiments are consistent with the hypothesis, at least for one institutional environment, that this structure directly affects outcomes.

Yet another series of experiments suggests that substantive applications of theorems such as 6.1 must proceed cautiously. Suppose that we modify the preceding procedures to permit discussion among subjects,

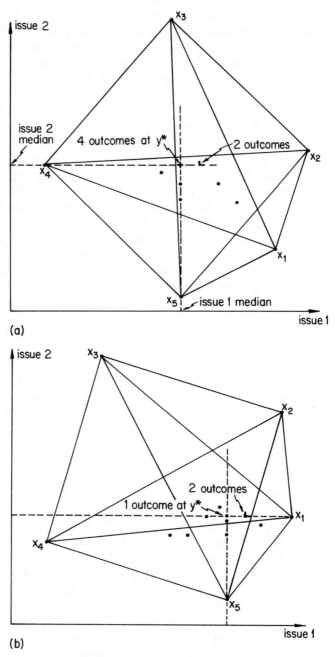

6.19(a) Issue-by-issue voting, no discussion. (b) Issue-by-issue voting, no discussion, rotated preferences.

as they make motions and vote on them. As before, we required issue-by-issue voting, and we imposed a germaneness requirement on the discussion (the subjects could only discuss the issue or dimension being voted on). Of course, we could not altogether prohibit subtle references to the second issue, and such references permit limited consideration of diagonal moves from a status quo, although the subjects could not vote on such moves directly. Figure 6.20a plots the outcomes from a series of experiments using this revised procedure. It reveals that the attractiveness of the issue-by-issue median, y^*, is considerably weakened. The rotation of ideal points, moreover, seems to affect outcomes little, as the distributions of outcomes in Figures 6.20a and 6.20b look quite similar. Correspondingly, the means of the two sets of outcomes from the two rotations of ideal points are considerably closer, and their difference is not statistically significant.

Although communication might affect the attractiveness of y^*, we are not prepared in this chapter to explain theoretically the patterns of outcomes in Figures 6.20a and 6.20b. With communication, some form of implicit and explicit cooperation occurs, and a phenomenon not yet considered emerges, coalitions. The study of coalitions, which we have thus far ignored and which will cause us to modify several of our conclusions about political processes, is a subject for later chapters.

These experiments suggest that procedures affect outcomes only if coordination and communication do not permit committee members to short-circuit the constraints that those procedures impose. Rather than short-circuiting procedures directly, however, legislators often can respond by voting in a sophisticated manner. And there are reasons for believing that sophisticated voting is pervasive in Congress. Riker's account of the DePew amendment in the Senate as a tactic for delaying the adoption of the Seventeenth Amendment and his analysis of the Powell amendment and sophisticated voting to defeat the 1957 school aid bill in the House are two examples. Enelow and Koehler offer two more contemporary examples. Consider their analysis and suppose that there are three alternatives: the status quo, Q; a bill that a committee reports out, B; and an amended bill, A. There are then six possible strict preference orders for legislators:

	Order no.					
Preference rank	1	2	3	4	5	6
1	B	B	A	A	Q	Q
2	A	Q	B	Q	B	A
3	Q	A	Q	B	A	B

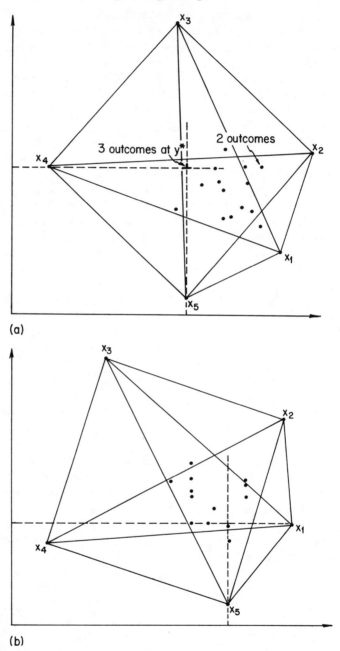

(a)

(b)

6.20(a) Issue-by-issue voting, some discussion. (b) Issue-by-issue voting, some discussion, rotated preferences.

Given the usual legislative procedure of voting on B versus A, and then the winner against Q, we can readily calculate legislators' sincere and sophisticated strategies if we know how many legislators hold each preference order. If supporters of B believe that B will lose to Q, then they might attempt a *saving amendment*, one that can secure a majority against both B and Q, so that under sincere voting the amendment prevails. To see what happens under sophisticated voting, let YN denote a vote for the amendment on the first ballot, and assuming that the amendment defeats B, against A in the final vote. For each of the preceding six preference orders, then, we have:

Preference order no:	1	2	3	4	5	6
Sincere vote:	NY	NN	YY	YY	NN	YN
Sophisticated vote:	YY	NN	YY	YY	NN	NN

To explain the votes that this table represents, consider a person who holds order 1 and votes sincerely, and keep in mind that we are assuming that B would lose to Q. Since he prefers B to A, on the first ballot he votes N, against the amendment A, whereas if the amendment beats B, he votes Y on the second ballot, since he prefers A to Q. Voting for B on the first ballot is not sophisticated, however. Because B loses to Q, voting for B against A is equivalent to a voting for Q against A. But legislators with preference order 1 prefer Q least among the alternatives.

Notice that sophisticated and sincere votes differ only for legislators who hold orders 1 and 6. With order 1, sophistication involves not being led to vote against the amendment and for a bill that ultimately fails. A sophisticated strategy consists of voting for a compromise bill that gains the support of some of those who oppose the original bill. With order 6, it requires not being suckered into supporting an amendment that increases the opposition's chances.

Since sophisticated strategies are easy to compute, their occurrence should not be uncommon. For example, consider the votes on the 1977 Common Sites Picketing Bill. The original bill, as reported out of the Education and Labor Committee, appeared in jeopardy. Representative Ronald A. Sarasin introduced an amendment that weakened several of the bill's provisions, thereby making it more acceptable to some of the opposition, but unacceptable to those who opposed the original bill's concept. Without observing preference orders directly, it is difficult to establish with certainty whether sophisticated voting occurred on this vote. But in this instance those with preference orders 1 and 6 are essentially the strong pro and anti labor members of Congress, respectively. Taking the support scores that the AFL-CIO reported, voting on the amendment for different values of this support appeared as follows:

		AFL-CIO support score				
		100–80%	79–60%	59–40%	49–20%	19–0%
Sophisticated:	YY	160	37	6	1	0
	NN	1	12	27	47	90
Sincere:	YN	7	12	10	8	3
	NY	0	0	0	0	0

The data in this table are striking, especially if we focus our attention on those with extreme support scores. If we assume that those with support scores in the range 100–80% have orders like 1, and that those with scores in the range 19–0% have orders like 6, then at least 250 legislators cast sophisticated votes.

Another common situation calling for sophistication involves killer amendments. If the members expect a bill reported out of committee, B, to defeat the status quo, Q, then those opposed to B might introduce an amendment A that defeats B but that loses to the status quo, Q. Riker's examples of the Seventeenth Amendment and the 1957 school-construction aid bill demonstrate strategic maneuvers of this sort: Opponents of B, ostensibly those with preferences like orders 5 or 6, introduce an amendment to create orders like 2 among the bill's original supporters. Interestingly, sophisticated voting should render killer-amendment strategies unworkable. Since if B and Q are paired in a final vote, B wins, and since the opponents designed A to lose to Q if they are paired in a final vote, the first ballot is not between A and B. A can never prevail, so the first ballot is between Q and B. Thus, with sophistication, the respective majorities that prefer B to A and B to Q should vote to defeat A and then Q.

Enelow and Koehler offer the ratification vote on the Panama Canal Treaty (B), guaranteeing the Canal's neutrality, as an example in which sophistication thwarted a succession of killer amendments (A). Treaty opponents designed such amendments to appeal to a broad spectrum of the Senate, but because their terms were not negotiated beforehand with then Panamanian President Omar Torrijos, opponents expected him to reject them, because they would open the door to further negotiations. Since the Senate subsequently adopted many of these amendments nearly unanimously as Senate resolutions, it is evident that most senators preferred A to B. Yet, B defeated A and the Senate then ratified B.

It is interesting to speculate about why killer amendments failed in the case of the Canal Treaty owing to sophisticated voting, but succeeded in Riker's examples. Didn't the proponents of school aid understand the Powell amendment and recognize the votes that it would cost among southern supporters of the original bill if adopted? Actually, Riker argues that both congressional leadership and the White House did understand the effect of this amendment. But the bill's supporters, especially those representing northern urban districts, feared that their constituents would not understand an apparent vote against desegregation. Thus, those who strategically supported the Powell amendment had their opponents trapped between voting strategically and losing electoral support and voting sincerely to maintain support within their constituencies but losing on the school-aid bill. It is also plausible, however, that the Powell amendment gave northern legislators an opportunity to appear liberal on a bill they did not want and knew could not pass.

6.6   Incomplete information and sophisticated voting

The principal limitation of the models in this chapter is the assumption that everyone's preferences are common knowledge. This condition corresponds to the usual game-theoretic assumption that everyone knows the extensive form of the game that they are playing. But committees often use agendas longer than two stages despite the theoretical result that, with sophisticated voting, any outcome that they can reach in n stages can be reached in two stages. We suspect that multiple-stage agendas occur because people have imperfect information about the preferences of others and because voting remains at least partly a mechanism for estimating preferences. If so, then the resulting game would be highly complex. Votes affect not only the path on a voting tree, but also beliefs about preferences, which in turn affect future votes. Voting, then, has a dual strategic character, to affect both paths and beliefs.

Although game theorists have only begun to look at such problems, we can indicate how research might proceed. Our approach is similar in some respects to the analyses of elections and voting with incomplete information presented in Chapters 4 and 5. There, as well as here, we look for strategies and beliefs that are simultaneously consistent. Unlike our analysis in Chapter 4, however, but like the analysis in Chapter 5, the model discussed here does not propose a mechanism through which

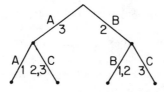

6.21 The agenda (A, B, C).

beliefs ultimately mirror reality. If beliefs are erroneous, then equilibria can prevail that are not complete-information equilibria.

Consider a three-person committee, and suppose that everyone knows that each person's preferences over three alternatives is one of the following three types:

Type 1	Type 2	Type 3
A	B	C
B	C	A
C	A	B

These are the same preference orders that illustrate the Condorcet paradox. Suppose, however, that although each person knows his own preference, his is uncertain about the other two committee members' preferences. Instead, each believes that another person is of Type i with probability p_i. Now consider the agenda (A, B, C) depicted in Figure 6.21.

As before, voting sincerely on the last ballot is a dominant strategy for everyone. Furthermore, Type 2 and Type 3 voters, regardless of their beliefs about other members' preference orders, have dominant strategies on the first ballot. Since people are unsure of the preference of others, choosing A or B on the first ballot is equivalent to choosing a lottery between A and C over a lottery between B and C. But Type 2 voters prefer any lottery between B and C to any lottery between A and C and, thus, they should vote for B. Type 3 voters must consider two possibilities: First, neither of the other two members is a Type 3. Second, at least one of the other members is a Type 3. In the first case, a Type 3 voter should choose A over B on the first ballot, since B wins if the final vote is B versus C (in this case, both of the other members prefer B to C), and B is a Type 3 voter's least preferred outcome. In the second case, a Type 3 voter does not care whether the final vote is between B and C or A and C, because the two or three Type 3 voters will choose C in the final vote. Thus, even if there is the slightest chance of the first case occurring, a Type 3 voter should vote for A on the first ballot. Figure 6.21 notes at the appropriate branches the dominant choices of members with each type of preference order.

The strategy that a Type 1 voter should choose is more difficult to compute. It depends on his beliefs, the p_i's, and his conjectures about what he and other members think Type 1 voters will do. To illustrate this choice problem, recall the normal-form game in Figure 5.1. There, information is complete, so that each person knows every other person's payoffs. Nevertheless, person 1's optimal strategy depends on what he thinks person 2 will do. If 1 conjectures that 2 will choose the strategy that yields 1 the greatest possible gain, b_1, then 1 should choose a_1. And this is the equilibrium that prevails if 2 conjectures that 1 will choose a_1 and 1 believes that 2 thinks this way. But consider another conjecture: All players choose the strategy that yields them the greatest possible gain, and suppose that all players share this conjecture. The problem with this conjecture is that it is inconsistent with the known payoffs and optimal choices. Specifically, 1 concludes that 2 will choose b_2 and that 2 thinks voter 1 will choose a_1. But b_2 is not a best response to a_1, nor is a_1 a best response to b_2. Thus, this conjecture, the information both persons have about payoffs, and the hypothesis that people maximize utility are inconsistent. Put simply, to identify equilibrium strategies we must find some conjectures for the players that remain consistent with their beliefs about preferences and with optimal choices.

We proceed first by normalizing the utility for any Type 1 voter so that the cardinal utility of his first choice, A, equals 1, that of his last choice, C, equals 0, and that of his second choice, B, equals v, where $0 < v < 1$. Second, notice that unless a Type 1 voter is decisive, he is indifferent between any pair of strategies. This means that in describing equilibrium strategies, we need only look at those instances in which the other two voters split on the first ballot; otherwise, our voter is not decisive and any strategy will do, including the strategy that he uses if he is decisive. A voter does not know beforehand, of course, whether he is decisive, but he can condition his choice of strategy on being decisive, since otherwise it does not matter what he does.

To search for possible equilibria, we must consider various conjectures that Type 1 players might entertain. We pick the two simplest possibilities:

Conjecture 1: All members vote sincerely, regardless of the preference order that each holds.

Conjecture 2: Type 2 and 3 voters choose their dominant strategies, but all Type 1 voters choose strategically.

Using one of these conjectures, we deduce the optimal strategy for a Type 1 voter, conditional on his beliefs, and we then check whether beliefs and conjectures are consistent.

Suppose that you are a Type 1 voter who thinks that others vote sincerely, regardless of their preferences. Thus, you think that Type 3 and all other Type 1 members vote for A on the first ballot, and Type 2 members vote for B. Hence, you will be decisive on the first ballot only if there is exactly one Type 2 member (the one who voted for B) and if the remaining member has preferences that match either Type 1 or Type 3 orders. Conditional on this third member not being a Type 2 member, the probability that he has Type 1 preferences is $p_1/(p_1 + p_3)$, and the probability that he has Type 3 preferences is $p_3/(p_1 + p_3)$. Thus, if you vote sincerely for A on the first ballot, then A defeats C with probability $p_1/(p_1 + p_3)$, the probability that another member is like you, and C defeats A with probability $p_3/(p_1 + p_3)$. If your utility for A is 1 and for C, 0, then your expected utility from voting sincerely is $p_1/(p_1 + p_3)$. But if you vote strategically for B on the first ballot, then B wins with certainty (your vote plus the vote of the Type 2 voter) and you get v. Hence, you prefer to vote sincerely if

$$\frac{p_1}{p_1 + p_3} > v. \tag{6.3}$$

Voting sincerely, then, is consistent with your beliefs (about preferences) and with the conjectures that you hold about the strategies of other Type 1 voters (conjecture 1). And since you also think that other voters share this conjecture, you have no reason to think that they will find their conjecture inconsistent with what you believe about them. That is, there is no infinite regress of conjectures. Hence, "all voters voting sincerely" is an equilibrium if expression (6.3) holds.

To see that equilibria resulting from this formulation are not unique, consider the second conjecture, namely, that all Type 1 persons vote strategically. Suppose again that you are decisive, that the other two voters split between A and B on the first ballot. Because you believe that other Type 1 voters are voting strategically for B, and since Type 2 voters also choose B as their dominant strategy, exactly one person must hold a Type 3 preference order. The third member of your committee (the one who votes for B) must have either Type 1 or Type 2 preferences. Given that he does not have Type 3 preferences, the probability that he is of Type 1 is $p_1/(p_1 + p_2)$ and the probability that he is of Type 2 is $p_2/(p_1 + p_2)$. By an argument similar to the one just given, it follows that you should also vote strategically if

$$\frac{p_1}{p_1 + p_2} < v. \tag{6.4}$$

As before, this preference is consistent with your initial belief about how other persons who hold your preference order will act. Thus, if expression (6.4) is true, then "everyone voting sincerely unless they hold a Type 1 preference order" is an equilibrium.

Notice that we can satisfy both of the preceding inequality constraints on v simultaneously. Hence, we can have two symmetric pure-strategy equilibria, one in which everyone votes sincerely and one in which all but Type 1 voters vote sincerely. And there may be other equilibria, as when some Type 1 members vote sincerely and others are strategic. Thus, although this discussion indicates how we might go about analyzing a voting game with incomplete information, it also reveals that generally there are multiple equilibria depending on what conjecture Type 1 voters hold about other Type 1 voters and that such conjectures are exogenous to the analysis. What we predict, then, depends as much on our measurement of beliefs and conjectures as it does on our understanding of people's preferences and the institutional structure in which they operate. This is the same problem that we encountered in making definitive predictions about outcomes for the normal-form game in Figure 5.1, and it demonstrates that in committees in which people are uncertain about the preferences of others, initial conjectures and beliefs are important determinants of final outcomes.

The preceding example explains a method of analysis. To illustrate the sorts of paradoxical conclusions that we can derive from this method, consider a committee of n persons, restricted to the same three preference orders as before, which uses the same agenda, (A, B, C). Hence, Type 2 and Type 3 voters have dominant strategies (persons with Type 2 preference orders choose B on the first ballot, and those with Type 3 orders choose A), but the strategies for Type 1 voters depend on initial conjectures and beliefs. Limiting the analysis to symmetric pure-strategy equilibria, let us take as an initial conjecture that all persons holding Type 1 preferences are sincere (and vote for A on the first ballot). But suppose that you are a Type 1 voter and that you are decisive on the first ballot. Then there must be exactly $(n - 1)/2$ Type 2 voters, because they are the only ones who vote for B. Hence, if you choose B, then B wins with certainty, because you and the other Type 2 voters will choose B over C on the second ballot. But if you vote sincerely for A on the first ballot, and if there is even one voter with a Type 3 preference, then the eventual outcome will be C, because this one Type 3 voter plus the $(n - 1)/2$ Type 2 voters make up a majority that prefers C to A. Given that he cannot be a Type 2 voter, the probability that a randomly chosen first-ballot A voter has Type 1 preferences (as against Type 3 preferences) is $p_1/(p_1 + p_2)$. The probability that all first-ballot A voters hold Type 1 preference orders – which is the probability that if you are decisive, then voting for A on the first ballot yields A over C – is

$$\left[\frac{p_1}{p_1 + p_3}\right]^{(n-1)/2}.$$

If alternative B's value to you is v, then you should vote sincerely for A on the first ballot, which is in accord with your conjecture about what voters hold your type of preference order do, only if

$$\left[\frac{p_1}{p_1 + p_3}\right]^{(n-1)/2} > v.$$

But since v is strictly greater than zero, for sufficiently large n this inequality cannot hold for any strictly positive p_1, p_2, and p_3. Thus, having all Type 1 members vote sincerely for A over B cannot be a symmetric equilibrium for sufficiently large committees.

Now consider the second possible symmetric equilibrium: All Type 1 members vote strategically and choose B over A on the first ballot. Paralleling the preceding argument, if you are pivotal on the first ballot, then there must be exactly $(n - 1)/2$ Type 3 voters, those who choose A. Thus, if you vote strategically for B on the first ballot, then you get A with certainty: You and all other Type 1 and Type 2 voters are a bare majority that prefers B to C. If you vote sincerely for A, however, whereas others who hold your preference order vote strategically, then you get A only if there are *no* Type 2 voters among the $(n - 1)/2$ members who choose B on the first ballot. Otherwise the one or more Type 2 voters will combine with the $(n - 1)/2$ Type 3 voters to choose C over A. The probability that all $(n - 1)/2$ first-ballot B voters hold Type 1 preference orders is

$$\left[\frac{p_1}{p_1 + p_2}\right]^{(n-1)/2}.$$

Hence, you should vote strategically for B on the first ballot only if

$$v > \left[\frac{p_1}{p_1 + b_2}\right]^{(n-1)/2}.$$

For sufficiently large n, this inequality necessarily holds, in which case the *only* pure strategy symmetric equilibrium is the following:

In the first ballot, Type 1 and 2 voters choose B, Type 3 voters choose A. If B wins, Type 1 and 2 voters choose B and Type 3 voters choose C. If A wins, Type 2 and 3 voters choose C, Type 1 voters choose A.

For sufficiently large committees the p_i's will approximate the actual proportions of Type 1, 2, and 3 voters (assuming that beliefs mirror the lottery that we use to generate preferences). Thus, if $p_1 + p_2 > \frac{1}{2}$, then Type 1 and 2 voters almost certainly make up a majority and B wins. But if $p_3 > \frac{1}{2}$, then C wins. Thus, for large n, and regardless of prior estimates of these probabilities, if a symmetric pure-strategy equilibrium exists, then alternative A can never prevail. Since this result holds true for p_1 arbitrarily close to 1, a range of situations occurs in which A is almost certainly the Condorcet winner, but the probability that the committee chooses A is zero!

The "trick" in this analysis that yields such a conclusion is that with just a little bit of uncertainty about preferences, Type 1 voters should condition their decisions on a very unlikely event, their being decisive. But if all Type 1 voters do this, then they all vote strategically, and thereby defeat an almost certain Condorcet outcome. There are doubtless a great many other equilibria in which some Type 1 voters vote strategically whereas others vote in a sincere manner. Nevertheless, this example illustrates the crucial role of information in our predictions about outcomes, and it demonstrates dramatically that sophisticated voting with incomplete information in binary voting systems is significantly different from voting under conditions of complete information.

The preceding examples hardly constitute a theory of agendas with incomplete information, and many avenues of research remain open. First, we should consider agendas with more alternatives. With two-stage agendas, voters can revise their beliefs after the first ballot, but by then it is too late to act on those revised beliefs. Second, recall the result that if information is complete and if everyone is sophisticated, then we require only two-stage amendment agendas to reach any "reachable" outcome from some status quo. We might then speculate that we sometimes observe multistage agendas because they provide one way for people to learn the preferences of others. Third, since strategies affect outcomes directly when the votes are tallied and indirectly as people adjust their beliefs to what they observe, an agenda setter's problem becomes complicated. The choice of an agenda may reveal something about his preferences before anyone else, since he must make this choice before anyone else reveals anything. Naturally, an agenda setter should take this problem into account, knowing that others members will take his actions into account. Complicating matters further, however, is the fact that these same members should assume that the agenda setter's choice anticipates their reactions, and so forth. Owing to the strategic complexity of such a scenario, it is evident that the outcome that prevails in a committee of this sort will depend on the members' sophistication and on the importance of the alternatives under consideration. That is, the final outcome will depend on the ability and the willingness of committee members to make complex strategic calculations. This underscores an earlier conclusion: Institutions are anything *but* black boxes into which we simply plug preferences and calculate outcomes in some straightforward, mechanical way.

6.7   Summary

This has been a long and perhaps tiresome chapter, especially for the reader who is encountering these ideas for the first time. Yet familiarity with the ideas offered here is essential to an understanding of an important approach to the study of institutions. Before proceeding, then, the reader should review the key concepts that this chapter discusses, some of which earlier chapters discuss and which include at least the following:

issue-by-issue voting	issue-by-issue median	structure-induced equilibrium
sincere voting	linearly independent	preference-induced
sophisticated voting	separability	equilibrium
elimination agenda	amendment agenda	sophisticated voting equivalent
voting tree	innocuous motion	binary agenda
cover	tournament	sophisticated agenda
killer amendment	saving amendment	uncovered set

An important lesson of this chapter is that cyclic social preferences do not preclude prediction, and as a corollary, that the existence of a Condorcet winner is not a necessary condition for prediction under majority rule. Instead, institutions and the rules that regulate individual choice can have as important an effect on final outcomes as the individual preferences that operate within them. This observation gives rise to a substantive, albeit imprecise, distinction between the two notions of preference-induced equilibria and structurally induced equilibria. The usual example of a preference-induced equilibrium is a Condorcet winner, an outcome whose prominence reflects a particular configuration of individual preferences and that we anticipate being chosen under a wide variety of institutional arrangements (although not necessarily all arrangements). The most evident example of a structurally induced equilibrium is the issue-by-issue median (Theorem 6.1) or its generalization reported in Theorem 6.2. There, if we suppose that people reveal their preferences sincerely, then we can identify a stable point even though no Condorcet winner exists.

This distinction between equilibria cannot be made precise, of course, because it is impossible for preferences to be revealed and cycles to become evident unless we implement *some* procedure. Nevertheless, this distinction helps to show that institutions constrain individual choice, and that a description of the social preference order under pairwise voting and simple majority rule alone is not a sufficient basis for prediction. That is, although a theorem such as 2.2 tells us that social preferences are wholly intransitive in a spatial context if no Condorcet winner exists, we cannot infer from this result that anything can happen, or that an agenda setter is dictator, or even that an agenda setter has inordinate power. Before we can affirm such conclusions we must establish the implications of cyclic social preferences in the context of the particular rules that people use to reach decisions. Do people make decisions directly in a committee, or indirectly with candidates competing in elections? In the case of committees, can people vote on issues in any order? Can they reconsider issues? Are preferences on the issues separable? If preferences are not separable, do people know the rules well enough so that they can vote sophisticatedly? If the committee abides by an agenda, what type of agenda must it use? And does everybody enjoy complete information?

Although terms such as *preference* and *structurally induced equilibria* are more journalistic conveniences that theoretically meaningful distinctions, the dependence of final outcomes on institutions that this chapter reviews is profoundly important. Accustomed as we are to digesting data gleaned from survey research instruments, we often forget that the attitudes people express in them are conditioned by and must operate through institutions. It matters little if one presidential candidate is ahead in the polls when his support is concentrated in the wrong states, and we will sometimes be disappointed if we believe that an outcome will prevail in Congress because a majority prefers it to the status quo. Proposals for reform of an institution, moreover, are likely to yield unintended consequences if their architects do not consider how people condition their decisions on those institutions.

This chapter examines two general procedural or institutional constraints, the first being issue-by-issue voting. We can summarize the effects of such voting on outcomes thus (keeping in mind our other assumptions, e.g., no communication and coordination, thin indifference contours, and so forth):

	All voters are sincere	All voters are sophisticated
Preferences are separable	A unique stable outcome exists, and that outcome is unaffected by voting order and whether issues are reconsidered. If an overall Condorcet winner exists, then it will be the stable point.	Same conclusions as when all voters are sincere.
Preferences are not separable	A unique stable outcome exists, but the outcome that is selected will depend on the voting order and other procedural details. But if an overall Condorcet winner exists, then it will be the stable point.	In general, no stable outcome will exist. The existence of a stable outcome may depend on the order in which the issues are considered.

Another way committees order complex decisions is to impose an agenda, a prespecified order for considering specific alternatives. Thus, congressional agendas require that a bill reported out of committee first be paired against any amendment, with the winner put to a final vote against the status quo. It might appear from Theorem 2.2, however, that agendas simply highlight the indeterminacy of social outcomes if social preferences are intransitive. If agendas can lead anywhere, and if the choice of an agenda setter or of a particular agenda simply pushes the cycle of social preference back one stage, then how can agendas induce stability? Theorem 2.2 assumes, however, that everyone votes sincerely, whereas Theorem 6.7 shows that by voting in a sophisticated manner in an amendment agenda, committee members can constrain the outcomes that they can reach from any status quo point. And the discussion that follows suggests that if amendment agendas are endogenous, then only outcomes in the uncovered set ultimately can prevail.

The relevance of the uncovered set here is interesting. In Chapter 4 we argued that if candidates in two-candidate majority-rule elections adopt undominated spatial strategies, then only strategies in the uncovered set prevail. Here we considered a different institution, majority-rule committees with endogenous amendment agendas, and we suggest that outcomes fall in the same set. Thus, two distinct institutions constrain outcomes in similar ways, and if one asks "Do institutions matter?" we must give two answers. First, they can matter in the sense of inducing stability or otherwise constraining outcomes. Second, they might not matter to the extent that different institutions constrain outcomes in the same way.

Of course, Chapter 4 looks only at a limited set of institutions, those that use majority rule, and in which everyone acts noncooperatively. We can explain the emphasis on majority rule and on formal mechanisms, in part, by the fact that such institutions are most susceptible to rigorous analysis. This analysis does not originate from the belief that these are the only institutions that matter. For example, we have not considered systems that combine elections with committee voting (as we must when examining the effects of electoral imperatives on congressional committee action) or less formal institutions, such as bureaucracies subject to congressional oversight. Instead, we focus on the simplest institutions, to illustrate fundamental principles and approaches.

But our analysis requires two important caveats. First, the discussion of voting with incomplete information suggests that initial beliefs are as crucial in determining outcomes as are institutions. Yet we must approach definitive conclusions cautiously. The analysis of games with incomplete information is a new area of inquiry, and we remain uncertain about what theoretical developments will arise in it. Second, our analysis ignores a possibility that Chapter 4 considers. There we present a model of incomplete information in elections in which voters and candidates, owing to the feedback that public opinion polls provide, eventually make the same decisions that they would make if their information was complete. Our analysis in this chapter, on the other hand, remains essentially static, and it does not model the possibility that equivalent mechanisms operate in other institutions. If such mechanisms exist, they will mitigate the effects of initial beliefs on outcomes.

The analysis in this and all previous chapters also assumes that action is noncooperative. This assumption precludes the possibility that people circumvent institutions or even change those institutions by communicating and coordinating their actions. Thus, it is impossible to understand fully the effects of institutions and their imperatives for individual choice until we weaken this assumption. Communication and coordination, then, are the subjects of the rest of this book.

CHAPTER 7

Cooperative games and the characteristic function

The preceding chapters have gone to extraordinary lengths to avoid the possibility that people cooperate and act in concert to achieve some outcome. Chapter 2 describes logrolling as an example of strategic voting without supposing that a decisive subset of the legislature such as a majority acts to thwart the vote trades that other members of the legislature might negotiate. Chapter 4 analyzes elections as noncooperative games between candidates; this approach seems appropriate for two-candidate elections, but it also assumes that voters are passive agents in the election and it thereby ignores the coalitions among voters that interest groups, such as organized labor, try to form. Chapter 5 applies the prisoners' dilemma to a variety of situations, but it does not allow people to try to avoid mutually distasteful outcomes by some form of prior, binding collusion. And Chapter 6 dispassionately analyzes issue-by-issue voting in committees by assuming that members cannot talk, negotiate, and coordinate their strategies, thereby precluding the important possibility that procedures affect outcomes only if members choose not to collude to bypass them.

The assumption that people reveal their strategies simultaneously and cannot communicate their choices beforehand is a useful theoretical abstraction. Many two-person situations are best modeled as zero-sum games, in which case communication serves no purpose, and it is useful in other circumstances to evaluate the consequences of the failure to communicate. It is probably rare, though, that communication among people, however imperfect, remains impossible. In response to this reality, cooperative game theory models those situations in which communication not only is possible, but also stands as a central feature of human interaction.

Communication achieves its significance because it admits of another profoundly new phenomenon, the formation of *coalitions*. Briefly, *a coalition is an agreement among two or more persons to coordinate their actions (choices or strategies)*. Coalitions include formal agreements, such as legally binding contractual arrangements involving two or more persons to act as a single entity, as well as agreements to avoid certain joint choices. Coalitions also include informal, perhaps even unobserv-

302

able, agreements, such as implicit understandings among legislators, to avoid mutually distasteful outcomes. Hence, we can interpret a lawyer and a client, a grocer and a customer, a business firm, a labor union, a political party, and so on, as coalitions.

Despite the importance of the concept of a coalition to politics, in a theoretical sense communication and the coordination that communication makes possible add nothing new to game theory. For example, we could decide that the opportunity to communicate merely augments the strategies available to people, thereby complicating both the extensive- and normal-form representations of situations. In this view, although cooperative games would yield a significant "bookkeeping" problem for the theorist, they should not cause any radical theoretical departures. For example, game theorists, and especially those who model the formation and maintenance of coalitions in substantive settings appreciate that the explanation for why people form and maintain coalitions in any specific situation often lies in an understanding of the larger game that contains that situation. For example, members of Congress honor vote trades verbally agreed to because they do not want other members to eliminate them from participating in future trades. Thus, to understand fully the logic of vote trading requires that we model the extensive form of one, or perhaps many, legislative sessions. Similarly, decisions by nations to form certain coalitions, to avoid others, and to honor agreements may reflect considerations of what international relations games national leaders will play subsequently as much as they mirror immediate advantages that a particular country might obtain. Stated simply, any equilibrium, cooperatively reached or not, must be a noncooperative equilibrium because a coalition itself is stable only if none of its members has an incentive unilaterally to seek some other agreement.

The logic of this argument, that communication merely augments the extensive and normal forms of games and that no new theory is required to study cooperative games, seems compelling. But there are advantages to ignoring it for the present. First, rather than concern ourselves with the issue of what mechanisms people use to enforce the agreements they might reach, we can achieve some simplification by directing our attention to the issue of what these agreements might look like and what coalitions would form to bring them about. Some simplification seems essential, furthermore, especially if we are concerned with specific bargaining processes. Even a brief introspection into the nuances of human interaction reveals the great number of strategies that communication permits: threats, the timing of actions, lying, and severing communication itself. Each possibility communicates something about intentions, each affects perceptions and beliefs, and thus, we can interpret each as a move in a monstrously complex game tree. The theorist's initial instinct is to abstract from such complexity and assume it out of existence, and to begin with a description that admits analysis.

The simplification that game theorists traditionally take is to let enforcement mechanisms lurk in the background, and instead, to refine a game's mathematical description so as to uncover the imperatives for specific coalitional agreements. That refinement, which we call the characteristic-function representation, focuses on the substance of conflict and negotiation, the resources that people are trying to divide among themselves, and the limits that technology places on alternative divisions. This chapter reviews that representation.

We begin by looking at a normal-form game, to see how we might extend the concept of a Nash equilibrium to treat certain problems. We then turn to the definition of the characteristic function. We illustrate this definition for games that model voting situations in committees, and for games in which people can transfer utility among themselves much like money. These games are especially useful for modeling various economic situations. We conclude with a discussion of this representation's strengths and weaknesses.

7.1   Strong equilibria

One approach to cooperative action is to make the concept of a Nash equilibrium a collective one without resorting to complicated extensive form representations. The Nash equilibrium concept is necessarily individualistic because it looks only at the incentives that people, taken one at a time, have to defect from specific strategy n-tuples. But a different casting of this concept models some of the incentives of people to cooperate and to defect from cooperative agreements, and it also illuminates one fundamental problem with deducing which cooperative agreements people are likely to choose. Specifically, instead of defining an equilibrium in terms of the incentives or disincentives of individuals to defect from a strategy n-tuple, suppose we look at the incentives for two or more people to defect simultaneously from a noncooperative equilibrium. Formally, let C be some subset of the set of all persons, N, and let c be the number of people in C. Next, let $s^* = (s_1^*, \ldots, s_n^*)$ be a Nash equilibrium. Third, let $(s_C, s_{\bar{C}}^*)$ be an n-tuple of strategies that denotes a defection from s^* by each member of the set C, but where everyone not in C (denoted $\bar{C}$) continues to choose their equilibrium strategies. Finally, let S_C denote the set of all strategy c-tuples available to C. Then,

Person 3

Person 2:	c_1		c_2	
	b_1	b_2	b_1	b_2
Person 1: a_1	57, 50, 15	0, 60, 35	55, 20, 0	30, 30, 40
a_2	70, 55, 30	0, 90, 60	50, 0, 80	20, 60, 70

7.1 Example of a strong equilibrium.

The n-tuple $s^* = (s_1^*, \ldots, s_n^*)$ is a *strong pure-strategy equilibrium* if for all subsets C of persons and for all $s_C \in S_C$,

$$u_i(s^*) \geq u_i(s_C, s_{\bar{C}}^*) \text{ for some } i \in C.$$

To illustrate this definition, consider the normal-form game in Figure 7.1. There, the unique Nash equilibrium is (a_1, b_2, c_2). To see if this is a strong equilibrium, we must first check whether any two persons have a unanimous incentive to defect and then whether all three persons have such an incentive. Holding 3's strategy fixed at c_2, 1 and 2 can move to any of the three utility outcomes (55, 20, 0), (50, 0, 80), and (20, 60, 70), but they jointly prefer none of these outcomes to (30, 30, 40). Next, holding person 1's strategy fixed at a_1, persons 2 and 3 can jointly shift to the outcomes (57, 50, 15), (0, 60, 35), and (55, 20, 0). Again, however, they jointly prefer none of these outcomes to (30, 30, 40). And, although 1 and 3 can shift unilaterally to (0, 60, 35), (0, 90, 60), and (20, 60, 70), holding 2 at b_2, 1 and 3 cannot jointly agree on any of these outcomes. Finally, although all three persons could move to any outcome in the game, no outcome Pareto-dominates (30, 30, 40), so unanimous agreement to move elsewhere is impossible. Hence, (a_1, b_2, c_2) is a strong pure-strategy equilibrium.

A two-person prisoners' dilemma, however, illustrates a Nash equilibrium that is not equivalent to a strong equilibrium. There, the Nash equilibrium is not Pareto-optimal, so the set of all persons, N, could agree to move elsewhere, and in particular, to the outcome that they unanimously prefer to the equilibrium. This simple example illustrates the fact that *strong equilibria are a subset of all Nash equilibria.*

The concept of a strong pure-strategy equilibrium is an attractive idea, but it has a serious drawback that we can most easily grasp in the context of mixed strategies. A mixed-strategy defection from s^* takes the form $(p_1, \ldots, p_n)$, where p_i denotes i's mixed strategy. Accordingly, suppose that in the previous example we fix person 3 at c_2, and assume that 1 chooses a_1 with probability p and 2 chooses b_1 with probability q. The weights placed on each of the four cells in the right-hand side of Figure 7.1 now become pq (for (a_1, b_1)), $p(1 - q)$ (for (a_1, b_2)), $(1 - p)q$ (for (a_2, b_1)), and $(1 - p)(1 - q)$ (for (a_2, b_2)). But these calculations do not permit *correlated mixed strategies* in which persons 1 and 2 agree, say, to mix their choices only between (a_1, b_1) and (a_2, b_2). Such an assignment of probabilities remains impossible if we restrict people to their "private" sets of strategies, as the preceding definition requires. Indeed, in the example, 1 and 2 prefer an even-chance lottery between (a_1, b_1, c_1) and (a_2, b_2, c_2) – which yields the expected 3-tuple (37.5, 40, 35) – to their strong equilibrium payoffs. But they cannot realize such a lottery without correlating their strategies. Hence, although a strong equilibrium requires that s^* be stable against "coalition" defections, it does not take full advantage of coordination possibilities if people can communicate and make binding agreements.

There is no reason that we cannot extend the definition of strong equilibria to accommodate correlated strategies. But before we do so, we must consider another issue. Suppose that persons 1 and 2 defect from (30, 30, 40) to the correlated lottery that yields the expected utility 3-tuple of payoffs (37.5, 40, 35). Now person 3 has an incentive to shift to c_1, in which case the expected utility 3-tuple (28.5, 70, 37.5) results. Although 2 is happy with this consequence, person 1 prefers the original outcome. What assumption should we make, then, about people's evaluations of defections? Should we rule out such secondary adjustments by 3, or should we take them into account?

Such questions arise whether or not we consider correlated mixed strategies. They are more apparent, though, with correlation, because through correlation people try to incorporate the coordination possible with communication. But if we permit coordination, why should we arbitrarily exclude consideration of secondary adjustments? Admittedly, answers to such questions are neither easy to formulate nor blessed with universal acceptance. We turn, though, to the traditional answer, which we offer in the form of the definition of the characteristic function, and we caution the reader that the following discussion describes a theory that research is constantly updating.

7.2   The characteristic function, $v(C)$

Let $N = \{1, 2, \ldots, n\}$ be the set of all relevant persons in a particular situation, and define a *partition of N* as a set $\mathbf{C} = \{C_1, C_2, \ldots\}$ that divides N into exhaustive and disjoint subsets. We call elements of C coalitions, and C is a *coalition structure*. For example, in noncooperative games the only admissible coalition structure is $\{\{1\}, \{2\}, \ldots, \{n\}\}$, but cooperative games consider all possible structures. We can interpret a decision to cooperate with one person and not another, then, as the choice of one set of coalition structures over another. If we conceptualize individual decisions now as involving the choice of one coalition structure over another, then our first task is to find out how people evaluate the attractiveness of alternative structures. To understand the nature of this task, refer again to the game in Figure 7.1. If the three persons play this game non-cooperatively, then we would predict the strategy 3-tuple (a_1, b_2, c_2), because it is the unique Nash equilibrium. But suppose that these persons can discuss their choice of strategies beforehand. Assume, further, that each has the means for granting either or both of the other persons the authority to choose a strategy for him, provided that they can form some absolutely binding contract about what joint strategies to choose. These agreements can include a decision to pick a specific strategy 2- or 3-tuple or *any* lottery over all of these tuples. Suppose, in particular, that persons 1 and 2 are evaluating the advantages of coordinated action. Ignoring probabilistic choices, they have four alternatives to evaluate, the four combinations of each of their two strategies. Thus, they can view the decision to form a coalition as a decision to expand their alternatives from their set of individual pure strategies to the set $\{(a_1, b_1), (a_1, b_2), (a_2, b_1), (a_2, b_2)\}$. Our task is to extend individual preferences over outcomes to preferences over this set.

This problem is similar to the one that we encountered in noncooperative game theory. There, in deciding which strategy a person should choose, we had to decide how to extend individual preferences over O to preferences over individual strategies. The approach we took was to evaluate each strategy in terms of its security value, and initially, to predict the strategy that maximizes this value. We then showed that at least for zero-sum games, such actions lead to a pure-strategy equilibrium if such an equilibrium exists. Cooperative game theory proceeds in a similar way. Ignoring some complications that mixed strategies occasion, notice that if persons 1 and 2 together choose (a_1, b_1), then regardless of how 3 responds, person 1 guarantees himself a payoff of at least 55 (if 3 chooses c_2), and person 2 guarantees himself at least 20 (again if 3 chooses c_2). Thus, the joint security value of (a_1, b_1) is (55, 20).

Proceeding through all four-strategy 2-tuples for the coalition $\{1, 2\}$, we have

Joint strategy	Security value
(a_1, b_1)	(55, 20)
(a_1, b_2)	(0, 30)
(a_2, b_1)	(50, 0)
(a_2, b_2)	(0, 60)

The value of coordination is evident. If 1 and 2 act alone, their individual security values are maximized at 0 and 30, respectively, if they choose a_2 and b_2, respectively. But, by coordinating to choose, say, an equal chance lottery between (a_1, b_1) and (a_2, b_2), 1 raises his security level to 28.5 and 2 raises his to 40.

The preceding example illustrates how people might evaluate joint actions with others in a game. What we require for tractable mathematical analysis, though, is a convenient means for representing these numbers. That representation is *the characteristic function*. Formally if $\mathbf{u} = (u_1, u_2, \ldots, u_n)$ is a vector of utilities, where u_i is i's payoff, then

> *The characteristic function*, $v(C)$, of a coalition C is a collection of utility vectors, such that if C can assure each of its members a payoff of at least u_i^* by their adopting some joint strategy, and if $w_i \leq u_i^*$ for all $i \in C$, then $\mathbf{w} = (w_1, \ldots, w_n)$ is in $v(C)$.

That is, if by coordinating their actions, the members of C can assure player $i \in C$ a payoff of at least u_i, then $\mathbf{u}$ is an element of $v(C)$. Much like the security value of individual players for noncooperative games, $v(C)$ summarizes as a set of payoff vectors the security values of the various joint strategies available to the set of persons, C.

To illustrate this definition with the previous example, Figure 7.2 graphs u_1 and u_2 and ignores u_3 since person 3 is not a member of the coalition that we are considering. This figure plots the security values of each of the four joint strategies of the coalition $\{1, 2\}$, and we connect these points with dashed straight lines to indicate that persons 1 and 2 might choose to consider lotteries between their four pure coordinated strategies. The characteristic function of $\{1, 2\}$ – $v(1, 2)$ – however, is the entire shaded region of this figure, including its boundary. To check this representation, consider $\mathbf{w} = (w_1, w_2) = (5, 20)$, which Figure 7.2 indicates is in $v(1, 2)$. Clearly, $(u_1, u_2) = (55, 20)$ is in $v(C)$. Thus, since $w_1 < u_1$ and $w_2 \leq u_2$, it follows from the preceding definition that $\mathbf{w}$ is in $v(1, 2)$.

Admittedly, the inclusion of $\mathbf{w}$ in $v(C)$ seems strange, because this point does not correspond to any real outcome in expected-utility terms or otherwise. Surely, we would not want to predict that $\mathbf{w}$ occurs, and to protect against this, it is useful to introduce the set U.

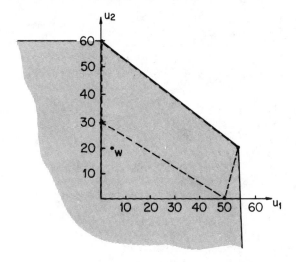

7.2 $v(1, 2)$.

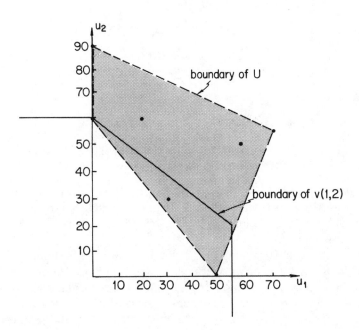

7.3 Feasible outcomes, U, for $\{1, 2\}$.

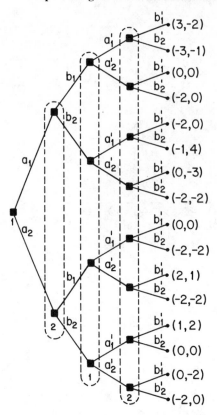

7.4 Extensive form.

U: The set U consists of all utility n-tuples that correspond to all outcomes O or lotteries over O.

Because the example in Figure 7.1 concerns three players, the set U is a collection of utility and expected utility 3-tuples. Figure 7.3, which reproduces $v(1, 2)$, represents by the shaded area the values of (u_1, u_2) that correspond to 3-tuples in U.

Notice that in calculating $v(1, 2)$ here, we assume nothing about the actions of person 3, except perhaps that person 3 acts to minimize the payoffs of players 1 and 2. In this sense, the calculation of $v(1, 2)$ provides a conservative assessment of a coalition's opportunities or value: Nothing precludes the possibility that a coalition's members cannot actually find themselves receiving more utility than $v(C)$ allows. Thus, U includes some outcomes not in $v(1, 2)$. This representation

	b_1b_1'	b_1b_2'	b_2b_1'	b_2b_2'
a_1a_1'	3, −2	−3, −1	−2, 0	−1, 4
a_1a_2'	0, 0	−2, 0	0, −3	−2, −2
a_2a_1'	0, 0	−2, −2	1, 2	0, 0
a_2a_2'	2, 1	−2, −2	0, −2	−2, 0

7.5 Normal form of the game in Figure 7.4.

suggests that in assessing $\{1, 2\}$'s value, we should take person 3's responses into account. Consideration of this possibility takes us ahead of ourselves, and we raise it here only to warn the reader not to interpret our definition of $v(C)$ rigidly or as the sole acceptable possibility.

Also, just because we label some subset of persons a coalition, we do not mean that these persons necessarily coordinate all of their actions. For example, consider the extensive form two-person game in Figure 7.4 and its normal form in Figure 7.5. Played noncooperatively, the players' maximum security levels from pure strategies are both −2. Furthermore, there are two pure-strategy equilibria, one of which – (a_1a_2', b_1b_2'), yielding (−2, 0) – both persons prefer to avoid in favor of the other – (a_2a_1', b_2b_1'), which yields (1, 2). Without some form of coordination, the players may not reach either equilibrium. Suppose that the players coalesce in the sense that before the start of the game, they coordinate their first moves and agree, for reasons that this game does not represent, to play noncooperatively thereafter. Notice that if they agree to (a_1, b_1), then the remaining subgame has the unique equilibrium (−2, 0). If they choose (a_1, b_2), then the remaining subgame has an equilibrium at (−1, 4). For (a_2, b_1) the equilibrium is (2, 1), and for (a_2, b_2) the equilibrium is (1, 2).

If both persons assume that an equilibrium prevails in the second stage, then the characteristic function, $v(1, 2)$, is the one that Figure 7.6 depicts. This figure also shows the characteristic function, $v'(1, 2)$, that results if persons 1 and 2 cooperate fully, and it also depicts by the shaded region the set of feasible outcomes, U. Full coordination expands 1's opportunities but it does little for 2. So it would not be unreasonable if 2 preferred partial coordination, owing to considerations that might be exogenous to the model. Hence, this example shows that by labeling C a coalition, we are not presuming necessarily that its members coordinate all of their actions. People often misinterpret the concept of a coalition to mean that its members act as a unit. Some

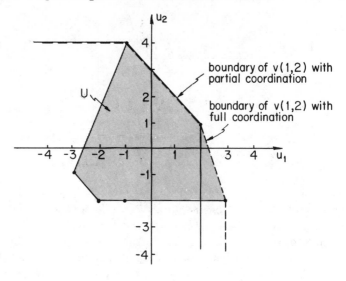

7.6 Partial and full coordination.

scholars debate whether the allied powers in World War II, for example, satisfied the game-theoretic definition of a coalition, because they failed to coordinate every aspect of their wartime strategies. The mistaken assumption, that coalitions must fully coordinate all strategies, probably originates from a common game-theoretic assumption that negotiations are costless. If they are so, then rational persons would realize all potential gains from coordination. Certainly, a game's extensive form can represent the costs of negotiation, and ultimately its characteristic function can do so also. If these costs are included in $v(C)$, then full coordination is a reasonable assumption. But often it is more convenient to treat these costs as exogenous constraints, in which case we need not require that a coalition's members coordinate their actions at every opportunity.

We can now define the primary representation of cooperative games:

> *Game in characteristic-function form*: We denote a game in c.f. form by (v, N, U), where v is a characteristic function defined over each subset C of N, the set of players, and U is the set of feasible utility outcomes.

Before proceeding to specific examples of characteristic-function form games, it is useful to state explicitly one common assumption about $v(C)$. The preceding example illustrates the gains from coordinated action between two persons, even if that coordination is limited. If negotiation is costless, then the players can lose nothing by cooperating, and they might gain something. We often state these gains as an explicit assumption:

> $v(C)$ is *superadditive* if and only if, when any two coalitions unite that have no members in common, the new coalition can ensure anything for its members that the two separate coalitions could ensure separately, and perhaps more.

Using the notation of sets, if C and C' are two disjoint coalitions (that is, $C \cap C' = \emptyset$), superadditivity requires that

$$v(C) \cap v(C') \subseteq v(C \cup C').$$

Superadditivity rules out negotiation costs that can exceed the benefits of coordinated action. This is often a convenient assumption to make, but cooperative game theory does not collapse without it. Furthermore, the solutions that we discuss later do not require it in their definitions. Instead, we use it to prove theorems about solutions, and the reasonableness of this assumption in different contexts simply affects the applicability of these theorems. Of course, if no two coalitions satisfy the superadditivity assumption, then we need not analyze the corresponding game using cooperative game theory, because no coalition has an incentive to form. Minimally, we should restrict ourselves to games that are essential:

> A game in characteristic function form is *essential* if at least two coalitions together can ensure more for their members than they can ensure for them separately. Formally, a game is essential if, for at least two coalitions,

$$v(C) \cap v(C') \subset v(C \cup C').$$

Hence, the intersection of $v(C)$ with U tells us what feasible utility outcomes a coalition can secure for its members. And to ensure that we are not modeling a game in which no coalition has an incentive to form, we suppose generally that $v(C)$ is both superadditive and essential.

7.3   Simple games

It is difficult at this point to appreciate the advantages and disadvantages of the characteristic-function representation. But given the emphasis of previous chapters, perhaps the easiest way to learn about $v(C)$ is through its application to committee games. Thus, we turn to *simple games*, because they model committees operating under formal voting procedures such as majority rule. *In a simple game, a coalition is either winning, losing, or – if ties are possible or if motions require extraordinary majorities to pass – blocking.* More formally,

> *Simple game*: A superadditive game in characteristic-function form is simple if, for all coalitions C,
>
> C is *winning*, in which case $U \subseteq v(c)$;
> or C is *losing*, in which case $v(C) = \cap_{i \in C} v(i)$;
> or C is *blocking*, in which case, if $\mathbf{u}^o = (u_1^o, \ldots, u_n^o)$ is the status quo, then $v(C) = \{u \mid u_i \leq u_i^o\}$.

Further,

> *Strong simple game*: A simple game is strong if all coalitions are either winning or losing.

The statement that $U \subseteq v(C)$ means that if a coalition is winning, then it can guarantee its members everything in U: All feasible outcomes are in its characteristic function. If a coalition is losing, then it cannot guarantee its members more than they can obtain individually. Thus, the outcomes in $v(C)$ for a losing coalition equal the outcomes that each member of C can secure individually. A blocking coalition cannot secure any motion in particular, but it can maintain the status quo and keep other motions from passing. Thus, if C is blocking, then the value that each member of C places on the status quo, u_i^o for $i \in C$, constrains the utility n-tuples in $v(C)$. Hence, if W is the set of all winning coalitions, and L the set of all losing coalitions, then the intersection of W and L is empty: A coalition cannot be both winning and losing. For strong simple games, a coalition is either in W or it is in L. Whether a game is strong or not, superadditivity implies that any expansion of a winning coalition is winning. This motivates us to identify those winning coalitions in which the defection of any member, i, renders the remaining members, $C\text{-}\{i\}$, losing or blocking.

> *Minimum winning coalitions*: W^*, which is a subset of W, is the set of minimum winning coalitions, where, for any C in W^*, and for all i in C, $C\text{-}\{i\}$ is not winning. Formally,

$$W^* = \{C \in W \mid \text{for all } i \in C, C\text{-}\{i\} \text{ is not in } W\}.$$

To illustrate the representation of a cooperative voting game in characteristic-function form, consider the three-bill, three-legislator vote-trading example last discussed in Chapter 1, and which we reproduce in Figure 7.7. Recall that $(1, 0, 0)$ denotes the outcome "bill 1

Legislator	(0, 0, 0)	(1, 0, 0)	(0, 1, 0)	(0, 0, 1)	(1, 1, 0)	(1, 0, 1)	(0, 1, 1)	(1, 1, 1)
1	0	2	4	−5	6	−3	−1	1
2	0	4	−5	2	−1	6	−3	1
3	0	−5	2	4	−3	−1	6	1

7.7 Three-person vote-trading game.

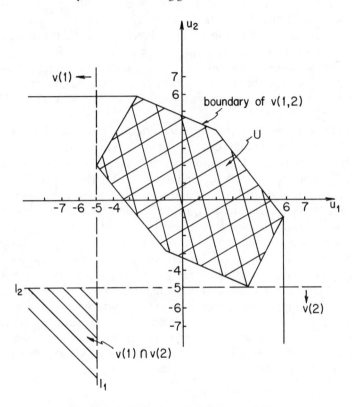

7.8 Superadditivity in a three-person vote-trading game.

passes and bills 2 and 3 fail," (1, 1, 0) denotes the outcome "bills 1 and 2 pass and bill 3 fails," and so on. If this committee makes its decisions by simple majority rule, then because n is odd, there are no blocking coalitions and the game is strong: The two- and three-person coalitions are winning; the two-person coalitions are minimal winning; and the one-person "coalitions" are losing. To compute $v(C)$, notice first that individual legislators cannot insure against the passage of anything. For example, the worst possible outcome for legislator 1 is that only bill 3 passes. Since he cannot block this outcome by himself, $v(1) = \{\mathbf{u} \mid u_1 \leq -5\}$. Of course, the same calculation holds for legislators 2 and 3. But a two-member coalition is winning, and it can thereby secure any outcome.

For this example, Figure 7.8 graphs $v(1, 2)$ and U, which appears as the cross-hatched part of $v(1, 2)$. To illustrate superadditivity, notice that all points on and to the left of the line l_1 do not give legislator 1 more than he can achieve by himself: They "satisfy" 1's individual security level and thus are in $v(1)$; similarly, all points on or below l_2 satisfy 2's individual security level. The intersection of these areas, the hatched area, corresponds to $v(1) \cap v(2)$, which in accordance with the definition of superadditivity, is a subset of the characteristic function of the union of these two one-person coalitions. That is, $v(1) \cap v(2) \subset v(\{1\} \cup \{2\}) = v(1, 2)$.

Alternatively, suppose that the committee requires unanimity to pass any bill. Since any member can veto any alternative, and assuming that the status quo is "no legislation passes," then $v(i) = \{\mathbf{u} \mid u_i \leq 0\}$. Similarly, since a two-person coalition is no longer sufficient to pass legislation, $v(i, j) = \{\mathbf{u} \mid u_i \leq 0 \text{ and } u_j \leq 0\}$. Only the coalition of all three members can secure a "positive" action, and the characteristic function of $\{1, 2, 3\}$ includes all possible payoffs.

Simple games are important because we use them to model cooperative committee processes with formal voting mechanisms. But for large games (those with more than three participants), it is inconvenient to represent $v(C)$ graphically or formally in terms of sets of utility n-tuples. Fortunately, a simple game admits of a more convenient representation. Let O denote the set of possible outcomes that can prevail as the result of the committee's actions, and let x_o denote the status quo outcome. Ignoring the clumsy notation of utility vectors and the like, if we do not consider lotteries, then we can redefine $v(C)$ thus:

$$v(C) = O \text{ if } C \text{ is winning;}$$
$$v(C) = \emptyset \text{ (the empty set) if } C \text{ is losing;}$$
$$v(C) = x_o \text{ if } C \text{ is blocking.}$$

That is, if a coalition is winning, then it can secure any outcome in O, and we set $v(C)$ equal to O; if C is losing, then $v(C)$ is the empty set; and if C is blocking, then the sole outcome in $v(C)$ is the status quo.

To appreciate the simplification that this representation entails, consider a five-person, majority-rule, one-man-one-vote game over two issues, and suppose preferences look like those that Figure 6.3 depicts. Suppose also that everyone is risk averse, so that we need not consider lotteries. Hence, to describe $v(C)$ in terms of utility, we must assign utility numbers to outcomes by using some function that is consistent with these indifference contours. At this point, however, we would stretch our graphical skills beyond the breaking point to represent $v(C)$ for three-, four-, and five-person coalitions, because such descriptions of $v(C)$ require as many dimensions as there are members of C. More conveniently, we can let each winning coalition's characteristic function be the set of all admissible outcomes, and we use a figure such as 6.3 to summarize individual preferences over those outcomes. In short, Figure 6.3 itself summarizes all information necessary to represent this majority game in characteristic-function form, so there is no reason to revert to additional diagrams. Thus, when applying the various solution concepts for n-person cooperative games to simple voting games (including vote trading and spatial games), we can use the representations of preference introduced in the previous chapters; there is no need to develop additional notation or special characteristic-function representations.

7.4 The special case of transferable utility

Thus far we have placed no specific constraints on $v(C)$ aside from superadditivity: Almost any function (actually, correspondence) that assigns sets of utility c-tuples to each subset of N will do. Owing to its initial proximity to economics, however, game theory first focused on a special case, *games with transferable utility*. To the extent that the substantive concerns of economics form a subpart of politics, this special case warrants emphasis. Nevertheless, as we show later, we cannot model many political processes using this special case without destroying the intent of the modeling exercise. Thus, although this section illustrates the meaning of $v(C)$, it also serves as a warning to those who might try simply to graft an economic model onto an analysis of political processes.

First, recall our discussion in Chapter 5 of the demand-revelation tax mechanism for soliciting honest preferences for public goods. The critical assumption there, which permits circumvention of Theorem 2.3, is that we can separate people's preferences for one commodity (money) from the preferences for other commodities, and that we can represent these preferences with a linear utility function. The appropriate context for imposing this same assumption here is to suppose that the game concerns the allocation of some set of m resources or goods, and that coalition members can trade these goods among themselves. Letting $\mathbf{x}_i = (x_{1i}, \ldots, x_{mi})$ denote the amount of each of the m goods that person i owns, and letting i's utility function, u_i, over $\mathbf{x}_i$ be linear and separable in the mth good, then these assumptions permit us to represent i's utility for x_i thus,

$$u_i(x_i) = u_i'(x_{1i}, \ldots, x_{m-1,i}) + x_{mi}.$$

Thus, i's utility for the mth good is independent of the amount of goods 1 through $m-1$ that this person has or consumes, and we can represent i's utility for the mth good as the amount of that good that he has. The transferable utility case in cooperative game theory also assumes that the mth good is separable in this fashion for all n persons, and that this good is perfectly divisible and transferable between and among people.

The simplest illustrations of this structure occur if this mth good is money, and if it is the only commodity exchanged, in which case we can suppress the subscript m and let $u_i(x_i) = x_i$. For a specific example consider the vote-trading scenario in Figure 7.7, but suppose that we are modeling graft in a legislature whose members are concerned only with direct monetary rewards. Individually, then, each person's security level is $-\$5$. But two people can form a majority coalition, and they can pass or fail any bill. Suppose that 1 and 2 coalesce. Looking across all feasible outcomes in Figure 7.7, we see that the outcome $(1, 0, 0)$ maximizes the *total* payoff to 1 and 2. Legislator 1 prefers the outcome $(1, 1, 0)$, of course, because it pays him $6. But choosing $(1, 0, 0)$ does not preclude this payoff for 1. If the only alternative is $(1, 1, 0)$, then player 2 should agree to $(1, 0, 0)$, even if he must give 1 most of his $4 payoff, because with $(1, 1, 0)$, player 2 loses $1. Moving from $(1, 1, 0)$ to $(1, 0, 0)$ is a Pareto improvement for 1 and 2, since the total amount of money that they now share increases. Of course, other features of the game might dictate that 1 makes a payment to 2. We do not know yet what payments will be necessary to form $\{1, 2\}$, but we do know that the total amount gained from coalescing constrains all such payments (that is, $x_1 + x_2 \leq 6$), and that $(1, 0, 0)$ yields the weakest constraint. Since utility and money are mathematically equivalent here, the set of all utility outcomes that the coalition $\{1, 2\}$ can achieve is

$$v(1, 2) = \{u \mid u_1 + u_2 = x_1 + x_2 \leq 6\}.$$

We traditionally simplify this notation to $v(1, 2) = 6$, and if we keep in mind that this shorthand means that persons 1 and 2 can attain any pair of payoffs whose sum does not exceed $6, then we can adopt this convention as well. Thus, if the vote-trading example in Figure 7.7 concerns only money, and if utility for money is linear and transferable, then a simple representation of this cooperative majority-rule model of legislative graft is

$$v(1) = v(2) = v(3) = -5$$
$$v(1, 2) = v(1, 3) = v(2, 3) = 6 \qquad (7.1)$$
$$v\,(1, 2, 3) = 3.$$

The transferable utility assumption makes several additional simplifications possible. First, we can express superadditivity for disjoint coalitions C and C' as

$$v(C) + v(C') \leq v(C \cup C'),$$

and a game is essential if strict inequality holds for at least one pair of disjoint coalitions. To see that the game that equation set (7.1) describes is superadditive and essential, notice that $v(1) + v(2) = -10 < v(1, 2) = 6$.

Characteristic-function form games with transferable utility also give rise to a class of cooperative games, called *zero or constant sum*. Specifically, any game satisfying

$$v(C) + v(N - C) = k \text{ (a constant)},$$

for all coalitions C, is constant sum. Of course, if $k = 0$, then the game is zero sum, and we can make any constant-sum game zero sum by subtracting k/n from each person's utility. These games model perfect conflict over divisible and transferable commodities such as money, and we use them to model governmental income-redistribution policies if collusive discussion and coalitional agreements, not some rigid voting scheme, decides those policies.

Another simplification that a transferable utility representation permits concerns the concept of an *imputation*, which is any payoff vector, $\mathbf{u} = (u_1, \ldots, u_n)$, that satisfies

$$u_i \geq v(i) \qquad (7.2)$$

and

$$\sum_{i=1}^{n} u_i = v(N). \qquad (7.3)$$

Later, we show how to normalize transferable utility games so that $v(i) = 0$ for all i and $v(N) = 1$, in which case expression (7.3) looks like a budget constraint. Thus, if the game is superadditive (that is, if it is impossible for some smaller coalition to win more for its members than what the coalition of the whole can give them), then all imputations must lie on the n-dimensional budget simplex. For example, if $n = 3$, then a triangle, such as the one in Figure 4.18a, describes the set of all feasible imputations. Furthermore, the formulation of utility functions that the transferable utility model assumes excludes the possibility that people care about the welfare of others. Because people care only about their own payoff, individual indifference contours over this triangle look like the ones that Figure 4.19a depicts.

It seems only reasonable that in searching for solutions – predictions about payoffs – for these games, we should limit our attention to n-tuples that satisfy (7.2) and (7.3). Expression (7.2) requires that no person receives less than he can secure by acting alone, and (7.3) asserts that the total amount that all players receive equals what the game is "worth." Although a search for a solution occurs in subsequent chapters, notice that if someone proposes an n-tuple, $\mathbf{w}$, such that $\sum_{i=1}^{n} w_i < v(N)$, then the coalition of the whole can form to redistribute the excess between what $\mathbf{w}$ offers and what $v(N)$ defines as feasible. And because we assume that communication and coordination are costless once we represent a game in characteristic-function form, we can assume that such a redistribution will occur. Furthermore, if $\mathbf{w}$ satisfies $\sum_{i=1}^{n} w_i > v(N)$, then $\mathbf{w}$ is infeasible.

That we can represent all feasible and reasonable outcomes to a cooperative game by a budget simplex simplifies analysis considerably and accounts for the attractiveness of this representation of cooperative situations. Its assumptions, though, warrant special emphasis. Many applications of cooperative game theory to politics, and especially to the study of parliamentary coalition formation, incorrectly presume that utility is transferable, and thus the resulting analyses are often wholly irrelevant to the problem under investigation. First, for a transferable utility representation to be appropriate, we must assume that there exists a perfectly divisible and unrestrictedly transferable commodity, such as money. Second, this commodity must be separable, so that each person's utility for the good does not depend on other characteristics of outcomes. Finally, each person's utility must be linear in this separable good. These assumptions may be more nearly acceptable in economics than they are in politics. We can model the vote-trading example in Figure 7.7 with transferable utility, for instance, but only if we are willing to impose some special assumptions that may require an unacceptable substantive interpretation. That legislators can exchange votes on bills does not mean that a transferable utility representation is appropriate. Instead, the appropriateness of this assumption would require that we can express the argument of each legislator's linear utility function in terms of the transferable commodity. One way to do this is to suppose that the bills under consideration concern money, and that legislators can exchange money freely among themselves. That is, with transferable utility we can model graft in a legislature.

But in the context of vote trading it is probably more useful if we assume that each bill concerns policies such as public works projects, with payoffs expressed in terms of votes won or lost in the next campaign for reelection. And although legislators can exchange votes on bills, the true and ultimate argument of each legislator's utility function, election votes, are not freely transferable under most electoral systems. In this context, a transferable utility representation is appropriate only if we can assume that legislators have sufficient monetary resources to compensate each other for election votes won or lost. But this, too, is a peculiar sort of legislature. Graft and the pecuniary rewards from passing certain bills instead of others provide interesting areas of inquiry, but we should not limit ourselves theoretically to an analysis of such legislatures.

Transferable utility is also an inappropriate assumption for spatial-committee games, such as the ones that Chapter 6 studies. There, presumably, the only negotiable commodities are the issues, which we represent as dimensions in the policy space. And again, unless we are willing to assume the existence of an additional issue – money that is freely and linearly transferable among legislators – it is more appropriate to model $v(C)$ as either the set of all outcomes if C is winning or as the empty set if C is losing. This comment applies especially to those who study parliamentary coalition formation. There, ministries might serve as the transferable resource (although perfect divisibility is a problem), but separability requires that there are no policy consequences associated with their allocation. This assumption provides a useful model of political parties that care only about providing members with employment, but it cannot aid in modeling coalitions if public policy intervenes as a concern.

The transferable utility formulation, then, is inappropriate for studying many political games. It is a powerful simplification, however, for those economic situations in which money is the principal measure of value. In addition, transferable utility is a useful device for illustrating definitions and the meaning of assumptions. With a number summarizing $v(C)$, and with a budget simplex in n dimensions representing all feasible outcomes, analysis proceeds with simple algebra instead of set-theoretic manipulations. But whenever possible, we should check whether the assumptions of this special case sufficiently reflect the conditions of the situation under study.

7.5   Some examples of characteristic-function games

Aside from considering voting games, the simplest way to see application of $v(C)$ is to review its form for specific examples.

The valley of the dump

For a primitive example of externalities, suppose that n people live in a valley, and that they cannot export out of the valley the garbage that each produces daily. Instead, people dump their garbage on other people's lawns. Suppose that each person produces exactly one bag of garbage per day, and that the utility that each associates with having x bags dumped on his lawn is $-x$. Thus, any one-person coalition cannot be certain that the rest of the residents will not decide one day to dump all of their garbage on that coalition's lawn. Since that one person can dump his garbage on someone else's lawn, $v(i) = -(n - 1)$. More generally, if a coalition, C, of size c forms, it can dump its garbage on the lawns of those not in the coalition, but it has no safeguard against the actions of the other $n - c$ persons not in the coalition. Hence, $v(C) = -(n - c)$. Here $v(C)$ depends solely on the size of C and not on the specific persons that make up this coalition. In this event, we call the game a symmetric game.

> *Symmetric game*: An n-person game in characteristic-function form is symmetric if $v(C)$ depends only on c, the size of C. That is, for all C and C' such that $c = c'$, $v(C) = v(C')$.

To complete the description of $v(C)$ in this example, we notice that a coalition of size N, called *the coalition of the whole*, cannot help but foul its own nest. Hence, $v(N) = -n$. Since n bags of garbage must be "distributed," no matter what coalitions form, the game is constant sum. That is, from the preceding expressions for $v(C)$,

$$v(C) + v(N - C) = -(n - c) - (n - (n - c)) = -n.$$

Thus, regardless of the size of C, the value of C plus the value of its complementary coalition sums to $-n$. This discussion, then, illustrates a constant-sum, symmetric, transferable utility game in characteristic-function form.

Normalized games

If we focus on games with transferable utility, we can often find games that look dissimilar but that are actually strategically equivalent. For example, consider these two games in characteristic-function form:

game v	*game v'*
$v(i) = 0,\ i = 1, 2, 3$	$v'(1) = -2,\ v'(2) = v'(3) = -4$
$v(1, 2) = v(2, 3) = v(1, 3) = 1$	$v'(1, 2) = v'(1, 3) = 4,\ v'(2, 3) = 2$
$v(1, 2, 3) = 1$	$v'(1, 2, 3) = 0$

At first glance, v and v' seem to represent entirely different games. But let us see what happens if we normalize v' with various linear transformations of utility, so that the value of all one-person coalitions is zero, and the value of the coalition of the whole is one. To normalize a characteristic function, recall from Theorem 1.2 that the information that a utility function conveys remains unchanged if we add or subtract any constant. Thus, for game v' we can add an amount, x_i, to each person's utility function, where we choose x_i so that

$$v''(i) = v'(i) + x_i = 0 \quad \text{for all } i.$$

This addition changes the value of the remaining coalitions, however, so that the new (temporary) value of C is $v'(C) + \Sigma_{i \in C} x_i$. That is, the new value of C is its old value, $v'(C)$, plus what we have added to each of its member's utility. In particular, the value of the coalition N is now

$$K = v'(N) + \sum_{i=1}^{n} x_i.$$

We can normalize the value of N to 1, then, by dividing each person's utility function by the positive constant K. Dividing each utility function by K does not affect the value of the one-person coalitions, of course, since zero divided by anything positive or negative remains zero. But larger coalitions are currently valued at $v'(C) + \Sigma_{i \in C} x_i$, and dividing individual utility functions by K changes this value to

$$v''(C) = \frac{v'(C) + \Sigma_{i \in C} x_i}{K}.$$

To illustrate this process using game v', let $x_1 = 2$, and $x_2 = x_3 = 4$, which sets $v''(1) = v''(2) = v''(3) = 0$. Hence, $K = 0 + (2 + 4 + 4) = 10$, so that

$$v''(2, 3) = \frac{2 + (4 + 4)}{10} = 1$$

$$v''(1, 3) = v''(1, 2) = \frac{4 + (2 + 4)}{10} = 1$$

$$v''(1, 2, 3) = \frac{K}{K} = 1.$$

But now game v'' is identical to game v for all C. Hence, game v' must be strategically equivalent to game v.

We refer to this particular normalization as *the (0, 1) normalization of transferable utility games*. It might seem strange to assert that games v and v' are strategically equivalent. The players' positions in game v are clearly symmetric; but in game v', person 2 appears to be at a distinct disadvantage with respect to person 1. Recall, though, that payoffs are utility numbers, and we cannot compare them across persons. A normalization is simply a linear transformation of each person's utility that does not alter the information that these numbers summarize. Normalizations, then, are useful devices for assessing whether there are strategic dissimilarities across games, as well as within a game, among players.

At this point we must warn the reader about some definitional confusion in the literature. Earlier, we defined simple games in terms of whether coalitions can secure any outcome or no outcome in the set O. In the context of transferable utility, however, some of the literature applying game theory to politics equates simple games with characteristic functions of the form,

$$v(C) = \begin{cases} 1 \text{ if } C \text{ is "winning"} \\ 0 \text{ if } C \text{ is "losing."} \end{cases} \tag{7.4}$$

However, consider the normalization of equation set (7.1), which models the characteristic function of the transferable-utility version of the vote-trading game in Figure 7.7. First, letting $x_i = 5$ for $i = 1, 2$, and 3, and adding x_i to each person's utility sets $v'(i) = 0$, $i = 1, 2$, and 3. And for these values of x_i we get $K = 3 + 15 = 18$. Dividing each person's utility by K, then, gives $v(1, 2, 3) = 1$. Finally,

$$v'(i, j) = \frac{[v(i, j) + x_i + x_j]}{K} = \frac{[6 + 10]}{18} = \frac{8}{9}.$$

But this characteristic function is not equivalent to (7.4), even though this vote-trading game has only winning and losing coalitions and is clearly simple.

The confusion and the unnecessarily restrictive definition that (7.4) offers, we suspect, originates from peoples' early preoccupation with transferable utility and from a desire to model political games in which coalitions are either winning or losing but in which payoffs are unknown. Instead of modeling payoffs, people adopted the simple expedient of assigning values of 0 and 1 to coalitions. This example shows, however, that this approach is inadequate.

Market games

Perhaps the most widely studied class of cooperative games are those in which people simply exchange resources, as in a market. To model a simple three-person exchange economy, then, we endow each person initially with a vector $x_i^* = (x_{i1}^*, x_{i2}^*, \ldots, x_{im}^*)$ of m resources or commodities, which i evaluates with the utility function $u_i(x_i^*)$. Because most models of exchange economies assume that people have property rights to their initial endowments, we assume that no trades can make a person worse off than the utility that he associates with this endowment. Thus,

$$v(i) = \{\mathbf{u} \mid u_i \leq u_i(x_i^*)\}.$$

Finally, assuming that people can trade all m commodities among themselves in an unrestricted way,

$$v(C) = \{\mathbf{u} \mid u_i \leq u_i(x_i), \text{ in which } \sum_{i \in C} x_{ij} \leq \sum_{i \in C} x_{ij}^*,$$
$$j = 1, \ldots, m\}.$$

The inequality involving summation signs states that no matter how the members of C distribute good j among themselves, after they make all trades, the total amount distributed cannot exceed C's initial endowment. Although C might get "lucky" and have people outside the coalition donate resources, property rights preclude C from expropriation. Hence, C's "security level," $v(C)$, reflects all of the various utility outcomes that it can achieve without exceeding the limits of its members' endowments.

Pollution, externalities, and the prisoners' dilemma

Lloyd Shapley and Martin Shubik offer a scenario to illustrate the problem of cooperation with economic externalities:

> There are n factories around a lake. It costs an amount A for a factory to treat its wastes before discharging them into the lake. It costs an amount cB for a factory to purify its own water supply if c is the number of factories that do not treat their wastes. Let $B < A < nB$.

Thus, factories both pollute and consume water. To derive the characteristic function for this game, we begin with an individual factory. For each of the factory's alternatives, we identify the worst that can happen to it with that choice. First, an individual factory can pollute. The worst possibility here is that all other factories pollute as well, in which case an individual factory must pay nB dollars to purify water for its own use.

But if a factory treats its own wastes, again the worst possibility is that all others pollute and that factory's total cost is $(n - 1)B$ to treat the water it must use for production, plus A to treat the water that it discharges into the lake, or $(n - 1)B + A$. From the assumption that $B < A$, it follows that $(n - 1)B + A > nB$. Because an individual factory acting alone cannot be certain that the remaining factories will not all fail to clean up their effluents, if it alone treats its waste it must pay $(n - 1)B + A$, so its security level is maximized if it also pollutes and pays nB instead. Hence, $v(i) = -nB$.

Turning now to the coalition of the whole, this coalition has several options, from some or all of its members polluting to some or all treating their wastes. But at one extreme, none of its members will treat their wastes, in which case each factory must pay nB to treat the water that it uses, yielding a total cost to the coalition of n^2B. At the other extreme, if all n factories agree to treat their wastes before discharging them into the lake, then it costs each of them nothing to purify the water for consumption but each of them must pay A for treatment, yielding a total cost to the coalition of nA. Because $nA < n^2B$ (from the assumption that $A < nB$), the coalition of the whole should choose this second option over the first, thereby setting $v(N) = -nA$.

Having thus described the characteristic function for one- and N-person coalitions, now suppose that a coalition C of size c forms. The members of C can do either of two things. First, they can pollute, and if everyone not in C pollutes, then the coalition incurs a total cost of cnB, the size of C times the cost to each member of C of cleaning its water if all n firms pollute. Second, C can treat its waste. In this case, the worst possibility is that those excluded from C pollute, so that C incurs a total cost of $c[(n - c)B + A]$ – the size of C times the cost to each member of cleaning the water it uses if $n - c$ firms pollute, plus the cost of treating its own waste. Comparing the two cases, we notice that $cnB > c[(n - c)B + A]$ if $cB > A$, or if $c > A/B$. Thus, if $c > A/B$, then the members of C minimize the total cost to the coalition by choosing the second option of not polluting. But if $c \leq A/B$, then the members of C should choose the first option and pollute. Summarizing,

$$v(C) = \begin{cases} -cnB & \text{if } 1 \leq c \leq A/B \\ -c[(n - c)B + A] & \text{if } n \geq c > A/B. \end{cases}$$

Figure 7.9 illustrates $v(C)$ for $n = 10$, $A = 4$, and $B = 1$. Notice that for $c \leq A/B = 4$, $v(C)$ is linear in c. Thus, small coalitions have no incentive to form, because the game is inessential in this range (for example, $v(1) + v(2) + v(3) = v(1, 2, 3)$). But larger coalitions have some incentives to form. For example, if $c = 5$, then $v(C) = -45$, but,

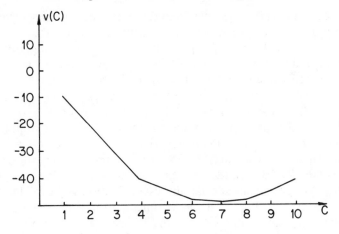

7.9 $v(C)$ for pollution example.

$v(i) = -10$, so that in this range the game is essential (for example, $v(1) + v(2) + v(3) + v(4) + v(5) = -50 < v(1, 2, 3, 4, 5) = -45$).

There are several political parallels to this game, because it is little more than a specific instance of a prisoners' dilemma in which the players can communicate before choosing strategies. If c factories pollute, then for the $c + 1$th factory the payoff from treating its waste is $-(cB + A)$, and if it pollutes as well, its payoff is $-(c + 1)B$. Thus, polluting is dominant in the noncooperative version of this game if $-(c + 1)B > -(cB + A)$, or if $B < A$, which is the model's assumption. If everyone pollutes, then each factory's payoff is $-nB$, while if no one pollutes, then their individual payoffs are $-A$. Thus, a prisoners' dilemma exists if $nB > A$, which is also one of our initial assumptions.

With this in mind, consider the n-person prisoners' dilemma that represents interest groups lobbying for particularized benefits, as Figure 5.6 depicts. (So as not to confuse the notation for a coalition and costs, we substitute K for C in Figure 5.6.) If any group opposes special-interest legislation, it has no assurance that the remaining $n - 1$ groups do not (successfully) seek such legislation, in which case its payoff is $-(n - 1)K/n$. The safest course for a single-member coalition, then, is to lobby for its own legislation, in which case the worst that can happen is that it receives nothing if a majority of other groups oppose such legislation, or $(B - K)$ if all other groups join it. If $B > K/n$, then a single-member coalition's security level is maximized if it lobbies, and, since everyone might lobby successfully, $v(i) = B - K < 0$.

A coalition of size c has two options: (1) oppose all special-interest legislation, and (2) seek special-interest legislation for its members. Suppose, first, that c is less than a majority. In case (1) the coalition's security level is defined by the possibility that the remaining groups all lobby for and get their bills, in which case the coalition's payoff is c times what each member must pay to fund what the members of the majority coalition, $N - C$, secure for themselves, $-(n - c)K/n$, or a total of $-c(n - c)K/n$. In case (2), if all members of $N - C$ oppose special-interest legislation, then each member of C gets and must pay nothing, but if $N - C$ also lobby for private benefits, then the members of C as well as everyone in $N - C$ secures the programs they seek. The worst possibility for case (2) from the coalition's perspective, assuming that $B < K$, is that all special-interest legislation passes. Then each person receives B, but must pay n times its share of each approved program, K/n, for a total of $B - K$. Hence, the total payoff to the coalition C is c times $B - K$. Notice that if $B > cK/n$, then C's security value from case (2), $c(B - K)$, exceeds its security value from case (1), $c(n - c)K/n$. Since we are assuming here that C is a minority (that is, $c < n/2$), then the inequality $B > cK/n$ is satisfied if $B > K/2$. That is, if $B > K/2$, then a minority coalition's security level is maximized if it seeks special-interest legislation for its members, and that security value is $c(B - K)$. Hence, $v(c) = c(B - K)$.

Finally, suppose that C is a majority. In case (1) its opposition blocks all such legislation, and $v(C) = 0$. In case (2) it cannot stop others outside of the coalition from also having their projects approved, so $v(C) = c(B - K)$. A majority coalition, then should choose (2) if $c(B - K) > 0$ or, equivalently, only if $B > K$. But since $B < K$, a majority coalition prefers case (1), in which case $v(C) = 0$. Summarizing, if $K/2 < B < K$, then

$$v(C) = \begin{cases} c(B - K) & \text{if } C \text{ is a minority} \\ 0 & \text{if } C \text{ is a majority.} \end{cases}$$

Thus, if winning coalitions can form, they will block any legislation since $B < K$. If only smaller than minimum winning coalitions can form as a consequence of exogenous constraints, then Pareto-inefficiencies prevail.

How might we incorporate such exogenous constraints into the description of $v(C)$? Recall Olson's argument from Chapter 5, that large groups are disadvantaged with respect to small groups in their ability to form and to maintain themselves, owing, presumably, to transaction costs, such as the costs of communication among members, capital costs associated with meeting facilities, and operating costs for clerical staff, and so on. For a simple assumption, let $(c - 1)T$ be the cost of maintaining a coalition of size c, in which case the previous characteristic function becomes

$$v(C) = \begin{cases} c(B - K) - (c - 1)T, & \text{if } C \text{ is a minority} \\ -(c - 1)T, & \text{if } C \text{ is a majority.} \end{cases}$$

This game is inessential for smaller than minimum-winning coalitions. It is inessential everywhere if $cv(i) > v(C)$ and C is a majority. Thus, the game is inessential if $c(B - K) > -(c - 1)T$, or equivalently, $K - B < (c - 1)T/c$. Since the coefficient of T, $(c - 1)/c$ increases as c increases, the most severe constraint on T occurs if C is a minimal majority, if $c = (n + 1)/2$. For large n (for $(n - 1)/(n + 1)$ approximately equal to 1), the condition for the game to be inessential becomes $T > (K - B)$. Thus, this game is inessential if transaction costs are greater than the differential between costs and benefits of individual programs. For example, if K is only slightly larger than B, but if transaction costs are "significant," then this society will reach a mildly Pareto-inefficient outcome in which no coalitions form and everyone lobbies for their (inefficient) special interest.

We are not arguing that this representation is an especially good model of interest-group activity. Larger groups enjoy net economies of scale and scope, so that the cost per member of maintaining a coalition declines as c grows (for example, we could model transaction costs by the term $(c - 1)^{1/2}T$). But groups might also face significant net diseconomies of scale, in which case the previous analysis provides a conservative assessment of the constraints on efficiency in interest-group lobbying. In either event, our intent is simply to show how to model transaction costs, which can be important in studying various committee processes. Theoretically, some coalition might circumvent the rules and procedures that Chapter 6 describes. Certainly, issue-by-issue voting can be the apparent procedure, but any winning coalition can circumvent a formal requirement precluding consideration of two or more issues simultaneously. And in most legislatures, a majority can change most formal procedures. Therefore, the collection of theorems in Chapter 6 may become irrelevant, to the extent that those rules are disadvantageous to a winning (majority) coalition. Circumventing or changing rules may be costly, though, and the costs of change may outweigh the gains from full cooperative action. Thus, the choice between modeling committee processes as a noncooperative game constrained by procedure, or as a cooperative game in characteristic function form in which winning coalitions can secure any outcome, depends on the relative importance of transaction costs.

Person 3

Person 2:		c_1			c_2	
		b_1	b_2		b_1	b_2
Person 1:	a_1	75, 75, 30	10, 10, 15		20, 20, 0	90, 10, 90
	a_2	10, 10, 15	80, 80, 35		10, 90, 90	30, 30, 80

7.10 Normal form for calculation of beta form.

	c_1	c_2	min u_1	min u_2
(a_1, b_1)	75, 75	20, 20	20	20
(a_1, b_2)	10, 10	90, 10	10	10
(a_2, b_1)	10, 10	10, 90	10	10
(a_2, b_2)	80, 80	30, 30	30	30

7.11 Computing alpha and beta forms of $v(C)$.

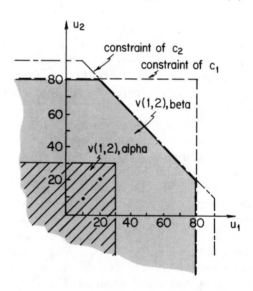

7.12 Alpha and beta versions of $v(1, 2)$.

7.6　Alternative forms of $v(C)$

Choosing between cooperative and noncooperative formulations is only one of the decisions that we must make in modeling political processes. If we wish to represent a game in characteristic-function form, then we must also decide on the exact definition of $v(C)$. Most applications of game theory rely on the definition that Section 7.1 offers, which some researchers regard as too conservative for most purposes. Recapitulating, this definition states that $\mathbf{u}$ is in $v(C)$ if and only if C has a (joint) strategy that assures $i \in C$ at least u_i, regardless of the choices of the other persons not in C. Game theorists refer to this as the *alpha* form of $v(C)$. An alternative definition is that $\mathbf{u}$ is in $v(C)$ if and only if the persons not in C cannot prevent $i \in C$ from receiving at least u_i. This is called the *beta* form of $v(C)$. Formally, C is beta-effective for $\mathbf{u}$ if for each strategy $(n - c)$-tuple that the players not in C might choose, there is a strategy for the members of C such that for all $i \in C$, the payoff to i is at least u_i. It is perhaps easiest to think of the difference between the alpha and beta forms in terms of who moves first, C in the alpha version and $N - C$ in the beta version.

To illustrate the differences between these two forms, consider the three-person game in normal form in Figure 7.10 and the table in Figure 7.11. Ignoring various complications occasioned by mixed strategies, Figure 7.12 presents the (alpha) characteristic function of $\{1, 2\}$ as the shaded hatched region, which we compute using the table in Figure 7.11. To compute the beta form of $v(1, 2)$, notice that if the excluded player, 3, chooses c_1, then players 1 and 2 can attain any outcome such that u_1 and u_2 do not exceed 80. But if 3 chooses c_2, then 1 and 2 can attain any outcome on or to the left of the dashed line: Either 1 gets 90 and 2 gets 10, or vice versa, or any lottery between these two possibilities may occur. If we assume that $\{1, 2\}$ must evaluate its opportunities without knowing what 3 will choose but knowing that it can respond afterward, then the minimum of the dashed and dotted lines, the shaded region in Figure 7.12, gives the beta form of $v(1, 2)$.

Generally, game theorists rely on the alpha form of $v(C)$. *For games with transferable utility, however, the alpha and beta forms are equivalent.* To see this identity, consider the following tableau, which we derive from Figure 7.11, assuming that utility is transferable:

	c_1	c_2	$(\frac{2}{9}, \frac{7}{9})$	min
(a_1, b_1)	150	40	64.4	40
(a_1, b_2)	20	100	122.2	20
(a_2, b_1)	20	100	122.2	20
(a_2, b_2)	160	60	82.2	60
$(0, 0, \frac{5}{9}, \frac{4}{9})$	82.2	82.2	82.2	82.2

To compute the alpha value, $\{1, 2\}$ approaches this table as though it were a zero-sum game, that is as though it is $\{1, 2\}$'s task to maximize its security value. Notice that if $\{1, 2\}$ uses the mixed strategy $(0, 0, \frac{5}{9}, \frac{4}{9})$, then $\{3\}$ cannot hold it to a payoff less than 82.2. This is the alpha value of $v(1, 2)$. But to compute the beta value of $v(1, 2)$ we assume that $\{3\}$ approaches this table as though it were a zero-sum game with its payoffs the negative of those shown. And if $\{3\}$ uses the mixed strategy $(\frac{2}{9}, \frac{7}{9})$, then $\{1, 2\}$ cannot get more than 82.2. This is the beta value of $v(1, 2)$. Hence, the alpha and beta values of $v(1, 2)$ are identical.

7.7   Ambiguities in $v(C)$

The preceding two definitions of $v(C)$ reveal some of the ambiguities in the characteristic-function representation of a game. Of course, we should anticipate some problems. When we transform a game from its extensive to its normal form, we may lose some information (which accounts for the discrepancies between Nash equilibria and subgame perfect equilibria). Similarly, moving to either the alpha or beta versions of the characteristic function can result in the loss of additional information, which loss, if we are not careful, can critically affect the description of a situation's strategic properties.

For a simple example, consider a three-person situation provided by Robert Rosenthal, which colorfully illustrates economic externalities. Suppose that three persons each live in a house with a large picture window that looks out over the resident's own garden, to the garden of his neighbor to the left. Assume that the homes are arranged in a triangle so that person 1 sees person 2's garden from his window, 2 sees 3's garden, and 3 sees 1's garden. Owing to the design of the windows, however, each person sees only one-fifth of his own garden but the entire garden of the neighbor to his left. Next, suppose that each person owns one bag of fertilizer, and that each person's utility is linear with the amount of fertilizer spread on the gardens or portions thereof that he views. For every bag of fertilizer spread on a garden in full view, a person gains 5 units of utility, while the utility that he associates with one bag spread on his own garden is 1, since he sees only a fifth of it. He derives no utility from fertilizer spread on the garden to his right, since it is out of sight. Suppose, also, that the value of fertilizer is additive, so that if two bags are spread on a garden to the left, the utility derived is 10. Finally, suppose that each person can keep others from fertilizing his garden without his consent.

Using the traditional notion of security value, a person can guarantee that at least his lawn is fertilized, in which case he is assured of receiving one unit of utility. Thus,

$v(i) = \{\mathbf{u} \mid u_i \leq 1\}.$

Since a two-person coalition cannot be certain that the excluded player does not leave his fertilizer in his garage, its best alternative is to fertilize the member's house on the extreme left with the two bags at its disposal. If $\{1, 2\}$ forms, for example, and if both bags are put on 2's lawn, then 1's payoff is 10 and 2's payoff is one-fifth of 10, or 2. Hence,

$$v(1, 2) = \{\mathbf{u} \mid u_1 \leq 10, u_2 \leq 2\};$$
$$v(1, 3) = \{\mathbf{u} \mid u_3 \leq 10, u_1 \leq 2\};$$
$$v(2, 3) = \{\mathbf{u} \mid u_2 \leq 10, u_3 \leq 2\}.$$

Finally, if the three-person coalition forms, then the three bags of fertilizer can be distributed in any way among the three gardens,

$$v(1, 2, 3) = \{\mathbf{u} \mid u_1 \leq 5b + a, u_2 \leq 5c + b, u_3 \leq 5a + c\},$$

such that $a + b + c = 3$, and a is the amount of fertilizer spread on 1's lawn; b, the amount on 2's lawn; and c, the amount on 3's lawn.

What makes this game especially interesting is that if 1 and 2 agree to use their combined supply of fertilizer on 2's garden and agree also as part of their bargain to exclude 3 from fertilizing 1's lawn, then 3's only *rational response* is to fertilize his own garden. This yields the utility vector $(10, 7, 1)$, which is not in $v(1, 2)$ as just defined. It is not in $v(1, 2)$ because 1 and 2 cannot strictly guarantee it: 3 might choose to discard the fertilizer and let his lawn turn brown out of spite (residents of suburbia have performed even more outlandish acts of retribution). Nevertheless, $(10, 7, 1)$ seems to be a reasonable outcome if we assume that we have adequately measured 3's utility function. (We should incorporate the benefits from spiteful behavior into u_3 if we believe that it is important.) After all, we are assuming that people are utility maximizers and that leaving fertilizer in the garage cannot maximize utility as measured. It seems, then, that there are good reasons for redefining $v(1, 2)$ to equal $\{\mathbf{u} \mid u_1 \leq 10, u_2 \leq 7\}$, which is why 1 and 2 agree to not let 3 fertilize 1's lawn.

Another possibility is that, after assessing the situation and the likely outcome $(10, 7, 1)$, 3 devises a doomsday mechanism that, if $\{1, 2\}$ forms, automatically destroys his bag of fertilizer. Person 3, then, can *threaten* 1 and 2 with this possibility, telling them that he has no choice in the matter: The only way that the bag of fertilizer will not self-destruct is if $\{1, 2\}$ agrees to form $\{1, 2, 3\}$ around a more equitable outcome, or if one of them joins him in a two-person coalition. This threat (or, more exactly, the device to implement the threat) alters the game's extensive form. But this example demonstrates that *the use of a security level in defining v(C) implicitly assumes that players expect others to actually carry out threats that may be injurious to their own welfare.* That is, excluding payoffs such as (10, 7, 1) from $v(1, 2)$ implicitly assumes that person 3 does not respond rationally to a coalition of his neighbors.

The normal-form game in Figure 7.1 shows a similar ambiguity. There, in calculating the characteristic function of $\{1, 2\}$, we assume that the security value of (a_1, b_1) is (55, 20). But if person 3 responds rationally to this joint strategy by 1 and 2, then he should choose c_1, in which case 1 and 2 receive 57 and 50, respectively.

Part of our problem here is that the extensive forms of these two games, the flower bed example and the normal form in Figure 7.1, remain unspecified. Should we assume that once coalitions form, the game is played strictly noncooperatively thereafter, or can we suppose that persons excluded from a coalition are privy to communications among that coalition's members? In answering this question it might seem reasonable to assume that excluded persons are not privy to such information. But what happens if the players discuss assumptions about responses during the tentative stages of negotiation and coalition formation? Such queries raise issues about the assumptions that we must make in specifying $v(C)$. If we cannot move readily from the normal to the characteristic-function form without specifying these details (without describing the extensive form of the negotiations process itself), what advantage do we gain by using the characteristic function, which we introduced to avoid the overly complex extensive and normal forms that communication engenders?

The particular difficulty is that no matter what we decide about rational responses after coalitions form, we can formulate an example that violates the rationale for our decision. Suppose that we decide that the concept of a rational response is meaningless, because we should view coalitions as playing a noncooperative game. McKinsey, though, offers a simple example in which player 1 has but one move and 2 has two alternatives:

$$(0, -1000) \quad (10, 0).$$

In traditional characteristic-function form, $v(1) = v(2) = 0$, although it is doubtful that 2 will choose his first strategy. We could revise this pessimistic evaluation of 1's prospects by deleting 2's dominated strategies. But to see the problems that arise if we pursue this logic, consider the four-person normal-form game in Figure 7.13.

Player 3

Player 4	Player 1	c_1 Player 2 b_1	b_2	c_2 Player 2 b_1	b_2
d_1	a_1	$-4, 3, -1, -1$	$1, 1, 1, 0$	$2, 3, 4, 5$	$2, 1, 2, 2$
	a_2	$0, 0, 0, 1$	$3, -2, -2, -3$	$-5, 2, 2, 2$	$3, 1, 1, 4$
d_2	a_1	$-2, 5, 2, 1$	$0, 3, 1, 2$	$3, 2, 3, 2$	$3, 2, 2, 1$
	a_2	$4, -6, 0, 3$	$5, 2, 0, 0$	$-2, 0, 1, 1$	$4, -5, -5, 2$
d_3	a_1	$4, 6, 3, 2$	$0, 3, 0, 3$	$-4, 4, 4, 4$	$1, 1, 1, 1$
	a_2	$-4, -5, 4, 4$	$3, 2, -2, 1$	$3, 5, 5, 0$	$4, -5, -5, 3$

7.13 Four-person normal-form game.

Figure 7.14 shows the characteristic function $v(1, 2)$ based simply on security levels. For example, if 1 and 2 jointly agree to (a_1, b_2), then 1 can do no worse than 0 and 2 can do no worse than 1. But this assessment of $v(1, 2)$ assumes nothing about the actions of persons 3 and 4, except that they might choose strategies in the worst interest of $\{1, 2\}$. This calculation of $v(1, 2)$, though, seems unreasonable, because it does not use any information about 3's and 4's preferences. For example, if 3 and 4 do not coalesce, so that the game is noncooperative among $\{1, 2\}$, $\{3\}$, and $\{4\}$, then no matter what actions persons 1 and 2 take, there is a unique pure-strategy equilibrium for players 3 and 4. No matter how the members of the coalition $\{1, 2\}$ coordinate their strategies, (c_2, d_1) is an equilibrium pair. In calculating $v(1, 2)$, then, should we assume that 3 and 4 abide by their equilibrium strategies? If it is reasonable to eliminate dominated strategies, should we not carry this reasoning further and assume that players excluded from a coalition choose noncooperative equilibrium strategies? To assume otherwise appears to contradict the predictions of part of game theory. If we make this assumption, then our prior assessment of $v(1, 2)$, clearly, is too conservative. Assuming that 3 and 4 choose (c_2, d_1) shifts the upper bound on $v(1, 2)$ to the dashed line in Figure 7.14.

Alternatively, we could assume that persons 3 and 4 coalesce to form

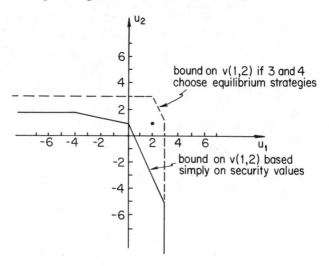

7.14 Alternative forms for $v(1, 2)$.

{3, 4}, in which case the concept of an equilibrium between them loses its relevance. Calculating $v(3, 4)$ using simple security values, though, reveals that the security level associated with (c_2, d_1) is the only outcome that is Pareto-optimal in terms of its security value, (1, 2), for {3, 4}. Hence, if {3, 4} forms, its members should choose (c_2, d_1), and the dotted line in Figure 7.14 depicts the appropriate upper bound on $v(1, 2)$. Continuing this line of reasoning further, however, we notice that if {1, 2} can reason that {3, 4} will choose its Pareto-optimal strategy n-tuple, then it should choose either (a_1, b_1) or (a_2, b_2), in which case {3, 4} realizes either (4, 5) or (1, 4). Thus, by thinking the matter through, {3, 4} can conclude that its security value is (1, 4) and not simply (1, 2).

This scenario appears to be a natural for the development of characteristic functions requiring consistency of beliefs about opposing coalitions. But suppose that we extend this example further, say, to five persons, in which case we could let $v(1, 2)$ depend on what coalitions players 3, 4, and 5 form, and on which strategies these coalitions choose. But this expansion would render the definition of $v(C)$ circular. We had originally defined $v(C)$ to simplify the representation of a game. Now, though, an analysis of coalition formation may be a precondition for defining $v(C)$. So, we have to ask what purpose $v(C)$ serves.

These ambiguities may signal that the analysis of coalitions should proceed directly from either the normal or the extensive forms. Any definition of $v(C)$ simply may obscure the strategic qualities of a game and produce a misleading analysis. For certain classes of games, though, such as simple games in which a coalition is either winning or losing, the definition of $v(C)$ is unambiguous. In sum, although difficulties with $v(C)$ should not detract from the characteristic-function's use as a modeling device, we should be aware of the fact that various definitions of $v(C)$ may be more or less profitable in different circumstances. This representation has facilitated much useful modeling, as our discussion in the remaining chapters will show.

7.8   Summary

The new concepts and ideas that this chapter introduces include the following:

coalition	strong equilibria	correlated mixed strategies
partition of N	coalition structure	characteristic function
game in $v(C)$ form	superadditive	essential
simple game	strong simple game	blocking coalition
winning coalition	losing coalition	minimum winning coalition
transferable utility	imputation	zero sum
normalized game	(0, 1)-normalization	symmetric game
market game	beta-$v(C)$	coalition of the whole
alpha-$v(C)$	rational response	threat

Although we can usefully model many political processes as noncooperative games, much of politics involves coordination, collusion, communication, and bargaining. The principal difficulty with extending the ideas of the previous chapters to situations in which cooperation is possible, though, is finding an acceptable way to extend individual preferences over strategies to preferences over strategy c-tuples. Formulating this extension is equivalent to learning how people evaluate alternative cooperative agreements and it is therefore an essential first step toward understanding why one coalition forms and not another.

Instead of modeling a situation's extensive form to identify the full range of strategies and responses, the characteristic function offers a mathematically tractable extension by imposing two assumptions. First, adequate enforcement mechanisms ensure the survival of any cooperative agreement. To appreciate this assumption, consider the two normal-form games in Figure 7.15. The first is simply a two-person prisoners' dilemma and the second is similar except that there is no dilemma, because each player's dominant strategy leads them jointly to the Pareto-optimal outcome (10, 10). Played noncooperatively, these

	b_1	b_2
a_1	10, 10	−5, 15
a_2	15, −5	−2, −2

(a)

	b_1	b_2
a_1	10, 10	15, −5
a_2	−5, 15	−2, −2

(b)

7.15 $v(C)$ and enforcement mechanisms.

games are quite dissimilar. Even if they communicate beforehand in the first, and agree to choose (a_1, b_1), nothing guarantees that (10, 10) prevails unless the players establish a mechanism for enforcing their agreement: Each person has a unilateral incentive to defect from his first strategy, regardless of prior commitments. We note with interest, however, that the characteristic function in one game for the coalition $\{1, 2\}$ is identical to the characteristic function for $\{1, 2\}$ in the other. That is, by treating the coalition $\{1, 2\}$ identically in both games, the definition of $v(C)$ ignores the problems that this coalition might have in maintaining any agreement in the first game.

The second assumption concerns the evaluation of the responses of players excluded from a coalition. Game theorists have focused on two alternative assumptions. The first, which yields the alpha form of $v(C)$, assumes the worst, that excluded players act in the worst interests of C, even if this formulation requires them to act against their own self-interest. The second alternative, which yields the beta form, assumes, in effect, that the excluded members choose their strategies first and that C responds with strategies that are best for its members. The alpha form gives a pessimistic view of a coalition's opportunities whereas the beta form gives, perhaps, an overly optimistic view. The discussion in the previous section indicates, though, that other options may include the assumption that the members of an opposing coalition respond in their own interest. Game theorists have not explored these possibilities fully, but their plausibility suggests that basic theoretical developments remain far from complete.

CHAPTER 8

The core

8.1 The core

Setting aside the potential ambiguities in the definition of the characteristic function, if we have a theoretical description that appropriately summarizes a situation's strategic properties, then we should ask two questions of it. First, is there anything in the feasible set of utility outcomes, U, that is stable against all other possibilities, that offers itself as a compelling prediction? Second, if we can identify such outcomes, then what coalitions bring them about? These questions parallel our concerns in noncooperative theory: the definition and search for Nash equilibrium strategies and the outcome that prevails if players choose them. But now we focus on the actions of coalitions and on people as coalition members instead of atomistic decision makers. Game theory, or more precisely cooperative solution theory, has concentrated more on the issue of predicting outcomes, but the theory has taken some important steps in respect to predicting coalitions. This chapter, however, discusses a solution concept, the *core*, that is legitimately concerned solely with predicting outcomes.

If the characteristic function is appropriate to the circumstance, then the core provides a compelling prediction about outcomes. But not all games have a core, and we should understand what this means for politics. Because the core is a compelling hypothesis about cooperative games, we interpret its existence or nonexistence in a particular circumstance as an indicator of whether the final outcome is sensitive to the details of bargaining, to the coalition that ultimately forms, and to the idiosyncracies of the participants' personalities, bargaining skills, and the like. Hence, we should not conclude that the core is an unsatisfactory concept simply because we cannot ensure its general existence: Nonexistence has a substantive interpretation. Games with no core are unlike those with one, and knowing whether a core exists in a particular instance tells us much about the relevance of bargaining and the stability of final outcomes. As dispassionate researchers, we are not disappointed to learn, for example, that a Condorcet winner does not always exist under majority rule. The absence of such a winner opens a situation to strategic voting and agenda control, and it makes the study of voting games interesting. The same is true for cooperative games

339

without a core. The definition of the core is a definition of a particular type of stability in cooperative social processes, and by examining the core's properties and the conditions for its existence, we can differentiate between those situations in which outcomes have a natural equil'b-rium and those for which we must invent new concepts to narrow the range of predicted outcomes.

8.1   The core

If we want to predict the utility n-tuples that we might expect to prevail in a game, then regardless of other possible considerations, it seems reasonable to limit our attention to those outcomes in the feasible set, U, that are *individually rational*. A payoff n-tuple is individually rational if no person, by unilaterally defecting from a coalition and by playing alone, can guarantee himself a more preferred result. In terms of the transferable utility representation of $v(C)$, for example, the n-tuple $\mathbf{u} \in U$ is individually rational if and only if

$$u_i \geq v(i) \text{ for all } i \in N.$$

A person can veto outcomes that are not individually rational for him, and we should expect everyone to do so.

The fact, however, that a single person is but a special limiting case of a coalition suggests that we should search for outcomes that are rational from *every* coalition's perspective. That is, if we interpret coalitions as players, then is it not reasonable to define a game's solution as those utility n-tuples from which no coalition has the means or the incentive for unilateral defection? The answer to this question is yes, and in raising the question we have essentially defined the core. Recall, however, that this is the approach we take in Chapter 7 to define strong equilibria (after accommodating for correlated mixed strategies). We ended that discussion, though, after we encountered the problem of assessing how coalitions judge the responses of excluded players. But such an evaluation is precisely what the characteristic function tries to provide. Hence, with an adequate representation of a situation in characteristic-function form, we can use the perspective of a strong equilibrium to define the core. We begin with the concept of domination.

> *Domination via C*: If $\mathbf{u}$ and $\mathbf{w}$ are any two payoff n-tuples in U, then $\mathbf{u}$ dominates $\mathbf{w}$ with respect to C if and only if $\mathbf{u}$ is in $v(C)$ and $u_i > w_i$ for all $i \in C$.

That is, if each and every member of C strictly prefers $\mathbf{u}$ to $\mathbf{w}$, and if C

Person 3

Person 2:		c_1		c_2	
		b_1	b_2	b_1	b_2
Person 1:	a_1	65, 50, 20	0, 60, 35	55, 20, 0	30, 30, 40
	a_2	70, 55, 30	10, 90, 60	50, 0, 80	20, 10, 70

8.1 A core at (30, 30, 40).

can secure **u**, then **u** dominates **w** with respect to C. Put differently, if **u** is in U, and if **u** dominates **w** with respect to C, then the members of C have a unanimous incentive to defect from **w**.

Domination: **u** dominates **w** if and only if there exists a coalition, C, such that **u** dominates **w** with respect to C.

Thus, **u** dominates **w** if we can find some coalition that can ensure **u** and whose members all strictly prefer **u** to **w**. With this concept of domination we can now define the core.

Core: The core of a game in characteristic-function form is the set of undominated payoff vectors in U.

Although we are concerned with cooperative games, it is nevertheless important, for appreciating the core's attractiveness as a hypothesis about outcomes, to understand how we can interpret this concept as a noncooperative equilibrium. Notice that for games in characteristic-function form, the essential players are the $2^n - 1$ nonempty coalitions, and that the strategies of each coalition, C, are the payoff vectors in $v(C)$. Elements of the core, then, are utility n-tuples that coalitions can attain and that are in equilibrium, because no coalition can unilaterally adopt some other strategy that is better for all of its members.

This interpretation of the core helps us see that its properties in a specific game depend on how we define $v(C)$, since the utility vectors in a coalition's characteristic function correspond to that coalition's strategies. To illustrate this relationship, as well as the definition of the core, consider the normal-form game that Figure 8.1 portrays, and in particular, consider the utility 3-tuple (30, 30, 40), corresponding to the strategy 3-tuple (a_1, b_2, c_2). Notice that this outcome is Pareto-optimal for the coalition of the whole, so (30, 30, 40) is undominated with respect to N. And since no person can guarantee himself more, the 3-tuple (30, 30, 40) is undominated with respect to the coalitions $\{1\}$,

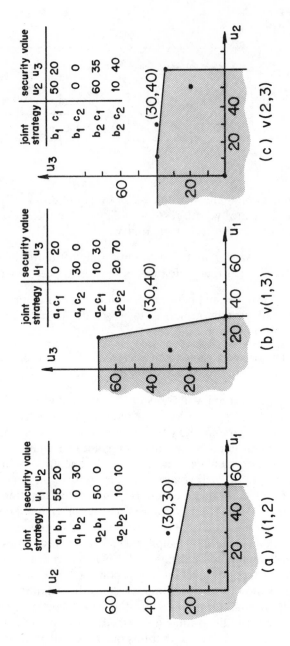

joint strategy	security value u_1 u_2
$a_1 b_1$	55 20
$a_1 b_2$	0 30
$a_2 b_1$	50 0
$a_2 b_2$	10 10

•(30,30)

(a) v(1,2)

joint strategy	security value u_1 u_3
$a_1 c_1$	0 20
$a_1 c_2$	30 0
$a_2 c_1$	10 30
$a_2 c_2$	20 70

•(30,40)

(b) v(1,3)

joint strategy	security value u_2 u_3
$b_1 c_1$	50 20
$b_1 c_2$	0 0
$b_2 c_1$	60 35
$b_2 c_2$	10 40

(30,40)

(c) v(2,3)

8.2 A core at $(30, 30, 40)$.

{2}, and {3}. Turning, finally, to the two-person coalitions, Figure 8.2 portrays the alpha versions of $v(1, 2)$, $v(1, 3)$, and $v(2, 3)$, in which the tables beside each figure give the security values that the figure plots to compute each characteristic function. Figure 8.2 also graphs the relevant components of the point (30, 30, 40), and it shows that this point is outside of each $v(C)$. But this means that no two-person coalition can ensure a utility vector that makes all of its members better off. Hence, (30, 30, 40) is in the alpha core.

If we use the beta version of $v(C)$, then the preceding definitions yield the beta core. Defining the core in terms of the characteristic function, however, may obscure how the definitions of $v(C)$ and of the core combine to resolve the issues that arose in Chapter 7 with respect to strong equilibria. It is useful, then, to restate the core's definition by referring only to the normal form. Letting S_i be player i's strategies in the normal form, and letting s_i denote a particular strategy in S_i then,

> *Alpha core*: For any C, let s_c be a c-tuple of strategies representing the joint choices of the members of C, and let s_{-c} be an n-c tuple of strategies representing the choices of all persons not in C. The payoff n-tuple $u(s^*) = (u_1(s^*), \ldots, u_n(s^*))$ is in the alpha core if and only if for any s_c there exists an s_{-c} such that,
>
> $$u_i(s_c, s_{-c}) \leq u_i(s^*),$$
>
> for all $i \in C$, or where the inequality is strict for at least one $i \in C$.

That is, $u(s^*)$ is in the alpha core if, whenever one or more members of C defect from s^*, the complementary coalition, $N - C$, can find a joint strategy that makes at least one member of C strictly worse off or that promises no improvement for everyone in C. This version of the core, then, corresponds to a characteristic function defined in terms of what C can guarantee its members if $N - C$ chooses in the worst interests of C, regardless of whether this choice is rational or irrational for $N - C$.

For the game in Figure 8.1, the threat to a two-person coalition is that the excluded player will continue to choose the strategy that corresponds to the core. That is, for the coalition {1, 2}, as long as {3} threatens to stay at c_2, no defection can raise 2's payoff above 30; for the coalition {1, 3}, no defection can raise 1's payoff above 30 as long as {2} stays at b_2; and for the coalition {2, 3}, no defection can raise 3's payoff above 40 as long as {1} stays at a_1. Thus, (30, 30, 40) is in the alpha core.

We can better appreciate the importance of threats in the definition of the alpha version of the core by referring to the game in Figure 7.1. To

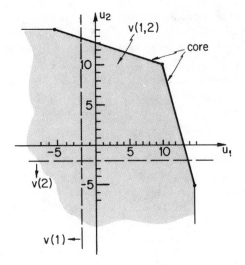

8.3 Characteristic function for prisoners' dilemma.

see that the utility n-tuple (30, 30, 40), corresponding to the strategy 3-tuple $s^* = (a_1, b_2, c_2)$, is in the alpha core, suppose that $C = \{1, 2\}$ defects from this 3-tuple by choosing an even-chance lottery between (a_1, b_1) and (a_2, b_2). If the excluded player, 3, continues to choose c_2, then this lottery improves 1 and 2's expected payoffs to 37.5 and 40, respectively. But players 1 and 2, in accordance with the definition of the alpha core, do not have a unanimous incentive to shift from s^* to the lottery, since 3 can respond by choosing c_1, thereby reducing 1's expected payoff to .5(57) + .5(0) = 28.5. More generally, the definition shows that the alpha core is an equilibrium in terms of the threats of complementary coalitions. If a coalition, C, proposes to defect to s_c from s^*, then $N - C$ threatens s_{-c}.

We can provide a parallel definition for the beta version of the core.

> *Beta core*: Using the preceding notation, $u(s^*)$ is in the beta core if and only if for any C, there exists a joint strategy, s_{-c}, for $N - C$, such that for all s_c,
>
> $$u_i(s_c, s_{-c}) \leq u_i(s^*)$$
>
> for all $i \in C$, or where strict inequality holds for at least one $i \in C$.

Thus, $u(s^*)$ is in the beta core if $N - C$ can threaten to commit itself to strategies that hold individual members of C to the payoffs that they associate with s^*. Referring again to the game in Figure 8.1, if player 3 commits to c_2, then no matter what strategy 1 and 2 choose, they cannot both get more than (30, 30); if player 1 commits to a_1, then 2 and 3 cannot both get more than (30, 40); and, if player 2 commits to b_2, then 1 and 3 cannot both do better than (30, 40). Hence, (30, 30, 40) is in the beta core.

Before we discuss other properties of the core, and before we apply it to specific models, we must note one of its features, in conjunction with the definition of $v(C)$. Figure 8.3 graphs $v(1, 2)$, $v(1)$, and $v(2)$, for the two-person prisoners' dilemma in Figure 7.15a. Notice that no one- or two-person coalition can dominate any of the Pareto-optimal outcomes on the dark solid line. Thus, all such outcomes are in the core, and only those outcomes are in the core. This means that the core predicts that the Pareto-inefficient outcome, $(-2, -2)$, which results from uncoordinated action, will not prevail. As we indicate in previous chapters, though, such a prediction supposes that persons 1 and 2 can enforce contracts between themselves; otherwise, each may have an incentive to defect from an agreement. The characteristic function assumes away the issue of enforceable contracts, and by defining the core in terms of $v(C)$ or, as in the previous two definitions, in terms of particular threats, the core assumes away the same problem. Thus, although the core is a compelling idea, it is perhaps most compelling for simple majority-rule games in which both the alpha and beta versions of $v(C)$ are identical and enforceable contracts seem less of an issue.

8.2   The core and Pareto-optimality

As an aid to applying the core to political games, especially simple majority-rule voting games, notice from the example of the two-person prisoners' dilemma that a payoff n-tuple cannot be in the core if it is an interior point of $v(C)$ for any C. Any point $\mathbf{w}$ that is interior to $v(C)$ is dominated by all points in $v(C)$ that are above and to the right of $\mathbf{w}$. This graphical notion of dominance gives us a perspective for defining the core in terms of Pareto-optimality. If

$$v^*(C) = \{\mathbf{u} \in U \mid \text{for no } \mathbf{w} \in v(C) \text{ is } w_i > u_i \text{ for all } i \in C\},$$

then,

$$\text{Core} = \bigcap_C v^*(C), \tag{8.1}$$

in which $\cap_C$ means the intersection of the $v^*(C)$'s for all coalitions. The set $v^*(C)$ consists of all feasible utility outcomes that are Pareto-optimal for the coalition C in the sense that C cannot ensure more for all of its members. The core corresponds to the intersection of all such sets, the intersection of all utility outcomes that are feasible and that are not interior to any coalition's characteristic function.

This restatement of the core's definition is useful for analyzing coalitions in formal voting bodies because of the following result:

8.1. *For strong simple games in characteristic-function form, if W^* is the set of all minimum-winning coalitions, then*

$$\text{Core} = \bigcap_{C \in W^*} v^*(C),$$

in which $\bigcap_{C \in W^*}$ means the intersection taken over all minimum winning coalitions. Hence, for strong simple games we need only look at the Pareto-optimal outcomes of minimum-winning coalitions to calculate the core and to ascertain its properties.

The proof of Theorem 8.1 provides a useful lesson in the mechanics of the core. Consider any n-tuple, $\mathbf{u}$, that is in the core, but suppose that $\mathbf{u}$ is not in $v^*(C)$ for some minimum-winning coalition C. But this means that there is an outcome in $v(C)$ that make all members of C better off. Hence, $\mathbf{u}$ is dominated with respect to C, and the assumption that $\mathbf{u}$ is in the core is contradicted. This contradiction establishes the relation,

$$\text{Core} \subseteq \bigcap_{C \in W^*} v^*(C). \tag{8.2}$$

Next, consider any n-tuple, $\mathbf{w}$, that is in the intersection of the Pareto-optimals of all minimum-winning coalitions. That is, let $\mathbf{w} \in \bigcap_{C \in W^*} v^*(C)$. But suppose that $\mathbf{w}$ is not in the core, in which case there must exist some winning coalition, C', such that $\mathbf{w}$ is not in $v^*(C')$. But if there is a winning coalition, C', such that $\mathbf{w}$ is not in $v^*(C')$, then there is a minimum-winning coalition, $C \subseteq C'$, such that $\mathbf{w}$ is not in $v^*(C)$. But then, $\mathbf{w}$ is not in $\bigcap_{C \in W^*} v^*(C)$, and thus $\mathbf{w}$ is not in the core. This reasoning proves that

$$\text{Core} \supseteq \bigcap_{C \in W^*} v^*(C). \tag{8.3}$$

We can satisfy expressions (8.2) and (8.3) simultaneously, though, only if expression (8.1) is true; that is, only if the core is identically equal to $\bigcap_{C \in W^*} v^*(C)$.

To illustrate this result, consider the three-member, three-bill vote-trading scenario in Figure 8.4. If we assume simple majority rule, then it follows that a two-person coalition can secure any outcome, including lotteries between outcomes. Figure 8.5, then, graphs $v(C)$ for each minimum-winning coalition. Notice that only the outcomes $(1, 0, 0)$ and

Legislator	(0, 0, 0)	(1, 0, 0)	(0, 1, 0)	(0, 0, 1)	(1, 1, 0)	(1, 0, 1)	(0, 1, 1)	(1, 1, 1)
1	0	5	1	−2	6	3	−1	4
2	0	6	−2	−1	4	5	−4	3
3	0	−2	4	1	2	−1	5	3

8.4 Vote trading with a core.

$(1, 1, 0)$ are Pareto-optimal for the coalition $\{1, 2\}$. Similarly, if lotteries are admitted, then the outcomes $(0, 1, 0)$, $(1, 1, 0)$, $(0, 1, 1)$, and $(1, 1, 1)$ are Pareto-optimal for $\{1, 3\}$. Finally, $(1, 0, 0)$, $(1, 1, 0)$, $(0, 1, 1)$, and $(1, 1, 1)$ are Pareto-optimal for $\{2, 3\}$. The only outcome that is Pareto-optimal for all three coalitions, then, is $(1, 1, 0)$, passing bills 1 and 2. Thus, the core to this game is the outcome $(1, 1, 0)$, or, equivalently, the payoff 3-tuple $(6, 4, 2)$.

Two aspects of this example warrant emphasis. First, although the core is unique and, thus, predicts a specific disposition of the bills, it provides no prediction about coalitions. Some students of politics may find this limitation troublesome, especially if they are familiar with legislative or parliamentary coalition-formation processes and if their research concerns predicting which legislators coalesce to pass a piece of legislation or which parties coalesce to form a government. But the core's failure to identify a specific coalition is reasonable in this example, since *any* winning coalition can bring $(1, 1, 0)$ about. Owing to the strict superadditivity that holds among the members of a minimum-winning coalition, some coalition will form. Unless we know something else about the situation, however, we have no reason to suppose that one coalition is more likely to form than another, and the silence of the core as to which coalition will form to produce $(1, 1, 0)$ seems appropriate. Thus, although there is an important literature that tries to predict legislative or parliamentary coalitions, our example shows that if a core exists, then such predictions may be impossible or they are at least irrelevant to the determination of final outcomes.

Our second observation about the example in Figure 8.4 is that the outcome $(1, 1, 0)$ is also a Condorcet winner. This correspondence between the core and a Condorcet winner is no accident. If we modify the definition of the core slightly and define a *strong core*, a utility vector that is not only undominated, but that dominates everything else, then,

8.2. *For strong simple majority-rule voting games, the Condorcet winner is the core, and the strong core and Condorcet winner are equivalent.*

This result follows almost immediately from Theorem 8.1. If coali-

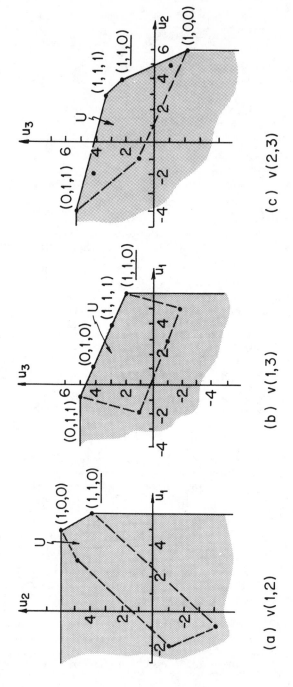

8.5 Characteristic function for vote-trading game.

tions are either winning or losing, and if x is a Condorcet winner, then not all members of any winning coalition can strictly prefer some other outcome to x. Thus, x must be Pareto-optimal for all winning coalitions, particularly for all minimum-winning coalitions; that is to say, x is in the core of the corresponding cooperative game in characteristic-function form. Conversely, if x is in the core, it must be Pareto-optimal for all minimum-winning coalitions, and correspondingly for all winning coalitions (this result follows from the easily established conclusion that the Pareto-optimal set for any greater than minimum-winning coalition, C, cannot be smaller than the Pareto-optimal set for any minimum-winning coalition made up of some members of C). Hence, nothing beats x in a majority vote. If, in addition, x dominates everything else, and if some winning coalition prefers x to any other feasible outcome, then x is a Condorcet winner.

This correspondence between cores and Condorcet winners is important. Recall from Chapter 4 that if a Condorcet winner exists, then it is an equilibrium strategy for each candidate in symmetric, two-candidate elections. Theorem 8.2 concerns a different procedure, cooperative-committee processes, and yet it also predicts the Condorcet winner as the outcome. Thus, it matters little if a two-candidate election or if all voters gathering together to negotiate a final outcome without candidates intervening in the process choose a final outcome. Of course, various transaction costs preclude implementing the New England–style town meeting in large societies, but Theorem 8.2 and our earlier results about elections establish an important principle of democracy (ignoring what we must pay candidates so that they run for office):

> *Since transaction costs preclude any reasonably sized electorate from engaging in a wholly cooperative majority-rule game, two-candidate elections are an indirect mechanism for achieving the same outcome that would prevail if transaction costs were zero and if there is a Condorcet winner in O.*

8.3   Transferable utility games

Transferable utility and the core

An alternative way to illustrate the core is to consider the special case of games with transferable utility. Assuming that a single number summarizes $v(C)$, we restate the definition of domination and the core. First, the utility n-tuple **u** dominates **w** with respect to C if, for all $i \in C$,

$u_i > w_i$, assuming that **u** is feasible for C. In the case of transferable utility, feasibility with respect to C means that,

$$\sum_{i \in C} u_i \leq v(C).$$

As before, **u** dominates **w** if there is a coalition, C, such that **u** dominates **w** with respect to C. Since the core consists of all feasible undominated payoff n-tuples, we can express its definition thus: **u** is in the core if and only if,

$$\sum_{i \in C} u_i \geq v(C) \text{ for all } C. \tag{8.4}$$

Expression (8.4) states that u must be Pareto-optimal for all coalitions, and it is the transferable utility equivalent of Theorem 8.1.

Recall from Chapter 7 that with transferable utility we can define an interesting class of games, constant-sum games, that usefully model the cooperative allocation of some fixed resource (bags of garbage in the valley of the dump example). But we can use expression (8.4) to show that the core is empty for this class of games:

8.3. *All essential superadditive constant-sum games in characteristic-function form have empty cores.*

To establish this result, recall that a cooperative game is constant sum if $v(C) + v(N - C) = k$, a constant, for all C. Contrary to the statement of Theorem 8.3, suppose that the game is superadditive and essential, and that the n-tuple, $\mathbf{u} = (u_1, u_2, \ldots, u_n)$, is in the core. Hence, from expression (8.4), it must be the case that

$$\sum_{i \in C} u_i \geq v(C) \quad \text{and} \quad \sum_{i \in N-C} u_i \geq v(N - C),$$

for all C. Summing these two expressions, we have

$$\sum_{i \in C} u_i + \sum_{i \in N-C} u_i = \sum_{i=1}^{n} u_i \geq v(C) + v(N - C). \tag{8.5}$$

But since the game is constant sum, $v(C) + v(n - C) = k$, so $\sum_{i=1}^{n} u_i \geq k$. On the other hand, the n-tuple **u** must be feasible and Pareto-optimal for the coalition of the whole, which requires that

$$\sum_{i=1}^{n} u_i = v(N) = k.$$

But this expression and expression (8.5) are satisfied only if

$$\sum_{i \in C} u_i = v(C) \quad \text{and} \quad \sum_{i \in N-C} u_i = v(N - C).$$

Since these equations must hold for every coalition C, it must hold for single-member coalitions, in which case $u_i = v(i)$ for all i, thereby implying,

$$\sum_{i=1}^{n} v(i) = v(N).$$

But this means that the game is inessential, so, as Theorem 8.3 asserts, the assumptions that a game is essential, superadditive, constant sum, and has a core are inconsistent.

For an application of this theorem, consider the valley-of-the-dump example from Section 7.5 of Chapter 7. Since the valley's residents can neither export nor otherwise dispose of their garbage except by dumping it on each other's lawns, the game is constant sum. Suppose that the specific utility n-tuple, $\mathbf{u}$, contrary to Theorem 8.3, is in the core of this game. Consider the n different coalitions of size $n - 1$, which each exclude one member, and for which the first coalition excludes person 1, the second, person 2, and so on. The value of $v(C)$ for each such coalition is -1 since an $n - 1$ person coalition cannot exclude the possibility that the excluded player will dump his garbage somewhere on the coalition's collective lawn. Hence, if $\mathbf{u}$ is in the core, then from expression (8.4)

$$u_2 + u_3 + \cdots + u_n \geq -1$$
$$u_1 + u_3 + \cdots + u_n \geq -1$$
$$\vdots$$
$$u_1 + u_2 + \cdots + u_{n-1} \geq -1. \tag{8.6}$$

Since there are n such equations, and since each person is excluded once, adding them yields,

$$(n - 1) \sum_{i=1}^{n} u_i \geq -n. \tag{8.7}$$

But the definition of $v(N)$ requires that the valley's residents cannot export their garbage, so

$$\sum_{i=1}^{n} u_i = v(N) = -n.$$

in which case (8.7) becomes $(n - 1)(-n) \geq -n$, or after rearranging terms, $n \leq 2$. Thus, if the valley has more than two residents, we have a contradiction and the core must be empty.

To see the logic of this result, suppose that the valley has three residents and that coalition $\{1, 2\}$ tentatively forms. If the members of $\{1, 2\}$ agree to split the one bag of garbage they receive from person 3 equally, then the payoff n-tuple $(-\frac{1}{2}, -\frac{1}{2}, -2)$ prevails. But person 3 can approach 1, offering to absorb all of 2's garbage in exchange for a coalition that gives 2 the two bags of garbage that $\{1, 3\}$ produces. In other words, 3 offers $(0, -2, -1)$. Persons 1 and 3 prefer this payoff vector to the original proposal, so it dominates the first proposal. But 2 can offer 3 $(-2, -1, 0)$. And at this point, 1 might resurrect his original agreement with 2, and thus a cycle will be completed. Indeed, in the absence of a core, the players can cycle endlessly throughout the set of outcomes.

Notice that this example is simply an extreme version of a game of redistribution. Instead of redistributing a desirable item such as income, the valley's residents are redistributing an undesirable commodity. But this difference is irrelevant, and we already know from the discussion in Chapter 4 that elections that concern redistribution do not have Condorcet winners. Hence, Theorem 8.2 implies that their cooperative counterparts do not have a core. Cooperative game theory and the concept of the core, then, give us an equivalent way to observe the instabilities in political processes occasioned by the problem of distributing scarce (constrained) resources.

To this point our results about the core seem mostly negative. Games of pure conflict, constant-sum games, have no cores, and majority-rule voting games have a (strong) core if and only if a Condorcet winner exists; this possibility, as we know, requires extremely fortuitous circumstances. An empty core, though, does not imply that prediction is impossible any more than the nonexistence of a Condorcet winner implies unpredictability. An empty core simply tells us that fundamental instabilities exist, that the game may be especially interesting when played, and that further prediction requires the development of additional ideas.

Nevertheless, some games have cores, and an analysis of them provides insight into public policy. Consider the example from Chapter 7 of business firms polluting a lake. This game is not constant sum, so Theorem 8.3 does not apply. Indeed, our problem here is that this game's core is large. For example, in addition to an equitable division of effluent-treatment costs, the payoff 10-tuple $\mathbf{u} = (-10, -10, -10, -10, 0, 0, 0, 0, 0, 0)$ and all similar permutations of payoffs are in the core. To see why this is so, notice that this 10-tuple gives no less than -40 to any coalition. But from Figure 7.9, which portrays this game's characteristic function, we see that the value of no coalition with more than four members exceeds -40. Hence coalitions with more than four members cannot upset $\mathbf{u}$. For smaller coalitions, $\mathbf{u}$ awards a minimum of $-10c$, $c = 1, 2, 3,$ or 4, and a maximum of 0. Again from Figure 7.9, we observe that no small coalition is worth more than $-10c$, and none can upset **u** either. So the 10-tuple **u** is in the core.

In this instance, the core tells us little about how people eventually distribute pollution costs. Shapley and Shubik point out in their analysis of this example, however, that if we modify the purification costs from cB to $\max[(c - 1)B, 0]$, if firms cannot "extort" profits from other firms (a lower bound of zero on costs precludes negative costs or profits), and if the government subsidizes each firm by an amount B if c firms pollute, then the core is a single point in which every firm pays for treating the pollution that it alone produces. The discovery of a large core, then, leads to the suggestion of a public policy (subsidizing cleanups) different from some other that might be attempted (regulatory edict).

The Coase theorem

For an especially important application of the core, consider the Coase theorem, which is the cornerstone of much contemporary theorizing about property rights, redistribution, and the economic analysis of law. This theorem asserts that if transaction costs are reasonably low, if an exogenous mechanism for enforcing contracts exists, if a freely transferable numeraire good such as money is available, and if prior property rights are established, then everyone involved in a situation in which the actions of some hurt or benefit others can reach mutually advantageous bargains without government intervention. Furthermore, if a unique outcome maximizes social wealth, then the parties will reach that outcome no matter what the prior assignment of property rights and liabilities might be. Recast in the terminology of Chapter 5, the Coase theorem counters the argument that the inefficiencies occasioned by public goods and externalities justify government intervention into markets, in particular, government regulation. Instead, governments ought simply to guarantee legally defensible and alienable property rights to ensure that markets function. If one can receive compensation for the benefits that one's actions bestow on others, and if an institutional arrangement could guarantee such a right, then the public goods problem dissolves. Political debate would then focus on the allocation of those rights rather than on decisions about what levels of various goods, services, and regulations to supply.

For example, an industry might have a property right to pollute, which the "victims" can buy away from it if they value clean air more than the industry values the use of the ambient air as a sink. Alternatively, the victims might have a property right in clean air, which the industry might buy away from them. And if victims (industries) are willing to pay more for clean (dirty) air, then the same outcome, clean (dirty) air, prevails no matter what the prior assignment of property rights has been.

To illustrate this proposition with a simple numerical example, consider this two-person situation. Suppose that 1 and 2 must choose one outcome from the set $O = \{o_1, \ldots, o_5\}$, where their dollar payoffs are as follows:

Outcome	Person 1	Person 2	Total
o_1	$0	$12	$12
o_2	4	10	14
o_3	6	6	12
o_4	7.5	4	11.5
o_5	12	0	12

Suppose, further, that person 1 has the legal right to decide which outcome is chosen, in which case the game in characteristic-function form becomes,

$$v(1) = 12; \ v(2) = 0; \ v(1, 2) = 14,$$

so that the core is the set of 2-tuples

$$\text{Core} = \{\mathbf{u} \mid u_1 \geq 12, \ u_2 \geq 0, \ u_1 + u_2 = 14\}.$$

Alternatively, suppose that 2 has the right to decide the outcome, so that the characteristic function becomes

$$v(1) = 0; \ v(2) = 12; \ v(1, 2) = 14,$$

and the core is

$$\text{Core} = \{\mathbf{u} \mid u_1 \geq 0, \ u_2 \geq 12, \ u_1 + u_2 = 14\}.$$

Thus, although the establishment of a property right dictates the eventual distribution of payoffs between 1 and 2, o_2 prevails no matter who controls the decision.

As we have tried to indicate, the policy implications of this example are profound. Instead of hiring labor to produce commodities such as mass transportation, mail delivery, education, and the like, instead of establishing specific clean-air and safety standards with which businesses must comply, and instead of centralizing decisions about the resources that people should allocate to each of these activities, the example suggests that governments ought simply to assist in reducing transaction costs among relevant parties, and ought to establish and enforce property rights in accordance with the directives of our electoral institutions. Unfortunately, an extended discussion about the proper functions of the state and about the difficulties of ensuring minimum transaction costs and defensible property rights is beyond the scope of this book. Our intent is simply to show how a simple game, in conjunction with the concept of the core, leads to a discussion of profoundly important political matters, and about how we can restate those matters in terms of simple abstract concepts.

8.4 Balanced games

A necessary and sufficient condition for existence

Theorem 8.2 states that a (strong) core to a majority-rule voting game exists if and only if a Condorcet winner exists, and Theorem 8.3 states that zero- (constant) sum games have no cores. The preceding examples, however, illustrate nonconstant-sum games with cores. Certainly, we would appreciate the parsimony that a general necessary and sufficient existence condition supplies. As a first step to such a result, consider these two transferable-utility games, in which the first but not the second has a core.

$v(i) = 0$, for all i $\qquad\qquad$ $v'(i) = 0$, for all i

$v(1, 2) = v(1, 3) = v(2, 3) = \frac{2}{3}$ $\qquad$ $v'(1, 2) = v'(1, 3) = v'(2, 3) = \frac{3}{4}$

$v(1, 2, 3) = 1$ $\qquad\qquad$ $v'(1, 2, 3) = 1$

Looking at the first game, if $\mathbf{u} = (u_1, u_2, u_3)$ is in the core, then it must be that

$$u_1 + u_2 \geq v(1, 2) = \tfrac{2}{3}$$
$$u_1 + u_3 \geq v(1, 3) = \tfrac{2}{3}$$
$$u_2 + u_3 \geq v(2, 3) = \tfrac{2}{3} \qquad\qquad (8.8)$$

Otherwise, $\mathbf{u}$ is not Pareto-optimal simultaneously for all three two-person coalitions. Summing these inequalities yields $2(u_1 + u_2 + u_3) \geq 2$, or after dividing by 2, $u_1 + u_2 + u_3 \geq 1$. A payoff 3-tuple that satisfies people's security values, which is feasible and Pareto-optimal for the coalition of the whole and which satisfies this inequality, exists. The core in this instance is the unique vector $(\frac{1}{3}, \frac{1}{3}, \frac{1}{3})$. But for the game that the characteristic function $v'(C)$ represents, if we substitute a value of $\frac{3}{4}$ for $\frac{2}{3}$ in equation set (8.8), then after summing the inequalities, we get, $u_1 + u_2 + u_3 \geq \frac{9}{8}$. Such a payoff 3-tuple, however, is infeasible. Hence, this second example has no core.

We can generalize the lesson of these two examples and establish a general necessary and sufficient condition for the existence of a core. First, let $C(i)$ denote all coalitions that contain person i, let $C(N)$ be the set of all coalitions, and let $\{w_c\}$ be a set of nonnegative weights that we assign to coalitions, where w_c is the weight that we assign to coalition C, with $0 \leq w_c \leq 1$. Then with respect to the set of weights $\{w_c\}$, the collection of coalitions, $\{C\}$, is a *balanced collection* if for every person, i, the coalitions in $\{C\}$ of which i is a member have weights that sum to 1. Formally, $\{C\}$ is a balanced collection if there exists a set of weights, $\{w_c\}$, such that

$$\sum_{C \in \{C\} \cap C(i)} w_c = 1 \quad \text{for all } i.$$

To illustrate this definition of a balanced collection, notice that for a three-person game, these three collections are each balanced:

Collection	Weights
Any individual coalition	$w_c = 1$ for all C
All 1-person coalitions	$w_{\{i\}} = 1$ for all i
All two-person coalitions	$w_{\{i,j\}} = \frac{1}{2}, i, j \in \{1, 2, 3\}$

Here, though, are examples of collections for a three-person game that can never be balanced, no matter what weights we choose:

a one-person coalition that overlaps a two-person coalition
a two-person coalition and the coalition of the whole
two two-person coalitions

Theorem 8.4, proved by Herbert Scarf, uses this notion of balanced collections to establish a necessary and sufficient condition for a game to have a nonempty core:

8.4.   (Scarf): *A game in characteristic function form has a nonempty core if and only if, for every balanced collection, $\{C\}$, if $\mathbf{u} \in v(C)$, $C \in \{C\}$, then $\mathbf{u} \in v(N)$.*

Instead of presenting a formal proof of this result, it is easier for us to demonstrate its logic by referring to the previous two examples of three-person, characteristic-function form games. First, consider the set of weights,

$$w_{\{1,2\}} = w_{\{1,3\}} = w_{\{2,3\}} = \frac{1}{2}, w_c = 0 \quad \text{otherwise.}$$

Since the sum of weights for the coalitions in which any given person is a member sums to 1, the collection $\{\{1, 2\}, \{1, 3\}, \{2, 3\}\}$ is balanced. Equation sets such as (8.8) and the subsequent check to see if the implied utility sums are no greater than the value $v(N)$ become equivalent to testing, for the special case of transferable utility, whether "**u** is in $v(C)$ for every C in a balanced collection" implies "**u** is in $v(N)$." Other weights yield different balanced collections that we must check. For example, if $w_{\{1\}} = 1$, and if all other weights are zero, then the coalition of 1 by himself is balanced. We know from superadditivity, however, that if **u** is in $v(1)$, then **u** is in $v(N)$. Indeed, a paper and pencil exercise shows that for three-person superadditive games, the only balanced collection that constrains the utility 3-tuples that might be in the core is the set of all two-person coalitions.

Our analysis of the garbage game in Section 8.3 proceeded in a similar fashion. Because this game involves more than three players, there are a great many balanced collections (for example, with weights $w_{\{1,2\}} = w_{\{5,6\}} = 1$, $w_c = 0$ otherwise, $\{\,\{1, 2\}, \{5, 6\}\,\}$ is balanced), but because we are trying to show that that game has no core if $n > 2$, all we must find is one balanced collection that violates the conditions of Theorem 8.4. This simplicity in testing is the advantage of a theorem that establishes a necessary condition. Hence, by assigning a weight of $1/(n - 1)$ to each coalition with $n - 1$ members, we find that the set of all such coalitions is a balanced set. Equation set (8.6) and its implication, (8.7), develop the consequence of **u** being in $v(C)$ for each member of a balanced collection. Subsequent algebra shows the consequence of the requirement that **u** be in $v(N)$, namely, that $n \le 2$.

Market games

For a deeper understanding of how we can apply Theorem 8.4, consider the example of a simple exchange economy, which we present in Chapter 7. This is more than a simple example, however, since we can cast much of modern microeconomic theory in terms of cooperative game theory, and it is important that we understand why an exchange economy has an equilibrium and why that equilibrium corresponds to the core. We begin by letting $\mathbf{x}_i^*$ denote person i's initial endowment of some set of goods. Thus, $\mathbf{x}_i^*$ is a vector, but in the following discussion we need not identify its specific elements. Let $v(C)$ be the set of utility payoffs that members of C can attain if they exchange the goods in their possession among themselves. That is,

$$v(C) = \{\mathbf{u} \mid u_i \le u_i(\mathbf{x}_i), i \in C, \text{ where } \sum_{i \in C} \mathbf{x}_i \le \sum_{i \in C} \mathbf{x}_i^*\}. \quad (8.9)$$

To see whether this game has a core, we limit discussion to a three-person example, so that the only balanced collection of interest to us is the set of two-person coalitions. Thus, the core exists if and only if, whenever $\mathbf{u}' = (u_1', u_2', u_3')$ is in $v(1, 2)$, $v(1, 3)$, and $v(2, 3)$, it is also in $v(1, 2, 3)$.

Our analysis of an exchange economy proceeds in three steps. First, we assume that $\mathbf{u}'$ is in $v(i, j)$. Second, we derive the consequences of this assumption in terms of what conditions must hold true about $\mathbf{u}'$. Finally, we show that these consequences imply that $\mathbf{u}'$ is in $v(N)$.

From the definition of $v(C)$, we know that $\mathbf{u}'$ is in $v(1, 2)$ if 1 and 2 alone can trade to the outcome $\mathbf{x} = (\mathbf{x}_1, \mathbf{x}_2)$, in which 1 gets $u_1(\mathbf{x}_1)$ and 2 gets $u_2(\mathbf{x}_2)$ and in which, from the definition of the characteristic function,

$$u_1' \leq u_1(\mathbf{x}_1) \text{ and } u_2' \leq u_2(\mathbf{x}_2). \tag{8.10a}$$

That is, $\mathbf{u}'$ is in $v(1, 2)$ if persons 1 and 2, by trading their initial endowments between themselves, can find a trade such that each of them gets u_1' and u_2', respectively. Similarly, $\mathbf{u}'$ is in $v(1, 3)$ and $v(2, 3)$ if there exists an allocation $(\mathbf{y}_1, \mathbf{y}_3)$ between 1 and 3, and an allocation $(\mathbf{z}_2, \mathbf{z}_3)$ between 2 and 3, that satisfy

$$u_1' \leq u_1(\mathbf{y}_1) \text{ and } u_3' \leq u_3(\mathbf{y}_3) \tag{8.10b}$$
$$u_2' \leq u_2(\mathbf{z}_2) \text{ and } u_3' \leq u_3(\mathbf{z}_3) \tag{8.10c}$$

To ensure that these pairwise trades are feasible, we must keep the following constraints in mind, since they ensure that no pair of players exceeds the summed initial endowments of that pair:

$$\mathbf{x}_1 + \mathbf{x}_2 \leq \mathbf{x}_1^* + \mathbf{x}_2^* \tag{8.11a}$$
$$\mathbf{y}_1 + \mathbf{y}_3 \leq \mathbf{x}_1^* + \mathbf{x}_3^* \tag{8.11b}$$
$$\mathbf{z}_2 + \mathbf{z}_3 \leq \mathbf{x}_2^* + \mathbf{x}_3^* \tag{8.11c}$$

But then,

$$\left(\frac{\mathbf{x}_1 + \mathbf{y}_1}{2}, \frac{\mathbf{x}_2 + \mathbf{z}_2}{2}, \frac{\mathbf{y}_3 + \mathbf{z}_3}{2} \right) \tag{8.12}$$

must be a feasible trade among all three persons, since the sum of the resources here is feasible. That is, if we divide both sides of equations (8.11a–c) by 2, then we get

$$\frac{(\mathbf{x}_1 + \mathbf{x}_2)}{2} \leq \frac{(\mathbf{x}_1^* + \mathbf{x}_2^*)}{2}$$

$$\frac{(\mathbf{y}_1 + \mathbf{y}_3)}{2} \leq \frac{(\mathbf{x}_1^* + \mathbf{x}_3^*)}{2}$$

$$\frac{(\mathbf{z}_2 + \mathbf{z}_3)}{2} \leq \frac{(\mathbf{x}_2^* + \mathbf{x}_3^*)}{2}$$

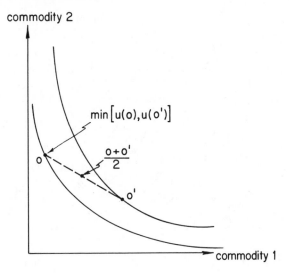

commodity 2

$\min\left[u(o), u(o')\right]$

$\dfrac{o + o'}{2}$

o

o'

commodity 1

8.6 Convex preference sets.

Summing these three inequalities yields

$$\frac{x_1 + x_2 + y_1 + y_3 + z_2 + z_3}{2} \leqslant x_1^* + x_2^* + x_3^*$$

Hence, $(x_1 + x_2 + y_1 + y_3 + z_2 + z_3)/2$ is less than the summed resources of the three players, and, therefore it is a feasible sum. Thus, an allocation that gives player 1 the amount $(x_1 + y_1)/2$, player 2 the amount $(x_2 + z_2)/2$, and player 3 the amount $(y_3 + z_3)/2$ is feasible.

That (8.11a–c) represent a feasible allocation among the three players tells us that its corresponding utility 3-tuple must be in $v(1, 2, 3)$, because $v(1, 2, 3)$ contains the utility 3-tuples from every feasible trade among the three persons. That is, the following 3-tuple is in $v(1, 2, 3)$:

$$(u_1, u_2, u_3) = \left(u_1\left(\frac{x_1 + y_1}{2}\right), u_2\left(\frac{x_2 + z_2}{2}\right), u_3\left(\frac{y_3 + z_3}{2}\right)\right).$$

Our final step is to show that $\mathbf{u}'$ is in $v(1, 2, 3)$. Whether or not this condition holds true depends on the shape of individual utility functions with respect to the commodities that people are trading. Suppose, then, that indifference contours all look like those in Figure 1.7b, so that preference sets are convex. Figure 8.6 shows that for preferences of this sort the following inequality must be satisfied whenever o is preferred or indifferent to o',

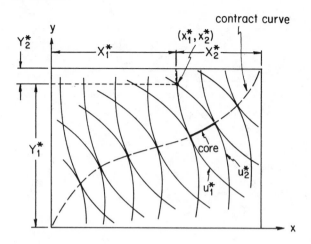

8.7 Edgeworth box.

$$u\left(\frac{o + o'}{2}\right) \geq \text{minimum of } [u(o), u(o')].$$

Hence, the first inequality in each of the following expressions follows from the convexity of preference sets. The second inequality follows from expressions (8.10a–c):

$$u_1\left(\frac{\mathbf{x}_1 + \mathbf{y}_1}{2}\right) \geq \min[u_1(\mathbf{x}_1), u_1(\mathbf{y}_1)] \geq u_1'$$

$$u_2\left(\frac{\mathbf{y}_2 + \mathbf{z}_2}{2}\right) \geq \min[u_2(\mathbf{x}_2), u_2(\mathbf{z}_2)] \geq u_2'$$

$$u_3\left(\frac{\mathbf{y}_3 + \mathbf{z}_3}{2}\right) \geq \min[u_3(\mathbf{y}_3), u_3(\mathbf{z}_3)] \geq u_3'$$

Thus, because we have already established that $\mathbf{u} = (u_1, u_2, u_3)$ is in $v(1, 2, 3)$, and because $u_i' \leq u_i$ for $i = 1$, 2, and 3, it follows from the definition of the characteristic function that $\mathbf{u}'$ is in $v(1, 2, 3)$. And since $\mathbf{u}'$ is an arbitrary utility n-tuple that is in $v(i, j)$ and $v(N)$, we conclude from Theorem 8.4 that this simple exchange economy has a core.

A graphic two-person example illustrates the core of an exchange economy. Figure 8.7, called an *Edgeworth box* diagram, represents the preferences of two traders. The horizontal length of the box equals the total amount of a good, say X, that the two persons jointly possess. Hence, $X_1^* + X_2^* = X^*$, where X_i^* is person i's initial holdings of good

X. The height of the box represents the total amount of a second good that they possess. Similarly, $Y_1^* + Y_2^* = Y^*$. In terms of our previous notation, then, $\mathbf{x}_1^* = (X_1^*, Y_1^*)$ and $\mathbf{x}_2^* = (X_2^*, Y_2^*)$. Moving from left to right in this box corresponds to 1 increasing the amount of good X that he holds at the expense of the amount of this good that 2 holds. Similarly, moving from bottom to top corresponds to 1 increasing his holdings of good Y at the expense of the amount of Y that 2 holds. Person 1, then, prefers to move to the upper right-hand corner of this box, whereas person 2 most prefers the lower left-hand corner (2's indifference contours are flipped over from their usual representation). The dashed line represents the contract curve between these two persons, the set of all Pareto-optimal outcomes.

This figure also graphs an initial endowment point, $(\mathbf{x}_1^*, \mathbf{x}_2^*)$. Thus, by refusing to trade, neither person can do worse than attain the utility associated with the indifference curve passing through this initial endowment point, so that u_1^* is 1's security value while u_2^* is 2's security value. The solid part of the contract curve denotes outcomes that satisfy the individual traders' security values (outcomes that are in $v(1)$ and $v(2)$) and that are Pareto-optimal for both of them (that are undominated with respect to the "coalition of the whole"). These outcomes correspond to the core.

8.5   Committees with a single issue

Despite their apparent dissimilarities, Theorems 8.1 and 8.4 are closely related. Theorem 8.1 equates the core of a strong simple majority-rule voting game with the intersection of the Pareto-optimals of all minimum-winning coalitions. If this intersection is empty (if no outcome lies in this intersection), then the core is empty. But since the coalition of the whole can guarantee everything, this condition is equivalent to saying that the core is nonempty if and only if the intersection of the Pareto-optimals of all minimum-winning coalitions is in $v(N)$. Notice, moreover, that for a simple voting game we can always find a set of voting weights such that the set of all minimum-winning coalitions is a balanced collection. Hence, for Theorem 8.4 to hold requires the same conditions as does 8.1. The only apparent discrepancy is that 8.4 requires looking at all possible balanced collections and not simply those that involve minimum-winning coalitions. For strong simple games, however, these additional requirements are redundant. Larger than minimum-winning coalitions cannot guarantee more than minimum-winning coalitions, so if we satisfy the conditions of Theorem 8.4 (or

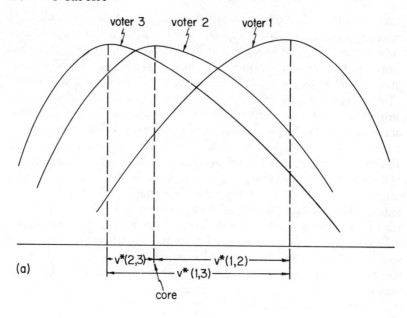

(a)

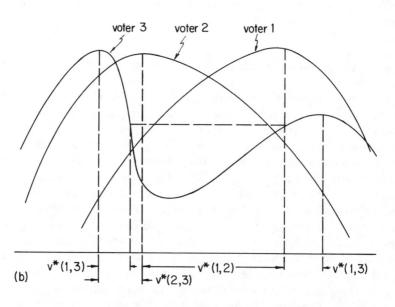

(b)

8.8(a) The core with single-peaked preferences. (b) No core with nonsingle-peaked preferences.

equivalently, of Theorem 8.1) for all minimum-winning coalitions, then we satisfy them for *any* other collections of coalitions. We can interpret Theorem 8.4, then, as a general result for all characteristic-function form games, such that Theorem 8.1 is its corollary for strong simple voting games.

To illustrate this discussion with a specific spatial voting example, consider Figure 8.8a, which depicts three voters' single-peaked preferences. Theorem 8.1 states that the core corresponds to the intersection of the outcomes that are Pareto-optimal for each of the minimum-winning coalitions. Figure 8.8a displays the three Pareto-optimal sets $v^*(1, 2)$, $v^*(1, 3)$, and $v^*(2, 3)$, and it is apparent that the median preference is the unique intersection of these three sets. Hence, the median and the core coincide. Figure 8.8b, on the other hand, shows preferences that are not simultaneously single peaked and for which a Condorcet winner does not exist (this figure reproduces Figure 6.12, and the reader should refer to our discussion of it in Chapter 6). This figure also portrays the Pareto-optimal outcome sets for each of the three minimum-winning coalitions, and it shows that there is no common intersection. Hence, there is no core.

We might ask at this point why we require the concept of the core to derive these conclusions. After all, Theorems 4.3 and 4.3′ establish the existence of a Condorcet winner and a transitive social preference without referring to characteristic functions and the like. But a Condorcet winner simply defines a property that an outcome may or may not satisfy with respect to the majority-rule relation. Yet satisfying that property provides no valid prediction about the game's outcome. Plurality-rule systems with three or more outcomes, or various configurations of preferences and the Borda count, for example, do not always yield Condorcet outcomes. Thus, to say that a Condorcet winner exists describes only in part the social preference order under majority-rule. We can no more use that description to assert a prediction than we can use, say, Theorem 2.2 to argue that if such a winner does not exist, then the universe is in chaos. Only the analysis of specific institutions can convert descriptions of social preference orders into predictions about eventual outcomes. Chapter 4 considers the specific institution of two-candidate, majority-rule elections, and Chapter 6 looks at committees in which various procedures such as agendas and a prohibition against communication constrain voting. We have learned thus far that those institutions yield Condorcet winners as final outcomes if such a winner exists. The institution that we consider here is simple majority rule in an otherwise procedure-free committee. Hence,

The correspondence of the core to the Condorcet winner is a prediction that this winner will emerge as the eventual outcome in "unfettered" majority-rule committees.

8.6   An extended example: the *Genossenschaften*

Alvin Klevorick and Gerald H. Kramer's analysis of a method of pollution control in the Rhine River basins of Germany illustrates a sophisticated application of the core to committees that are concerned with a single issue. These basins are divided into *Genossenschaften,* committees that regulate effluent taxes. Each committee is responsible for setting a tax for a specific geographical region, to be paid by firms and communities alike on the basis of how much pollution each disposes into the region's rivers. To simplify presentation, we focus on a specific firm, a specific household, and the pollution control board that sets the effluent tax. First, suppose that a firm uses an amount L of labor (measured, for example, in terms of hours worked per week or number of employees), and as a byproduct of its operations, it produces G units of garbage that it dumps into a river. The function $F(L, G)$ describes the firm's total product output, in which F increases as L increases. Suppose, further, that if we hold L constant, then production, F, declines if the firm emits less garbage, G, and if G increases, then for a fixed amount of labor, L, it can produce a higher level of output, F. If the marginal cost of labor is w dollars per unit, and if the firm's product sells at p dollars per unit, then the firm's profits are

$$\text{profits} = pF(L, G) - wL - tG,$$

where t is the tax that the pollution control board imposes. Firms, then, derive no intrinsic pleasure from clean water. They are concerned about pollution only because of the tax that they must pay if they pollute at a given level. Assuming that firms maximize profits, a firm prefers the smallest tax possible, zero. We should keep in mind for our subsequent arguments, though, that the amount of garbage it emits declines as its tax, t, increases; so we can express G as a function of t, $G(t)$, that decreases as t increases.

The second key actor in this scenario is the household. Unlike firms, we suppose that households cannot control the amount of garbage that they dump into the rivers. Instead, they care about the quality of the rivers for recreation. We represent a household's utility function, then, by the function $u(I, S)$, where I is its disposable after-tax income, S is the total waste discharged into the rivers, and u increases with I but decreases with S. The household's after-tax income, I, is simply its pretax income, Y, minus the effluent tax rate, t, times the pollution that the household produces, H. If $tH > Y$, then we set $I = 0$.

We now model the third key actor, the pollution control board. Letting n denote the total number of households, the total waste discharged into the water becomes

$$J(t) = \sum G(t) + nH,$$

and the revenue that the board collects equals

$$R(t) = tJ(t) = t\sum G(t) + ntH.$$

We take the sum in each expression over all firms. The pollution control board uses its revenue, $R(t)$, to reduce the pollution in the rivers by treating the water. The amount of waste that the board can remove from the rivers is $Q(R(t))$, where Q is a function that represents water-treatment technology, and Q increases with R. Hence, the river's pollution level equals the amount that firms and households emit, J, less the amount that the pollution board removes, Q, or,

$$S(t) = \begin{cases} J(t) - Q(R(t)) = J(t) - Q(tJ(t)) \text{ if } J(t) > Q(tJ(t)) \\ 0 \text{ otherwise.} \end{cases}$$

With a fixed treatment technology, the sole variable that the board can control, then, is the tax t.

To this point, this description of the Genossenschaften seems unexceptional, but it is here that the institution's designers have become imaginative. The firms and households themselves, acting as a committee under majority rule, comprise the board. And each member's relative voting weight is an increasing function of the effluent charges that it pays. Ignoring the complications occasioned if $t = 0$ or if $J(t) = 0$, a representative household's voting weight equals its tax charge, tH, divided by the board's total revenues, $tJ(t)$, or $tH/(tJ(t)) = H/J(t)$. A particular firm's voting weight is $G(t)/J(t)$.

This definition of voting weights establishes an interesting problem for the board. First, suppose that the preceding assumptions are sufficient to yield single-peaked preferences over the issue of what effluent tax to set, so that for a specific distribution of voting weights, t' is the median ideal point and, hence, the core. The interesting twist here is that the selection of t' depends on the current distribution of voting weights, but t' may cause a shift in these weights. That is, t' may cause firms to adjust their outputs of pollution and, therefore, the fees that they pay, which in turn changes firms' and households' relative voting weights. This adjustment can result in a new (weighted) majority-preferred tax, say t'', that causes further adjustments. The question now

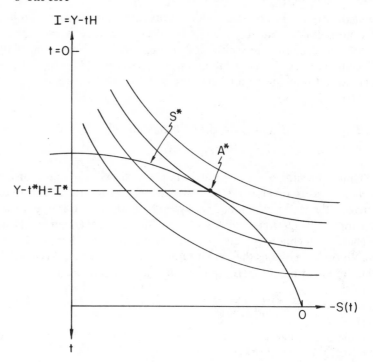

8.9 Household tradeoffs between water quality and disposable income.

becomes: Is there an equilibrium value of t that causes no further changes in voting weights? There are, then, two research questions. First, are preferences single peaked? Second, does a fixed-point tax exist, one that a specific distribution of voting weights implies and that, in turn, implies those same weights?

With respect to the first question, we know that firms' ideal points are $t = 0$, and that their profits, and thus their preferences, decline monotonically as t increases. The critical issue, then, is ascertaining a household's preferences for different tax charges. We proceed in two steps. First, recall that we define a household's utility function over two commodities, its after-tax income, $I = Y - tH$, and the quality of rivers for recreation, $-S$. Suppose that indifference curves for these two variables look like the ones that we posit in Chapter 1 in modeling tradeoffs between two commodities (see Figure 1.7b, for example). Figure 8.9 graphs an equivalent set of indifference curves. Notice that both axes of this figure, after-tax income and water quality, are functions of t. If $t = 0$, then water quality is minimized but I is maximized, whereas if t is set to expropriate all household wealth, then water quality, $-S$, is maximized. A particular tax rate, then, yields a specific amount of disposable income and a specific level of water quality. Thus, if we graph $I = Y - tH$ (and by implication, $-t$) against $-S(t)$, then we can identify the feasible outcome that places the household on its highest possible indifference curve.

Bypassing some mathematics, we note simply that by carefully selecting our assumptions (specifically, that $G(t)$ is a strictly decreasing convex function of t for each firm and that the board's water-treatment technology, $Q(R)$, is a strictly concave and increasing function of R), we are assured that the relationship between feasible I and feasible $-S(t)$ looks like the concave function that Figure 8.9 graphs. Hence, the region bounded above by the curve S^*, denotes the technologically feasible levels of water quality for the various values of t.

Notice that our household has a unique ideal point on S^* at A^*, and that its preferences decline on S^* as we move along this curve in either direction from A^*. Furthermore, movement along S^* corresponds to moving up or down on the I axis, which implies a single-peaked preference over I. But $I = Y - tH$, which also implies a single-peaked preference over t. Hence, if I^* is the disposable income level that corresponds to A^*, then t^*H is the amount that it prefers to spend on cleaning the water, and $t^* = (Y - I^*)/H$ is its ideal effluent tax.

Households probably differ in the tradeoffs that each is willing to make between consumption and clean water, in which case each will prefer different effluent charges. But if our assumptions adequately describe each household, then all household preferences are single peaked over the issue t. Overall, then, preferences over t for, say, a six-firm, seven-household Genossenschaften look like the preferences in Figure 8.10.

If voting weights did not change as effluent charges change, then we would be done. We could apply Theorem 8.2 to assert the existence of a core that we know must be at the ideal point of the median voter. Since votes are weighted, the core might not equal t_1^* in Figure 8.10. If households outweigh firms three votes to one, for example, so that with 27 total votes, the core corresponds to the preference of whoever holds the 14th vote, then if the voting weights of all households are identical, household 3's ideal, t_3^*, is the core. The difficulty here, though, is that if the board chooses t_3^*, then a new voting-weight distribution may arise after firms adjust their discharge for which t_3^* is not the core. Thus, we can imagine a scenario in which the board chooses t', firms then adjust their production levels and effluent discharges so that a new set of voting

preference

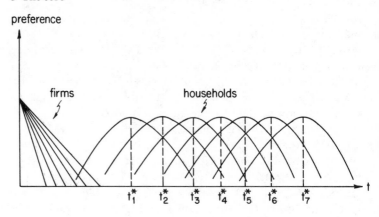

8.10 Preference in the *Genossenschaften.*

weights prevails, after which the board adjusts the tax to t'', whereupon firms readjust their production levels and effluent discharges, and so on.

To show that there is an overall equilibrium effluent tax, suppose that the tax increases from t' to t''. The single-peaked preference functions of firms and households remain unchanged, since neither depends on t. But to see that the relative voting weight of households increases, notice that the fraction of all votes that households control equals $\Sigma H/J(t)$, but the total waste discharged into the water, $J(t)$, decreases as t increases, since only firms can adjust their pollution, and they will decrease their output as t rises. Hence, as t increases, voting power shifts from firms to households. Furthermore, each household's voting weight rises in the same proportion as the total voting strength of all households. If $v_i(t)$ is the voting weight of the ith household when the tax rate is t, where $v_i(t) = H_i/J(t)$ and H_i is the pollution that household i produces, then the ratio of i's weight to j's weight, $v_i(t)/v_j(t)$, equals the constant H_i/H_j. That is, regardless of t, household i's voting weight relative to household j's is constant. Thus, if at t' all households control 60% of the vote, with, say, the 3d household controlling 10% of this total (6% of all votes), and if t' rises to t'' and all households now control 75% of the vote, then household 3's share of this total remains 10% (or 7.5% of all votes).

We use these calculations to graph t against the tax that it implies, T, after the firms adjust their decisions to t and the board assigns appropriate voting weights, to identify a fixed-point tax, a tax that implies itself. Suppose that we increase t from t' to t'', and that at t', household 2 in Figure 8.10 is the median voter. From the previous argument, we know that, after firms adjust their production, the new

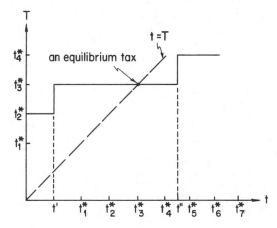

8.11 Fixed-point effluent charge.

tax t'' increases the weight of all households, with household 2's share of this total remaining constant. Thus, either household 2 remains the median voter or the median shifts upward to a household with a preference for less pollution and a greater tax. The upper bound on t occurs when households control 100% of the vote. If each household in our example emits the same amount of pollution and controls an equal share of this total, then t can be no-greater than t_4^*, the median household preference.

To see what the final equilibrium effluent tax rate looks like, consider Figure 8.11, which graphs a specific tax charge, t, on the horizontal dimension and the tax charge that t implies, T, on the vertical dimension. Thus, as t increases, the households' voting weights increase, so that, if households control more than 50% of the vote when $t = 0$, then we assume in this example that in the range $0 \leqslant t < t'$, household 2 is the median weighted preference; in the range $t' \leqslant t < t''$, household 3's preference is the median; and for $t > t''$, t_4^* is the median. The final equilibrium, though, corresponds to a fixed point, a tax that implies itself. Thus, paralleling our discussion of fixed points with reference to Figure 3.16, Figure 8.11 includes a diagonal line representing $T = t$. So the equilibrium tax rate in our example is t_3^*, which means that after all adjustments occur, household 3's preference is the agreed-upon effluent tax charge. Depending on the pollution board's technology and other parameters, of course, different equilibria might prevail, including multiple equilibria and an equilibrium at $t = 0$.

What is interesting about this model is its sophisticated use of the

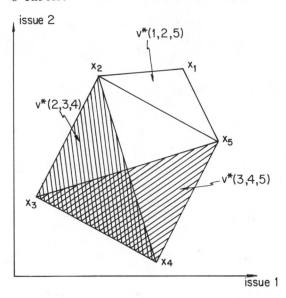

8.12 Majority-rule spatial game without a core.

concept of the core in unidimensional committees. The result that the core corresponds to a committee's median preference appears straight-forward. But when we combine this result with a specific and complex institutional structure, we derive a nonobvious result about the exist-ence of an equilibrium tax even though we permit firms to adjust their production decisions after the board announces the tax. An analyst could now begin to fill in the model by estimating the actual parameters and functions that pertain to the several Genossenschaften, to assess whether such an arrangement will work. This analysis also gives us some insight into possible strategies for firms (for example, pollute to the maximum the first year to gain control of the board if possible, then set effluent charges to zero the next year), and into policy proscriptions for avoiding unintended consequences (for example, outlawing a zero tax or requiring that initial voting weights give households more than one-half the vote).

8.7   Multiple dimensions and some experimental evidence

The analysis of the Genossenschaften is tractable because the situation concerns a single policy variable, the effluent tax charge. Thus, carefully selected assumptions ensure single-peaked preferences on the tax dimension, which in turn ensures a core's existence for any set of voting weights. For the more general case of multiple issues, however, ensuring the existence of the core is problematical. Indeed, we can interpret Theorem 8.2 as a negative result, since with m issues and circular indifference contours, a Condorcet point, and hence a core, exists only if a median in all directions exists. To see the application of this theorem in terms of the concepts that we use to define the core, consider the five-person committee in Figure 8.12. Assuming simple majority rule and equal voting weights, three-person coalitions are minimum winning. The core, then, corresponds to the outcomes that are Pareto-optimal simultaneously for all minimum-winning coalitions. For circular indifference contours, the Pareto-optimal outcomes for the coalition {1, 2, 5} coincide with the triangle formed by the ideal points of persons 1, 2, and 5, plus this triangle's interior. Similarly, the Pareto-optimal outcomes for {2, 3, 4} are bounded by the triangle formed by the ideal points of persons 2, 3, and 4, and the Pareto-optimals of {3, 4, 5} are similarly constructed. As Figure 8.12 shows, however, these three triangles do not have a common intersection, no outcome is simultaneously Pareto-optimal for these three minimum-winning coalitions. Hence, the core is empty.

A nonempty core in this situation requires a median in all directions. That is, with circular indifference contours, Theorem 4.4 establishes not only a necessary and sufficient condition for an equilibrium in two-candidate elections, but also a core in simple majority-rule committees. To see this, notice that if a median in all directions exists, then by definition every line through such a point is a median. That is, at least one-half of the ideal points lie on and to either side of the line. Hence, the Pareto-optimal portion of $v(C)$ for any minimum-winning coalition necessarily includes the median in all directions. The core is that median. (If the committee has an even number of members, then this median can be a region, in which case the core is not unique.)

To illustrate this correspondence between the core and the intersections of the $v^*(C)$'s for minimum-winning coalitions in a slightly different context, consider again the city-block indifference contours in Figure 1.8. Figure 8.13 depicts city-block indifference contours for three voters in two dimensions. Notice that if indifference contours are circles, then this arrangement of ideal points does not yield a Condorcet winner, and the cooperative majority-rule committee game has an empty core. With the square indifference contours in Figure 8.13, however, a core exists. For the minimum-winning coalition {1, 2}, $v^*(12)$ is the locus of all tangencies of 1 and 2's indifference contours. With circular indifference contours, $v^*(12)$ is necessarily the straight

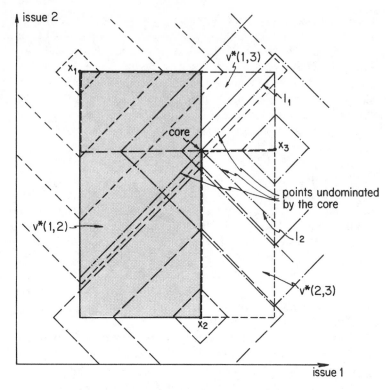

8.13 City-block preferences and the core.

line connecting the ideal points of 1 and 2, but here indifference contours are flat, and the squares all share a common rotation. Thus, $v^*(12)$ is the shaded rectangle in Figure 8.13. Similarly, $v^*(13)$ and $v^*(23)$ are rectangles, as illustrated. These three rectangles have a common intersection, and this intersection is the core.

Although we can generalize this example's implications to larger committees, provided that the indifference contours of all voters have the same orientation, we cannot generalize it to more than two issues. Nevertheless, the example illustrates a somewhat different core than the one that we get if indifference contours are circles or ellipses. If the number of voters is odd, then with circular indifference contours, the core is necessarily strong: It dominates all other feasible outcomes. For majority voting games, then, committee members can reach an outcome from any other outcome that is a strong core in a single vote. But in Figure 8.13 the core is not strong. Points along the lines l_1 and l_2, while defeated in a majority vote by points off the lines, are not defeated by

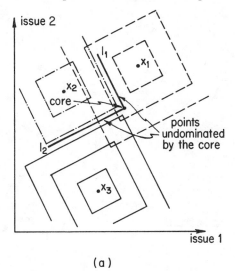

(a)

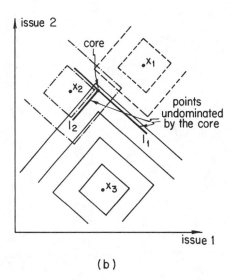

(b)

8.14 Effect of rotation on the core.

the core. Thus, it takes two votes to reach the core from any point on l_1 or l_2.

Aside from illustrating our definitions with a different kind of preference structure, the city-block metric is interesting from the perspective of the experimenter who tests the core as a predictive

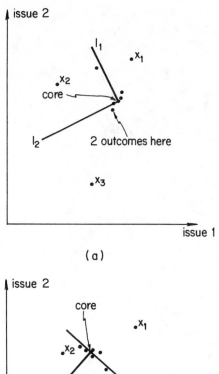

(a)

(b)

8.15 Experimental outcomes.

concept. One difficulty with circular contours is that, because the core corresponds to a median in all directions, it is central, perhaps even "fair," if everyone knows everyone else's preferences. In other words, there are several hypotheses about people's motives that are consistent with a core prevailing as the final outcome, which means that subjects of an experiment designed to test the core as a predictive concept might

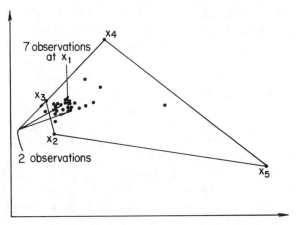

8.16 Experimental outcomes.

choose it for reasons other than the purely game–theoretic properties it satisfies. But consider Figures 8.14a–b, which differ from each other only in the rotation of peoples' indifference contours. Yet they yield quite distinct points as cores, neither of which appears to satisfy any criteria of fairness.

Reviewing some experimental evidence about the core in these circumstances, Figures 8.15a–b each summarize a series of seven experiments in which undergraduates negotiated over points on a grid, with no constraints on discussion. No formal votes were taken, and the experiment ended when a majority indicated that it had arrived at an agreeable outcome. Although subjects knew each other's (ordinal) preferences, no one knew how much money others would receive for specific points. Hence, the sole items for negotiation were positions on the two issues. As these figures indicate, all but two outcomes fall close to the core, even though subjects knew nothing about this concept. The two outcomes that are furthest from the core, however, lie near points that the core does not dominate and thus there is little incentive for at least two persons to move from them to the core.

Now consider the experimental outcomes that Figure 8.16 reports. In this instance the undergraduates' indifference contours were circles, so person i's ideal point is the core. Outcomes are dispersed somewhat more widely than are the outcomes in Figures 8.15a–b, but there is a reasonable explanation for this greater dispersion. The city-block experiments try to model as closely as possible a characteristic-function form game. Hence, to minimize transaction costs, there were no impediments to bargaining, and discussion proceeded unconstrained.

With respect to the data in Figure 8.16, however, each experiment began with a status quo point that a formal parliamentary process could alter: After a subject made a motion (which can be any point in the space), another person had to second it. But unlike the spatial experiments that Chapter 6 reports, voting here did not proceed issue by issue. Instead, the seconded motion was put against the status quo and voted on directly. If it won, then it became the new status quo, and the experiment proceeded until there was a formal motion to end the game that was seconded and approved by a majority. Despite the slight friction that this process generates in bargaining, as Figure 8.16 shows, the core accounted for much of the experimental committees' bias.

These experiments increase our confidence that similar predictions are possible in "real" committees. Admittedly, it is unlikely that our information about preferences will be as good there as it is if preferences are induced in an experimental environment. But the core, in conjunction with our theoretical explorations into strategic revelation of preference, suggest that as our theory develops, our problems should concern measurement, not measurement and theory simultaneously.

8.8   Logrolling and some experimental ambiguities

The preceding discussion does not imply that all experimental evidence supports the core's predictions. Indeed, some experimental vote-trading games offer an important note of caution about the formal modeling of political processes. Recall from the analysis of Figure 8.3 that the core allows a full cooperative analysis of vote trading, and that Theorem 8.2 tells us that if preferences yield a Condorcet winner, which corresponds to the sincere voting outcome with separable preferences (see Theorem 2.4), then that winner is the cooperative equilibrium that the core predicts. To check whether a vote-trading game has a core, we simply check whether the sincere voting outcome is undominated. The sincere voting outcome for the game in Figure 8.3 is (1, 1, 0), and as the characteristic functions in Figure 8.4 reveal, this outcome is undominated. Hence, (1, 1, 0) is the core.

Experimental vote-trading games with a core, however, reveal some instructive empirical irregularities. Figure 8.17 presents the preferences of five legislators across five bills, in which if preferences are separable across bills, and if the utility that a person associates with the defeat of any specific bill is zero [for example, the utility that legislator 1 associates with the outcome (1, 0, 1, 0, 1) is $10 + 0 + 5 + 0 + 5 = 20$], then the outcome from sincere voting, (0, 0, 1, 0, 1), is a core. But in a series of 42 experiments using undergraduates, in which bargaining

Legislator	Bill 1	Bill 2	Bill 3	Bill 4	Bill 5
1	10	−2	5	4	5
2	−2	10	5	−5	−4
3	4	−8	5	3	8
4	−8	4	−3	−5	8
5	−5	−5	−4	−10	−4
Sincere outcome	Fail	Fail	Pass	Fail	Pass

8.17 Utilities of legislators from passage of five bills.

again remained unrestricted and in which the only negotiable commodity is one's vote on a specific bill (that is, the experiment did not permit discussion of the money used to induce preferences), the core prevailed only 47% of the time.

Although a variety of outcomes prevails, the most common event associated with the core's failure is that persons 1 and 2 trade votes on bills 1 and 2. Persons 3 and 4 could trade to negate this agreement between 1 and 2, but this second trade did not always occur. Other trades sometimes occurred, but the trade between persons 1 and 2 commonly initiated an experiment in which something other than the outcome from sincere voting prevailed.

Suppose that we conduct this experiment differently. There are $2^5 =$ 32 distinct pass–fail permutations (outcomes) with five bills. If we eliminate 6 of the more obviously unreasonable permutations, then we are left with 26 possibilities, which we can label A through Z. And instead of inducing preferences indirectly over these 26 possibilities by assigning payoffs to bills, suppose that we induce preferences over alternatives A through Z directly. Specifically, we can let these preferences correspond to the preferences that we derive from Figure 8.17. For example, if the core $(0, 0, 1, 0, 1)$ is associated with the letter G, then we assign subject 1 a payoff of $0 + 0 + 5 + 0 + 5 = 10$ for G. In this way we can formulate a game over "prepackaged" bills that is equivalent mathematically to a game with the preference structure in Figure 8.17.

Despite this mathematical equivalence, subjects now choose the core 100% of the time, compared with 47% of the time in the vote-trading scenario. This difference shows dramatically that although two situations may be mathematically equivalent, people may not respond to this equivalence. Instead, they may use various heuristics to simplify decision tasks, and these heuristics yield different outcomes in different situations. In a vote-trading experiment, calculating payoffs and learning the preferences of others is costly. To reduce such costs, a simple heuristic such as "trade votes on bills of little personal consequence for votes on bills of primary significance" seems reasonable. If the players take all advantageous trades, then outcomes should converge or otherwise cycle back to the core. But if two or more persons fail to negotiate such a trade, then an outcome other than the core may prevail. With prepackaged bills, a simplifying heuristic is unnecessary, and the players can agree to the core directly.

Although we can explain away various experimental anomalies, we should be prepared for the likelihood that the real world will present even more complex anomalies. In analyzing a legislature, for example, we should be aware that people might deviate from theoretical predictions because of the way in which alternatives are posed. Different structures, although equivalent mathematically, may present legislators with different analytic tasks, which impose different costs. If these costs are high, then procedural impediments to unrestricted bargaining, such as issue-by-issue voting, a formal committee structure, and rules of agenda formation, may affect outcomes. And when a committee considers complex legislation, the final outcome, even with the possibility of coalitions, may depend on whether the members vote on the components of legislation one at a time or together, packaged into complete bills.

8.9   Extraordinary majorities

Failing to observe the negotiation of all advantageous trades in a vote-trading experiment should make us aware of another possibility, namely, that various coalitions may be unable to form either because of formal or informal rules, or because of the costs of maintaining, say, large coalitions. If we ask what effect such prohibitions have on the core, our answer would depend generally on the situation being modeled. There is one kind of exclusion, however, that admits of a general analysis. Notice that most of the formal voting examples in this volume assume simple majority rule. But some committees use extraordinary majorities when making unusually important decisions, such as when Congress uses a two-thirds rule to vote on constitutional amendments. Extraordinary majorities have the effect of restricting the set of winning coalitions while simultaneously admitting a new kind of coalition, a blocking coalition (assuming that n is odd since, if n is even, blocking coalitions exist even under simple majority rule).

The inherent bias of extraordinary majorities is that they make it more difficult to upset the status quo, because blocking coalitions seem easier to form than winning coalitions. We know that the status quo figures prominently in committee voting if full cooperation is precluded and if agendas or other committee procedures constrain voting. The status quo loses its importance in strong simple games, however, because winning coalitions can do anything. But if the committee uses an extraordinary majority, then the character of the core changes, and the status quo again is relevant.

To state a general theorem about extraordinary majorities that reveals some of the effects of these majorities on the core, let X be a subset of O, and suppose that the elements of X share the following property: For every $o \in X$, no $o' \in O$ exists such that o' dominates o with respect to any winning coalition, C. For simple games, Theorem 8.1 identifies X, except that now the set of minimum-winning coalitions, W^*, contains all coalitions larger than the minimum majority required to pass a motion, say k'. With mathematics that are too complex to review here, Joseph Greenberg proves Theorem 8.5 about X: Letting $k = k'/n$,

8.5 (Greenberg) *Assuming a simple n-person game with k-majority rule, X is nonempty for every set of preferences over O if and only if*
    1. $k > m/(m + 1)$ if O is a closed, convex, and bounded subset of m-dimensional Euclidean space, and if preference sets are convex,
    2. $k > (m - 1)/m$ if O is a finite set of m alternatives.

Hence, with two issues, if we use anything greater than two-thirds rule, then X is necessarily nonempty, while with three issues, X is guaranteed to be nonempty if $k > \frac{3}{4}$. And in the case of a finite number of alternatives, then with but five alternatives we require something more than a four-fifths rule to guarantee a nonempty set, X. This does not mean that X is nonempty with weaker rules. But Theorem 8.5 states that if we want to *ensure* that X is nonempty for *every* possible preference profile, then we must satisfy condition (1) or (2).

To illustrate X, consider Figure 8.18, which has no core if $k = \frac{1}{2}$ (simple majority rule). For $k = \frac{4}{5}$, however, X consists of all points in the vertically hatched interior pentagon formed by the appropriate contract lines. These are the points that are Pareto-optimal for every four-person coalition.

Although the definition of the set X is similar to that of the core, X is not a core, because its definition takes account of winning but not

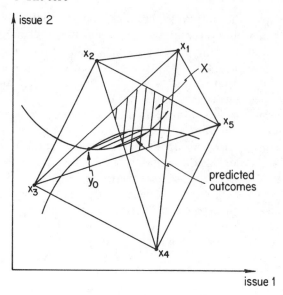

issue 2

issue 1

8.18 A core with 4/5 rule.

blocking coalitions. There may be elements of X that, although undominated with respect to any winning coalition, nevertheless are
dominated with respect to a blocking coalition. Keeping in mind that
blocking coalitions can ensure only the status quo, consider Figure 8.18
again and suppose that the point y_o is the status quo. Assuming
four-fifths rule, notice first that player 3 should vote against any move
from y_o into X, whereas players 1 and 5 prefer everything in X to y_o.
Thus, 3 prefers to block outcomes in X, while 1 and 5 would not join
such a coalition. The critical players are 2 and 4, and if a proposed
change makes *either* of them worse off, then they can coalesce with 3
and form a two-person blocking coalition to maintain y_o. Hence, the
only outcomes that can defeat y_o in the direction of X are those in the
hatched region formed by the indifference contours of persons 2 and 4
through y_o, since no other points can gain the unanimous support of 2
and 4. Thus, the intersection of these points with X, the cross-hatched
region in the figure, are the outcomes that will not be blocked and that a
minimum-winning coalition unanimously prefers to y_o. These outcomes
are the core.

Figure 8.19 reproduces these five ideal points, but it illustrates a status
quo, y_o, that is especially unattractive. In this instance the members of
some four-person coalition unanimously prefer a great many points to

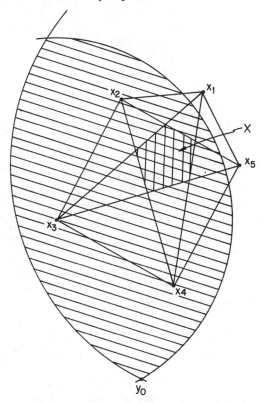

8.19 An irrelevant status quo.

y_o, and X is a subset of these points. The horizontally hatched region, for example, denotes all points that $\{1, 2, 3, 5\}$ prefers unanimously to y_o. The status quo, then, is irrelevant to the final outcome, and all elements of X form the core.

This discussion offers a perspective for studying the effects of the status quo in committees that do not have impediments to cooperation, aside from the constraints that the definition of winning imposes. For example, if y_o is in X, then y_o is the unique prediction, which is what makes constitutional amendments difficult to pass. Of course, we might want to study other kinds of constraints, such as the historical prohibition (until recently) against including the French Communist party in government coalitions. We can incorporate such constraints readily into any analysis, even if we must add them in a less than general way to model a specific circumstance.

	c_1		c_2		
	b_1	b_2	b_1	b_2	
a_1	2	1	0	3	← core
a_2	4	4	1	1	

8.20 Frequency of experimental outcomes.

8.10   The core and ambiguities with $v(C)$

The preceding sections assume that the characteristic function provides an appropriate summary of strategic possibilities and imperatives. Indeed, with the exception of the simple exchange economy in Section 8.4, our focus has been on simple games, in which case the definition of $v(C)$ (and correspondingly, of the core) remains unambiguous. But consider the game in Figure 8.1, and recall our discussion in Chapter 7 (Section 7.7) of the flower bed example, where the definition of $v(C)$ becomes ambiguous if we consider the rational responses of players excluded from a coalition. The same ambiguity holds true with the game in Figure 8.1. If we use either of the alpha or beta forms of $v(C)$, then the corresponding game has a unique core, the 3-tuple (30, 30, 40). But if persons 1 and 2 announce that they will choose a_1 and b_1, respectively, then 3's only rational response is c_1, which yields the utility outcome (65, 50, 20). Thus, it appears that 1 and 2 can force outcomes away from the core. And by shifting subsequently to a_2, player 1 can secure (70, 55, 30), which Pareto dominates (65, 50, 20). But now 3 has an incentive to shift to c_2, and 2 has an incentive to shift to b_2. Hence, although the core predicts a unique outcome, there remains ample reason to believe that bargaining will be more than perfunctory in this game.

In an experimental application of this game, in which subjects bargain freely for a predetermined length of time without the opportunity to engage in sidepayments, the core prevails only 3 times out of 16 trials. Figure 8.20 reports the actual distribution of outcomes, and the deviations from the core suggest that subjects do take rational responses into consideration. A review of the discussions among subjects confirms that we can attribute the two outcomes at (a_1, b_1, c_1) directly to the realization that 3 is unlikely to choose c_2 if 1 and 2 coalesce. Similarly, the four outcomes at (a_2, b_1, c_1) represent agreements among all three persons to shift to a Pareto-preferred outcome after the coalition $\{1, 2\}$ seems a fait accompli to 3. Similarly, the four outcomes at (a_2, b_2, c_1) result from the coalition $\{2, 3\}$ and the rational response of person 1, which 2 and 3 seem perfectly willing to bring to 1's attention. In short, then, the core fails to predict these experimental outcomes adequately because a characteristic function based on security values fails to model the game's strategic features adequately.

For another example, consider a three-person game in which two small persons, 1 and 2, and one large strong person, 3, are negotiating over how to divide \$100 that they would be paid if they move a refrigerator. No person alone can move the refrigerator, so $v(i) = 0$ for $i = 1, 2$, and 3. Indeed, the only coalitions that can complete the job successfully are those in which 3 is a member. Hence, $v(1, 2) = 0$, but $v(2, 3) = v(1, 3) = v(1, 2, 3) = \$100$. Player 3 is uniquely positioned with this characteristic function, since only those coalitions of which he is a member have any value. The unique outcome in the core to this game, the 3-tuple $(0, 0, 100)$, reflects 3's monopoly position.

Suppose, however, that 1 and 2 agree not to negotiate separately, but instead they agree to divide equally anything that they can get. If they can change the game in this manner, then the new core becomes,

$$\{(u_1, u_2, u_3) \mid u_1 = u_2 = c,\ u_3 = 100 - 2c,\ 0 \leqslant c \leqslant 50\},$$

which includes outcomes that are more advantageous for 1 and 2 then did the original core.

Experiments and examples such as these warn us that we must treat our analytic representations with care. Although the preceding example illustrates a market with two buyers and one seller (of labor), an economist might argue that as the numbers of buyers and sellers increase, competitive forces dissolve *proto-coalitions* such as $\{1, 2\}$, coalitions that people form to affect the coalitions that finally prevail. Despite this appeal to competitive forces, the preceding example does disturb some game theorists and economists, although some mistakenly believe that it indicates a problem with the core, rather than with $v(C)$. Their argument is that the core is not self-enforcing, in the sense that as people learn the implications of the core, some might change their actions to avoid the outcomes that it predicts. Actually, what is changing is the game being played. But if players can change the game to some other form, then it must be that the original form misrepresented or inadequately captured strategic possibilities.

Political scientists should be sensitive to this problem because politics often concerns the disruption or revision of competitive forces by changes in the game. Coalitions such as oil cartels, consumer lobbies, and labor unions form, receive government sanction, and benefit from the actions of political entrepreneurs, not because they have intrinsic value as represented by what they can secure their members, but because they are proto-coalitions that change the strategic character of a situation.Theorizing about politics can never be immune from the need to be vigilant about such possibilities. Unless we model all feasible future histories, any representation must exclude strategic possibilities. And such exclusions place us at the mercy of decision makers who choose a branch that is not included in our description of the extensive form or who invent a branch we had not anticipated.

The preceding examples suggest, though, that the ambiguities associated with $v(C)$, if combined with a particular solution concept such as the core, can guide the study of how people prefer to structure the games that they play. Laws prohibiting cartels and closed shops, for example, support one characteristic function, while the repeal of such laws or the establishment of regulatory agencies to maintain a cartel establish a different one. Exploring the ambiguities of a situation's strategic representation, then, is very much part of theorizing about politics.

8.11   Summary

Although we cannot guarantee the general existence of the core (indeed, it is certainly or almost certainly empty for large classes of games), we should not interpret this problem to mean that the core is flawed. Such a conclusion might be appropriate if we adhered to some rigid criterion of precise predictability of outcomes, but nonexistence also tells that the situation being modeled displays a fundamental instability that makes precise prediction inherently difficult or impossible. Whether we must add finer details to our description of the situation or whether we must invent auxiliary solution hypotheses to predict action, the nonexistence of the core tells us that the political process in question does not follow the path of a simple journalistic accounting of events.

Actually, though, the core is not a single concept. Its meaning and interpretation depend on the implicit or explicit characteristic function that we use to model a game. We discuss two alternatives, the alpha and the beta core. But there are other alternatives, such as the use of a characteristic function derived from assumptions about how opposing coalitions act in their own interests. Some games have a core under one definition of $v(C)$ but not under another definition. For example, the three-person flower bed game from Chapter 7 (section 7.7) has a core if we define $v(C)$ in terms of security values. In this instance the value of the coalition $\{1, 2\}$, for example, is $\{\mathbf{u} \mid u_1 \leqslant 10, u_2 \leqslant 2\}$, and the equitable outcome $(6, 6, 6)$ is included in the core's elements. But if persons 1 and 2 assume that 3 responds rationally to the actions of their coalition and fertilizes his lawn, then $v(1, 2)$ becomes $\{\mathbf{u} \mid u_1 \leq 10, u_2 \leq 7\}$. If the coalitions $\{2, 3\}$ and $\{1, 3\}$ change their perspective in a similar way, then the game has no core.

One possible area of future research, then, is what we might call solution-consistent characteristic functions. Such representations would require, in a somewhat circular fashion, that $v(C)$ be consistent with whatever solution hypothesis we apply to it, including the assumption that people act in their own interests. But developing such an analysis may require greater attention to the extensive-form details of a situation, and any ultimate theoretical resolution remains unclear. This indeterminacy demonstrates, however, that the fundamental structures of game theory and political theory are now hardly complete.

For simple voting games, these problems seem less of an issue because the various definitions of $v(C)$ are equivalent and, therefore, the core is a singular concept. In the case of majority rule, the core generally corresponds to the Condorcet winner, and thus it justifies that winner as a prediction for committees in which bargaining is unrestricted or in which the transaction costs of circumventing procedural constraints are low relative to the importance of the outcomes under consideration.

Once again, though, we must acknowledge that Condorcet winners probably are rare in most circumstances. In two-candidate elections without a Condorcet winner, we introduced the uncovered set to narrow the range of possible predictions. In committees for which bargaining is precluded, we considered issue-by-issue voting and sophisticated voting with agendas. Chapter 9 introduces some solution hypotheses that are designed explicitly to treat cooperative committee games without cores, and to proceed the reader must be familiar with the following concepts and definitions:

individually rational	domination via C	domination
core	alpha core	beta core
strong core	Coase theorem	balanced collection
market games	Edgeworth box	proto-coalition
k majority rule		

Before we review those concepts that treat games without cores, however, we should address an apparent misconception that many political scientists hold concerning the interpretation of these concepts and the core in the context of specific institutional constraints. Often, people believe that the core and other solution hypotheses are designed to treat games without constraints, such as agendas and issue-by-issue voting. Owing perhaps to the fact that the most straightforward illustrations of these concepts arise in committees that function with an otherwise unconstrained majority rule procedure (and thus are the dominant examples in the literature), some researchers mistakenly take individual preference configurations, solve for a core after assuming that the dominance relations among elements of U are defined by majority rule, and then try to incorporate specific institutional details by assessing whether or not those details preclude the selection of the core or an outcome predicted by some other cooperative game-theoretic hypothesis. But the problem with this approach is that, if such details do indeed constrain individual choice, if transaction costs render institutions real impediments to coordinated action, then those details ought to be represented in the extensive-, normal-, or characteristic-function forms of the game. If, for example, the status quo holds special significance because the committee uses a particular type of agenda, then the game cannot be modeled as simple, and a blind application of cooperative solution hypotheses yields meaningless results. The core and the several hypotheses we are about to review are not limited to simple games, market games, or to committees lacking procedural constraints, and their application assumes that we possess beforehand an adequate representation of the game, including a consideration of relevant institutional constraints.

CHAPTER 9

Solution theory

That cooperative situations might not have a natural equilibrium such as the one that the core defines came as no surprise to game theorists. The first solution concept proposed, the V-set or stable set, was designed specifically to treat games without a natural equilibrium. Although it is easy to criticize this concept today with the more than 30 years of hindsight since its invention, it adopted a perspective that departs significantly from the way in which we commonly think about political outcomes. Much of political science explains events as the result of specific actions and prior conditions. Nature may intervene to cast a probabilistic sheet over events, and our predictions may not be unique, as when two equally good alternatives exist, but each prediction contains its own logic. A specific n-tuple of strategies is a Nash equilibrium because no one has a unilateral incentive to defect from *it*, and an outcome is in the core because *it* is undominated. And although nonunique equilibrium strategies or core outcomes imply some degree of predictive indeterminacy, it is tempting to infer that we can reduce all other indeterminacies by supplying more detail to a situation's description. The toss of a coin, analogously, might not be random if we measure precisely its initial velocity, spin, and so forth. Hence, we might suppose that if we refine the measurement of people's preferences and perceptions, and if we describe strategic alternatives in more detail, then political events would seem to follow a more deterministic logic as well.

The theory that this chapter examines questions this view and suggests that it may be impossible often to narrow predictions about outcomes beyond subsets of possibilities. This theory, however, goes further than simply raising the possibility that our predictions cannot be unique. After all, the core is not always a unique element and a noncooperative game may have several Nash equilibrium strategy n-tuples. But an element of a core, for example, constitutes a prediction, because, by itself, it satisfies certain properties in relation to all other outcomes or n-tuples. Cooperative solution theory, on the other hand, focuses on subsets of outcomes and on the properties that those subsets satisfy. Thus, we predict a specific outcome not because of its properties alone, but because this outcome is a member of some set and that set satisfies certain properties, properties that it might not satisfy if that outcome is excluded. Our predictions no longer take the form "the outcome o will prevail because o ... [defeats, ties, is greater than, or dominates] ... all other possibilities," but rather "o is a possible outcome because o is a member of the set X, and X satisfies the properties ..., which X would not satisfy if o is excluded from it." Hence, this approach admits of the possibility that we can only limit predictions to sets of outcomes. And to the extent that such a set contains more than one element, this approach admits of social processes that we can characterize by fundamental indeterminacies.

As social scientists, we should be comfortable with this perspective. Balance-of-power theories in international relations, for example, do not assume necessarily that a single winning coalition forms to dictate events. Instead, the notion of balance implies a form of instability, a willingness on the part of nations to shift alliances, to keep certain coalitions from winning over an extended period or at least from becoming dominant. And although this set of motives implies that we cannot predict exactly what coalition will form in any specific period, it does not imply that events remain wholly unpredictable. Balance-of-power theories exclude certain coalitions, militarily dominant ones, and this prediction may be sufficient for predicting other events, such as peace or war. Similarly, legislative alliances between liberal Democrats, conservative Democrats, liberal Republicans and conservative Republicans may be in constant flux, as various bills emerge from committee and meet adoption or defeat on the floor. Again, the political scientist does not regard such possibilities as an impediment to prediction. The ephemeral nature of legislative coalitions or international alliances breathes a life into politics that makes its practice artful and its study interesting. But it does not make it unpredictable, for we still exclude the possibility that certain coalitions form on a given issue, and often that is enough to know whether some centrist position will emerge or whether extremist policies can prevail.

This chapter reviews three solution hypotheses, the V-set, the bargaining set, and the competitive solution, that adopt this perspective of predicting sets of outcomes. But we should not infer from such a list that game theory enjoys a surplus of solution hypotheses. Each hypothesis builds on the previous one, and our argument is that this sequence points to some as yet undeveloped general hypothesis. This chapter, then, proceeds historically. It begins with the V-set, which established the theoretical perspective for solution theory. We then turn to the bargaining set, which provided the V-set with some behavioral motivation. And we conclude with the competitive solution, which finds motivation in some of the bargaining set's difficulties with situations that are especially relevant to politics. This review should convey not only a grasp of solution theory's present accomplishments and limitations, but also a sense of the outlines of some as yet undiscovered general synthesis.

9.1   The stable set

The reason that the core cannot serve as a general solution hypothesis is not that its conceptualization is fundamentally flawed, but, instead, that the strong form of stability that it defines does not exist universally. Hence, we must define a more general, albeit weaker, kind of stability. Clearly, though, we must relate any new definition or generalization to the definition of the core. Otherwise, we cannot take advantage of the insight into the meaning of stability that the core offers, and we are likely to produce a disconnected theory that proliferates hypotheses like gerbils, without hope of parsimony. With this desired connection in mind, notice that the core is most compelling as a hypothesis about outcomes when it is a unique payoff n-tuple, when it dominates every other n-tuple. Of course, such an outcome's existence is rare in voting games, and its nonexistence is guaranteed for all zero-sum, transferable-utility games. To generalize the concept of the core, then, a natural first step is to forgo any uniqueness requirement; after all, we do not even require that equilibria to noncooperative games be unique. In so doing, though, we must answer these two questions: What relationship should exist among the (nonunique) outcomes in this new solution set (called X for the present)? What relationship should exist between specific outcomes in X and outcomes not in X?

A reasonable answer to the first question would suppose that no element of X dominates another element of X. This answer, of course, is consistent with the definition of the core, because whenever the core contains more than one element, no element of the core can dominate another element in it. Answering the second question is more difficult. If we require every outcome in X to dominate every outcome not in X, then we have simply defined a rather strong version of the core, and if we require that every element in X not be dominated by feasible elements outside of X, then we have simply defined the core anew. But consider another answer to the second question. Suppose that for every point not in X, there is a point in X that dominates it. If we accept this answer, which is admittedly ad hoc and not the only possibility, then we have defined the von Neumann-Morgenstern V-set (also called the stable set). Recalling that U denotes the set of all feasible utility n-tuples,

> *Stable set V*: The stable set for a game in characteristic-function form is any set $V \subseteq U$ satisfying (1) *internal stability*: if **u** and **w** are any two elements of V, then **u** does not dominate **w** and **w** does not dominate **u**, and (2) *external stability*: for all **w** $\in U$ that are not in V, there exists at least one **u** $\in V$ that dominates **w**.

Notice the difference between this solution concept and the core. An n-tuple, **u**, is in V not only because of its properties (no other elements in V dominate it) but also because of the properties of the other elements in V (although an element not in V may dominate **u** itself, something else in V dominates whatever dominates **u**). If **u** prevails, it does so not because of properties that it alone satisfies, but because it is a member of a set with certain properties. If a legislature approves a particular bill, x, then this approval may not reflect merely the properties only of bill x, but may reflect that x is a member of some class of bills, which, although only one member of this class can prevail, together they satisfy certain criteria.

This perspective might seem alien to the social sciences. When debating why a particular parliamentary coalition forms to establish a government, for example, we are likely to evaluate the preferences of key actors, their skills and political strengths, the relative sizes of competing parties, contemporary circumstances, behind-the-scenes deals, and chance events. If we look at other outcomes, we do so with the view that they are competitors of the outcome that prevailed. Rarely, if ever, do we argue that a particular outcome prevails because it, in conjunction with some other outcomes, satisfies properties that sets of other possibilities do not satisfy. This is not to say that our earlier explanations are wrong; solution theory identifies sets of outcomes that can prevail because of the properties that those sets satisfy. Explaining why one outcome in a set prevails rather than another may still require an appeal to bargaining skills, chance events, and the like.

To appreciate this perspective better, consider the three-person, three-bill vote-trading situation in Figure 9.1. The sincere voting outcome with simple majority rule is for all bills to fail, but passing all three bills dominates this outcome. Hence, we know from Theorems 2.4 and 8.3 that the core is empty. But consider the three outcomes $(1, 1, 0)$, $(1, 0, 1)$, and $(0, 1, 1)$ and the corresponding set of payoff 3-tuples, $\{(-4, 3, 3), (3, 3, -4), (3, -4, 3)\}$. None of these three outcomes dominates the other. Comparing $(1, 1, 0)$ and $(1, 0, 1)$, person 2 is indifferent, 3 prefers the first, and 1 prefers the second. Hence, the

Legislator	(0, 0, 0)	(1, 0, 0)	(0, 1, 0)	(0, 0, 1)	(1, 1, 0)	(1, 0, 1)	(0, 1, 1)	(1, 1, 1)
1	0	−2	−2	5	−4	3	3	1
2	0	5	−2	−2	3	3	−4	1
3	0	−2	5	−2	3	−4	3	1

9.1 Vote trading without a core.

outcomes tie in a majority vote. Yet at least one outcome in this set dominates every outcome not in it. So these three outcomes (or, equivalently, the corresponding three payoff n-tuples) constitute a stable set to this vote-trading game.

This example illustrates a situation in which we predict that a set of outcomes will occur because of that set's properties, not because of the properties of the outcomes in it, taken one at a time. The core looks for outcomes that are stable against *all* possibilities, those that no coalition can upset. But many games, like the one that this example provides, do not have such outcomes. The idea behind the V-set (as well as nearly all other alternatives to the core) is that we must settle for isolating stable *sets* of outcomes. External stability, together with internal stability, provides one definition of a set's stability. In the example, the outcome (1, 0, 0) dominates the outcome (1, 1, 0), with associated payoffs $(-4, 3, 3)$ in V, with respect to the coalition {1, 2}. Hence, although this outcome by itself is unstable, it is a prediction because it is in a stable set.

We emphasize that we should not interpret V in this example to mean that the three outcomes in it are equally likely. We can make no probabilistic inferences. Instead, we simply predict that one of the outcomes (1, 1, 0), (1, 0, 1), and (0, 1, 1), will occur. Narrowing the prediction further with reference to probabilities requires an appeal to considerations exogenous to the analysis, such as the participants' bargaining skills and the timing of offers and counteroffers. Because these exogenous details seem more apparent to the observer of actual events than do the abstract properties that sets of alternatives satisfy, journalistic accounts of legislation are likely to focus on these secondary considerations, and not on any complete causal explanation.

Predicting outcomes that are not stable by themselves might seem unsatisfactory. But because we cannot escape such instabilities in games without cores, and thus we cannot eliminate them in considering political events, the best that we can do is to isolate the sets of outcomes that, for theoretical reasons, seem most likely to prevail. We do not claim that the V-set is the best generalization of the core. It has serious shortcomings that render it suspect. Indeed, the reader should ask: What reasons do we have to suppose that the definition of V constitutes a valid prediction? What assumptions about human action could substantiate the idea that V-sets model instabilities in coalition formation? Is the V-set anything more than an ad hoc notion that is related to game theory only because it uses game-theoretic notation and terminology? Legitimate and defensible answers to these question are none, none, and yes. By itself, the V-set is a mathematician's invention, with properties that make it attractive to those who value parsimony above all other criteria. But the political scientist must balance parsimony with meaning, and the arguments for supposing that people choose outcomes in V-sets often remain obscure or simply constitute appeals to the conciseness of its definition and (perhaps illustrating the fallacy of sunk costs) the significant literature of game theory devoted to it.

One of the V-set's attractions, however, is the plausibility of its predictions in some games. For example, consider a simple three-person game of "divide the dollar" in which any two persons can dictate the outcome. The characteristic function for this game, after we equate money and utility, becomes

$$v(i) = 0 \quad \text{for } i = 1, 2, 3$$
$$v(1, 2) = v(2, 3) = v(1, 3) = v(1, 2, 3) = 1.$$

One possible prediction is that, given the obvious symmetry to this game, the three persons settle on the fair outcome $(\frac{1}{3}, \frac{1}{3}, \frac{1}{3})$. But this equal division finds motivation in considerations exogenous to the analysis, namely, equity. If such considerations exist, then they should find reflection in the utility numbers that we attach to outcomes: We should not incorporate them into the analysis in an ad hoc fashion. Assuming more Machiavellian players, the following set of outcomes seems more plausible: $\{(\frac{1}{2}, \frac{1}{2}, 0), (\frac{1}{2}, 0, \frac{1}{2}), (0, \frac{1}{2}, \frac{1}{2})\}$. That is, two persons will gang up on the third and divide the dollar between them. The V-set's attractiveness now is that these three payoff 3-tuples constitute such a set: They do not dominate each other, and at least one of them dominates every alternative proposal for dividing the dollar. For example, all three dominate the fair division point $(\frac{1}{3}, \frac{1}{3}, \frac{1}{3})$. More generally, to see that one of these three vectors dominates any other feasible outcome, consider any 3-tuple of the form $\mathbf{w} = (x, y, 1 - x - y)$. Suppose that $\mathbf{w}$ dominates $(\frac{1}{2}, 0, \frac{1}{2})$ with respect to the coalition $\{1, 2\}$, so that $x > \frac{1}{2}$ and $y > 0$. But y must not equal or exceed $\frac{1}{2}$, or $1 - x - y$ will be less than 0, which person 3 alone can preclude. Hence, it must be that $0 \leqslant 1 - x - y < \frac{1}{2}$. But then $(0, \frac{1}{2}, \frac{1}{2})$ dominates $\mathbf{w}$ with respect to the coalition $\{2, 3\}$.

The mathematical game theorist's fascination with V has less to do with the plausibility of its predictions. We can attribute the social scientist's interest in V, by contrast, in large part to the plausibility of its predictions in such games as divide the dollar. But there are at least four questions that we should ask and answer about V-sets. First, if a core and a V-set exist simultaneously, how are they related? Second, do all games have nonempty V-sets? Third, are V-sets unique? The fourth and final question may be the most important. Because stable sets consist of outcomes that are not individually stable as defined by the concept of domination, its predictions are necessarily more fragile. Hence, we must counter with good arguments the claim that outcomes in V might not occur. The claim that V-sets have "nice" mathematical properties will not suffice. Our fourth question, then is, What assumptions about human motivation and action lead us to suppose that people only choose outcomes in V?

9.2　Some properties of V-sets

Relationship of V to the core

Part of the answer to the first question is that if the core and V both exist, then they are not necessarily identical. The core is the set of undominated outcomes in U. Hence, an outcome $\mathbf{u} \in U$ may exist that is not in the core because something in U dominates it but that is not dominated by anything in the core. But such a point cannot exist outside of V: By external stability, V must contain at least one point that dominates $\mathbf{u}$. What we can say, though, is that *if the core and V are both nonempty for a game in characteristic-function form, then V must contain the core*. To see that this claim is true, suppose, to the contrary, that not all of the core is included in V. There is an n-tuple $\mathbf{u}$, then, that is in the core but is not in V, that remains undominated. Unless the V-set is empty, then from the definition of external stability there must exist a point in V that dominates $\mathbf{u}$, which is a contradiction. It is possible, of course, that the core not only exists, but also that outcomes in it strictly dominate all other feasible outcomes, in which case the core and V are equivalent. Since, in that event, the core satisfies both internal stability and external stability, the core is a V-set. The reader should be able to confirm that it is the unique V-set.

For an example in which the core is a strict subset of V, consider the three-person, majority-rule game with city-block preferences in Figure 8.13. There, although the core does not dominate points on the line l_1

and l_2, those points are not in the core because points elsewhere can defeat them. The core plus all points on these two lines, though, constitute a V-set. Because the core does not dominate points on l_1 and l_2, because they do not dominate each other, and because nothing dominates the core, internal stability is satisfied: Nothing in the set $\{l_1\}$ $\cup$ $\{l_2\}$ $\cup$ $\{core\}$ dominates anything else in that set. A simple pencil exercise should convince the reader that the core dominates all remaining points. Hence, the set $\{l_1\}$ $\cup$ $\{l_2\}$ $\cup$ $\{core\}$ satisfies external stability and constitutes a V-set.

Existence and uniqueness of V

The second question posed earlier concerns existence. It is not difficult to contrive games with finite numbers of feasible outcomes that do not have a V-set. For example, a three-person, majority-rule game that excludes lotteries and that we restrict to the three payoffs, (2, 1, 0), (1, 0, 2), and (0, 2, 1), has no V-set. But if we impose some structure on outcomes and preferences – for example, by requiring that preferences be spatial, by admitting transferable utility, or by permitting lotteries – it becomes more difficult to find a counterexample to existence. William F. Lucas, however, has formulated a 10-person game in characteristic-function form, with transferable utility, with a (weak) core, but without a V-set. And Lucas and Rabie have designed a 14-person game with an empty core and an empty V-set, thus ending years of speculation. But these games seem sufficiently rare and abnormal that the possible nonexistence of V may be a minor difficulty.

Perhaps the major problem with the V-set is not that it is empty, but, to answer our third question – that games generally have too many V-sets. Consider again a three-person spatial voting example with circular indifference contours, such as the ones in Figure 9.2a. With simple majority rule and one vote per person, we already know that this game has an empty core, because there is no Condorcet winner. The set of outcomes $V = \{o_1, o_2, o_3\}$, though, constitutes a V-set. For example, person 1 is indifferent between o_1 and o_2, while the preferences of persons 2 and 3 are divided. Thus o_1 and o_2 do not dominate each other. Continuation of this argument establishes V's internal stability. Figure 9.2b demonstrates external stability. The vertically hatched region (including its boundary) corresponds to all points that o_2 does not dominate, the cross-hatched area denotes all outcomes that o_3 does not dominate, and the corresponding regions formed by the three indifference curves through o_1 correspond to all points that o_1 does not dominate. As this figure shows, there is no point that is common to all

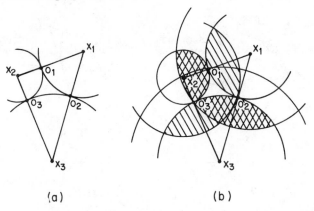

9.2 *V*-set for three-person spatial game.

three areas: At least one outcome in V dominates every outcome outside of V. Hence, V is externally stable and a V-set.

As predictions, these three outcomes seem plausible. First, we can associate each with a specific minimum-winning coalition. That is, the coalition $\{1, 2\}$ probably would choose o_1, $\{1, 3\}$ would choose o_2, and $\{2, 3\}$ would choose o_3. Second, the outcomes appear to reflect the relative positioning of each person. For example, person 3's ideal point is far from the ideal points of the other two persons. Roughly speaking, it would appear that 3 must "give up the most" to ensure membership in a winning coalition. Hence, the outcomes that we associate with coalitions of which he is a member are further from his ideal point than they are from those of the other two persons. Conversely, person 2, on the average, is closer to the other two members of the committee, so he must give up the least.

If these three points were the unique V-set for this game, then the definitions that establish stable sets would gain in appeal. Unfortunately, this game has V-sets that make little or no sense. Consider the arc in Figure 9.3a, which corresponds to a segment of one of 3's indifference contours, and the two points o' and o'' on the arc. By construction, person 3 is indifferent between o' and o'' while the preferences of persons 1 and 2 are divided. Hence, all points on the arc satisfy internal stability. Establishing external stability is more difficult and it might tax our geometric intuition. Briefly, the shaded regions in Figure 9.3a correspond to the points that o' does not dominate. As o' moves from left to right along the arc, region A shifts, so that all points above the arc lie outside of it for some positioning of o'. Correspondingly, region B

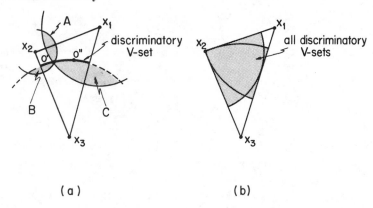

(a)   (b)

9.3 Discriminatory solutions.

expands while C shrinks, so that again, any point below the arc eventually lies outside of B and C for some value of o'. Stated simply, all points on this arc constitute a V-set. But that is not all. Many arcs have these properties, and if we find them all, then we can show that all points in the shaded region of Figure 9.3b are in some V-set.

We call sets such as the arcs in the preceding example *discriminatory V-sets*, because they appear to discriminate for or against a particular person (depending on one's point of view). Mathematically, they satisfy the definition of V-sets. Their relevance to politics is another issue entirely. For a considerable time, social scientists sought to make sense of them by arguing that in certain processes, societal norms prohibit the ruination of one member. Alternatively, contracts negotiated before a game is played might establish certain minimum payoffs to particular persons. But all such rationalizations are unsatisfactory. If these constraints are real, then they should be reflected beforehand in the description of $v(C)$. For example, if person i can receive no more or less than c, then we should set $u_i = c$ and remove him from the game.

The existence of discriminatory V-sets is a major irritant. For the simple voting game in Figure 9.3a, the V-set excludes few Pareto-optimal outcomes as predictions. And far from being a rare pathology, discriminatory V-sets pervade almost every major class of cooperative games in characteristic-function form. The response of game theorists to so many solutions, has been, in part, to narrow the scope of V-sets to more behaviorally palatable predictions. For political scientists concerned with voting institutions, the most interesting restriction defines a main-simple V-set.

Main-simple V-set: For simple (voting) games, a V-set is main-simple if and only if, for each minimum-winning coalition C, there is an outcome, $\mathbf{u} \in V$, such that for all other $\mathbf{w} \in V$, $u_i \geqslant w_i$, for all members $i \in C$.

This notion's attractiveness is twofold. First, it begins to incorporate coalition predictions into the analysis. Although it assumes that minimum-winning coalitions form, it formalizes the reasonable conjecture that they do so by providing their members with outcomes that are at least as good as what other coalitions offer. We can best grasp the second reason's attraction by referring to the three-person voting game in Figure 9.2a. We know that $\{o_1, o_2, o_3\}$ is a V set. But notice also that,

o_1 is preferred or indifferent to o_2 and o_3 by 1 and 2;
o_2 is preferred or indifferent to o_1 and o_3 by 1 and 3;
o_3 is preferred or indifferent to o_1 and o_2 by 2 and 3.

Hence, $\{o_1, o_2, o_3\}$ is a main-simple V-set. Indeed, it is the unique main-simple V-set for this game, which thereby illustrates one way to eliminate discriminatory V-sets from consideration.

For some time game theorists believed that the definition of main-simple V-sets resolved, at least for simple games, the problem of discriminatory solutions. Indeed, *for strong simple symmetric games in characteristic-function form with transferable utility, a nonempty and unique main-simple V-set necessarily exists.* Unfortunately, this result does not extend to games without transferable utility (a result that we establish later). And we may remain unwilling to assume that all political situations concern money that is freely transferable among all persons and that each person's utility for money is separable and linear. Because, as we argue earlier, these assumptions can be especially problematic in politics, the main-simple restriction on V does not resolve our problems.

In spite of such problems, the V-set remains an important part of our understanding of cooperative committee processes. Much of the early thinking about how to model such situations relied on the V-set as the sole descriptive supplement to the core as a solution hypothesis. Its most important contribution is the view that social scientists may have to remain satisfied with limiting their predictions about outcomes only to sets of possibilities. This perspective, once we explore it in such examples as "divide the dollar" or in a cooperative vote-trading game without a core, now seems natural and appears in nearly all modern revisions of the V-set and alternatives.

9.3   Bargaining sets for simple voting games

That the core is concerned with payoffs but not with coalition outcomes is understandable. Games with nonempty cores (even large ones) enjoy a kind of stability that renders payoff outcomes insensitive to the specific coalitions that form and vice versa. For example, we might hypothesize that if the n-tuple $\mathbf{u}$ is in the core and in $v(C)$ as well, then C is a potentially observable coalition. Although for simple games this prediction tells us only that some winning coalition will form, in this instance some indeterminacy in the relationship between outcomes and coalitions is natural. But if a core does not exist, then our ability to predict payoffs may depend on our ability to predict which coalitions form. The attractiveness of the main-simple V-set, for example, is that if, for exogenous reasons, we believe that only minimum-winning coalitions form, then we can use this belief to eliminate discriminatory V-sets from consideration. More generally, however, the V-set remains problematical because it is unconcerned with coalition predictions. We may appeal to the reasonableness of the V-set's predictions in specific examples. But the absence of "corelike" stability means that we can evaluate the V-set as a prediction about political processes only if we can give its definition some behavioral justification. But its silence on coalitions seems to preclude that justification. As a result, its definition yields peculiar results, such as discriminatory solutions that we can eliminate only with ad hoc restrictions.

This section reviews an approach that models the coalition-formation process directly. Recognizing that the V-set is not rooted in any conceptualization of the bargaining process, and that without such a model, we cannot refine it in accordance with any hypothesis about how people negotiate and choose, Aumann and Maschler offer an alternative solution hypothesis, which they call the *bargaining set*. Although their solution's predictions often do not differ much from those of V-sets, it is motivated by some specific behavioral considerations.

Actually, bargaining-set theory encompasses several alternative but closely related solution concepts, and to define them we require some additional notation and definitions. To simplify this notation, we limit discussion temporarily to strong simple games, voting games, so we need only be concerned with winning coalitions. First, recalling that the set $v^*(C)$ consists of all outcomes that are Pareto-optimal for the coalition C, define

> A *proposal* is a pair $(\mathbf{u}, C)$, in which $\mathbf{u}$ is in $v^*(C)$ and C is a winning coalition.

Now consider the logrolling example in Figure 9.1 and the proposal,

$$q = (\mathbf{u} = (-2, -2, 5), C = \{2, 3\}).$$

Suppose that 2 is contemplating the desirability of q, and after scanning Figure 9.1, he turns to 3 and says: "You're getting too much and I'm not getting enough. Either we move to a better outcome from my point of view or I'll form a coalition with 1. In particular, I'll offer 1 the proposal,

$$q' = ((3, 3, -4), \{1, 2\})."$$

After looking over Figure 9.1 himself, 3 admits that perhaps he is getting too much. He cannot make an equally attractive offer to 1 and, simultaneously receive a payoff of 5. So he turns to 2 and submits to the proposal,

$$p = ((-4, 3, 3), \{2, 3\}).$$

Person 2, though, has now become greedy, and heady with success, he demands more. Specifically, he threatens,

$$p' = ((-2, 5, -2), \{1, 2\}).$$

But this time 3 calls 2's bluff and makes a counterthreat that is at least as good for 1 and 3 as the offer that he originally accepted to placate 2; namely, he counters with

$$p'' = ((3, -4, 3), \{1, 3\}).$$

Naturally enough, 2 backs off from his demands and accedes to p.

It is reasonable to suppose that bargaining proceeds in this way if everyone is aware of all possibilities and all payoffs. And it is this process that Aumann and Maschler model and call bargaining-set theory. To tract their analysis, let C be a winning coalition with i and j two of its members, and let C' be an alternative winning coalition,

> An *objection* of i against j with respect to the proposal $p = (\mathbf{u}, C)$ is an alternative proposal $p' = (\mathbf{w}, C')$, such that p' is feasible (that is, $\mathbf{w} \in v(C')$) and:
>
> 1a. i but not j is in C' (that is, i threatens to exclude j from a new coalition).
> 2a. $w_k \geq u_k$ for all k in C' (that is, all members of the new coalition C' like the new proposal, p', at least as much as they like the old, p).
> 3a. $w_i > u_i$ (that is, the objection is an improvement for the objector, i).

Referring to the logrolling example, with $p = (\mathbf{u} = (-4, 3, 3), C = \{2, 3\})$, then $p' = (\mathbf{w} = (-2, 5, -2), C' = \{1, 2\})$ satisfies the definition of an objection. First, it is a legitimate proposal. Second, for $C' = \{1, 2\}$, $i = 2$ is in C' but $j = 3$ is not. Third, persons 1 and 2 like p' at least as much as they like p, and in particular, the objector, 2, strictly prefers p' to p.

Next, using the notation that defines an objection, we define a counterobjection by j against i thus:

> A *counterobjection* by j against i is a feasible proposal $p'' = (\mathbf{z}, C'')$ that satisfies these three conditions:
>
> 1b. j but not i is in C'' (that is, the person who is being objected against, j, now counters with the threat of excluding the objector, i).
>
> 2b. $z_k \geqslant w_k$, for all $k \in C' \cap C''$ (that is, all persons involved in the objection and counterobjection like the counter at least as much as they like the objection).
>
> 3b. $z_k \geqslant u_k$, for all $k \in C''$ (that is, all members of the coalition involved in the counter like the counter at least as much as they like the original proposal).

In the logrolling example, after greedy person 2 objects to $((-4, 3, 3), \{2, 3\})$ with $((-2, 5, -2), \{1, 2\})$, person 3 has a legitimate counter, $p'' = ((3, -4, 3), \{1, 3\})$.

Using these simplified notions of objection and counterobjection, we define the bargaining set, $\mathcal{M}'_1$, thus:

> *Bargaining set $\mathcal{M}'_1$:* $\mathcal{M}'_1$ is the set of all proposals of the form $(\mathbf{u}, C)$, such that for every objection of any person $i \in C$ against another person $j \in C$, j has a counterobjection against i.

Aside from formal definitions and notation, the essential idea behind the bargaining set is simple. Suppose that i and j are members of the same coalition. Then a particular coalition and outcome is "stable" with respect to i and j if both persons can defend their payoffs against objections by the other. Regardless of i's threat, j can counter with a proposal that gives him at least as much as he receives in the original proposal and that makes his new coalition partners at least as well off as they are with i's threat. If i and j can defend the proposal against all objections, then it is in the bargaining set.

To illustrate these ideas further, consider again the logrolling example in Figure 9.1 and the proposal $p = ((-4, 3, 3), \{2, 3\})$. There is only one objection that person 2 can raise against 3, the unique proposal that

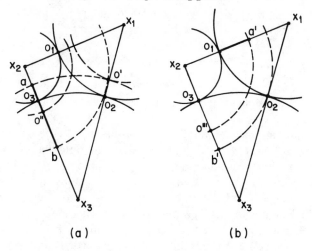

(a)   (b)

9.4 Bargaining set for three-person spatial game.

gives 2 more than a payoff of 3, $p' = ((-2, 5, -2), \{1, 2\})$. But as we have already seen, 3 can counter with p''. An equivalent argument holds for any objections by 3 against 2, so p is in $\mathcal{M}_1'$. Similarly, $((3, 3, -4), \{1, 2\})$, and $((3, -4, 3), \{1, 3\})$ are in $\mathcal{M}_1'$ as well. It is interesting to note now that the only proposals in $\mathcal{M}_1'$ that entail minimum-winning coalitions are those that correspond to this game's main-simple V-set. That is,

$$V = \{(3, 3, -4), (3, -4, 3), (-4, 3, 3)\},$$
$$\mathcal{M}_1' = \{((3, 3, -4), \{1, 2\}), ((3, -4, 3), \{1, 3\}), ((-4, 3, 3), \{2, 3\}).$$

In this way we can interpret bargaining-set theory as a behavioral rationalization for predicting outcomes in V-sets.

Since it sometimes helps to see things geometrically, consider again the three-person spatial example from Figure 9.2 and the three V-set points o_1, o_2, and o_3, which we draw again in Figure 9.4a. Suppose initially, though, that the coalition $\{1, 3\}$ is considering o'. Player 3 can object against 1 and o', using player 2 and any outcome in the interval (a, b) (player 3 prefers any point in this interval to o', and any such point makes 2 no worse off than he is at o'). In particular, suppose that 3 uses o''. Now, however, 1 cannot counter, because he cannot find an outcome that 2 likes as much as o'' and that he likes at least as much as o': The indifference curve for 2 that includes o'' does not intersect the indifference curve for 1 that includes o'. Thus, since person 3 can find an

Legislator	Bill 1	Bill 2	Bill 3	Bill 4	Bill 5	Bill 6
1	3	3	2	−4	−4	2
2	2	−4	−4	2	3	3
3	−4	2	3	3	2	−4

9.5 Riker–Brams vote-trading scenario.

objection that cannot be countered, proposal $(o', \{1, 3\})$ is not in $\mathcal{M}'_1$.

Referring to Figure 9.4b, though, consider the outcome o_2, which is one of the outcomes in this game's main-simple V-set. Although 3 prefers this outcome to o', suppose that he remains unsatisfied. As before, he can object with 2 and any outcome in the interval (o_3, b'), such as o'''. But now 1 can counter with any outcome in the interval $[o_1, a']$, because 2 likes every outcome here at least as much as o''', and because outcomes in this interval are no worse and sometimes better for 1 than is the original agreement with 3, o_2. This argument is wholly general, and it establishes that the outcomes o_1, o_2, and o_3, in conjunction with the appropriate coalitions, are in the bargaining set.

9.4   Cooperative vs. noncooperative analysis of committees

At this point in the development of solution theory it may be useful to see how a cooperative analysis of committee processes differs from a noncooperative approach, and to see the sorts of conclusions that each approach engenders. Consider the logrolling example from Figure 5.13, which Riker and Brams use to illustrate the prisoners' dilemma occasioned by uncoordinated vote trading. We reproduce it here in Figure 9.5.

The sincere voting outcome is $(1, 1, 1, 1, 1, 1)$: All bills pass. But this outcome yields the payoff 3-tuple $(2, 2, 2)$, which, for example, $(5, 5, -8)$ dominates with respect to the coalition $\{1, 2\}$, passing bills 1 and 6 only. Hence, because sincere voting with separable preferences must yield the core if the core exists (Theorem 2.4), the core to this game is empty. But consider these three proposals:

$p_1 = ((5, 5, -8), \{1, 2\})$, only bills 1 and 6 pass;
$p_2 = ((5, -8, 5), \{1, 3\})$, only bills 2 and 3 pass;
$p_3 = ((-8, 5, 5), \{2, 3\})$, only bills 4 and 5 pass.

We intend to show that these three proposals are in $\mathcal{M}'_1$ and that, together, they form a main simple V-set. Figure 9.6 lists all 64 outcomes and payoffs. With respect to the specific proposal $p_1 = ((5, 5, -8), \{1, 2\})$,

Bills passed	Payoff	Bills passed	Payoff	Bills passed	Payoff
none	(0, 0, 0)	5, 6	(−2, 6, −2)	1, 2, 3, 6	(10, 3, −3)
1	(3, 2, −4)	1, 2, 3	(8, −6, 1)	1, 2, 4, 5	(−2, 3, 3)
2	(3, −4, 2)	1, 2, 4	(2, 0, 1)	1, 2, 4, 6	(4, 3, −4)
3	(2, −4, 3)	1, 2, 5	(2, 1, 0)	1, 2, 5, 6	(4, 4, −4)
4	(−4, 2, 3)	1, 2, 6	(8, 1, −6)	1, 3, 4, 5	(−3, 3, 4)
5	(−4, 3, 2)	1, 3, 4	(1, 0, 2)	1, 3, 4, 6	(3, 3, −2)
6	(2, 3, −4)	1, 3, 5	(1, 1, 1)	1, 3, 5, 6	(4, 4, −3)
1, 2	(6, −2, −2)	1, 3, 6	(7, 1, −5)	1, 4, 5, 6	(−3, 10, −3)
1, 3	(5, −2, −1)	1, 4, 5	(−5, 7, 1)	2, 3, 4, 5	(−3, −3, 10)
1, 4	(−1, 4, −1)	1, 4, 6	(1, 7, −5)	2, 3, 4, 6	(3, −3, 4)
1, 5	(−1, 5, −2)	1, 5, 6	(1, 8, −6)	2, 3, 5, 6	(3, −2, 3)
1, 6	(5, 5, −8)	2, 3, 4	(1, −6, 8)	2, 4, 5, 6	(−3, 4, 3)
2, 3	(5, −8, 5)	2, 3, 5	(1, −5, 7)	3, 4, 5, 6	(−4, 4, 4)
2, 4	(−1, −2, 5)	2, 3, 6	(7, −5, 1)	1, 2, 3, 4, 5	(0, −1, 6)
2, 5	(−1, −1, 4)	2, 4, 5	(1, −5, 7)	1, 2, 3, 4, 6	(6, −1, 0)
2, 6	(5, −1, −2)	2, 4, 6	(1, 1, 1)	1, 2, 3, 5, 6	(6, 0, −1)
3, 4	(−2, −2, 6)	2, 5, 6	(1, 2, 0)	1, 2, 4, 5, 6	(0, 6, −1)
3, 5	(−2, −1, 5)	3, 4, 5	(−6, 1, 8)	1, 3, 4, 5, 6	(−1, 6, −1)
3, 6	(4, −1, −1)	3, 4, 6	(0, 1, 2)	2, 3, 4, 5, 6	(−1, 0, 6)
4, 5	(−8, 5, 5)	4, 6	(−2, 5, −1)	3, 5, 6	(0, 2, 1)
4, 5, 6	(−6, 8, 1)	1, 2, 3, 4	(4, −4, 4)	1, 2, 3, 5	(4, −3, 3)
		all	(2, 2, 2)		

9.6 Outcomes for six-bill vote-trading game.

the eight outcomes underlined in this table can all serve as the basis for legitimate objections by person 1 against 2, using the coalition {1, 3}. Each of these outcomes gives person 1 more than 5 and his potential new coalition partner, 3, at least −8. But against each of these, the proposal p_3 serves as a legitimate counter by person 2, because 2 receives as much with p_3 as with the original proposal p_1, while person 3 likes the counter as much as any objection. Hence, p_1 is in $\mathcal{M}_1'$. A similar argument holds for p_2 and p_3.

To illustrate that no other proposal involving minimum-winning coalitions is in $\mathcal{M}_1'$, consider $p_o = ((-5, 7, 1), \{2, 3\})$. If 3 objects against p_o with the proposal $((3, -4, 2), \{1, 3\})$, for example, 2 has no legitimate counter, because no outcome exists that gives 2 a payoff of 7 and that attracts 1 away from the objection. Hence, p_o is not in $\mathcal{M}_1'$. A relationship between the V-set and $\mathcal{M}_1'$ is apparent, moreover, since the three outcomes (5, 5, −8), (5, −8, 5), and (−8, 5, 5) also constitute a main-simple V-set.

Recall Riker and Brams's argument that vote trading in this game should yield the utility 3-tuple (0, 0, 0) defeat of all bills. Their analysis assumes, however, that vote trading is uncoordinated and that each legislator seeks the best outcome without negotiating an overall plan with others. The V-set and bargaining-set theory predict significantly different outcomes. We should expect this difference because coalition formation with complete information and no transaction costs precludes Pareto-dominated outcomes: If an outcome is Pareto-dominated, then cooperative theory assumes, minimally, that the coalition of the whole should form to eliminate such possibilities. In the preceding example, minimum-winning coalitions are sufficient to eliminate them.

But we cannot assert that one conclusion is right and the other is wrong. In certain legislative situations – for example, when the relevant issues are of sufficiently low salience that the transaction costs of coordination outweigh corresponding gains – perfect coordination and communication may not occur. Conversely, on bills in which importance outweighs transaction costs, coalitions and effective bargaining seem more likely to occur. Legislative leaders can also be crucial to any process, because they are often instrumental in forming coalitions, disseminating information about forthcoming legislation, and planning the sequence of floor votes. The size of the legislative body and the amount of legislation considered can also affect matters. The gains from coordinated action are like public goods. All members of a winning coalition share them, so prisoners' dilemmas of the sort that Olson describes may emerge, as people seek to avoid the entrepreneurial costs of making effective coalition agreements. Riker and Brams's prediction and the predictions of the V-set and of $\mathcal{M}'_1$, then, establish two extreme possibilities. Researchers ultimately must establish in a particular context whether a cooperative or a noncooperative approach is appropriate.

9.5   Extensions of the bargaining set

The preceding definition of a bargaining set is actually a special case of a wide class of such sets. First, we find no reason to limit definitions to simple games. Second, we can also consider games in which sets of persons can object against other sets, and in which the number of persons allowed to object simultaneously and the number that must have legitimate counterobjections varies. To generalize $\mathcal{M}'_1$, let $\mathbf{C} = (C_1, C_2, \ldots)$ be a *coalition structure* that partitions all persons into disjoint and exhaustive subsets. Next, let $\mathbf{u}$ be a utility n-tuple such that $\mathbf{u}$ is Pareto-optimal for each coalition in $\mathbf{C}$: that is, $\mathbf{u}$ is in $v^*(C_i)$ for all $C_i \in \mathbf{C}$. Then $p = (\mathbf{u}, \mathbf{C})$ is a proposal. Finally, let K and L be disjoint subsets of some coalition in $\mathbf{C}$, say C. That is, let $K \subset C$, $L \subset C$

with $K \cap L$ equal to the empty set, and $C \in \mathbf{C}$. The general definitions of objection and counterobjection proceed as follows:

> *Objection*: An objection of K against L with respect to $p = (\mathbf{u}, \mathbf{C})$ is a proposal, $p' = (\mathbf{w}, \mathbf{C}')$, such that for some $C' \in \mathbf{C}'$,
>
> 1a. $K \cap C' = K$
> 2a. $L \cap C' = \emptyset$, the empty set
> 3a. $w_i \geq u_i$ for all $i \in C'$
> 4a. $w_i > u_i$ for all $i \in K$

With this definition, a set of persons, K, can object against another set, L. If they do so, they must exclude all members of L from their new coalition, their new coalition partners must be at least as well off with the new proposal as with the original, and they in particular must be strictly better off.

> *Counterobjection*: A counterobjection of L against K is a proposal $p'' = (\mathbf{y}, \mathbf{C}'')$, satisfying, for some $C'' \in \mathbf{C}''$,
>
> 1b. $L \cap C'' = L$
> 2b. $K \cap C'' \neq K$
> 3b. $y_i \geq w_i$ for all $i \in C' \cap C''$
> 4b. $y_i \geq u_i$ for all $i \in C''$

Thus, in a counterobjection L responds with the coalition C'', which includes all of L, and if necessary, some but not all of K. L's response, p'', makes everyone in C'' at least as well off with the counterobjection as with the original proposal, p (condition 4b). And all members who are in C' and C'' prefer p'' to the counter, p' (condition 3b). Game theorists have proposed several variants of the bargaining set using these definitions. They include

> The bargaining set $\mathcal{M}$: $\mathcal{M}$ is the set of all proposals, such that for every objection of a K against an L, L has at least one counterobjection.

Often, though, $\mathcal{M}$ fails to exist. To secure existence we must make certain modifications on who must counter and who can object. The most commonly studied variants of the bargaining set are the following:

> *Bargaining set $\mathcal{M}_1$*: $\mathcal{M}_1$ is the set of all proposals, such that for every objection of a K against L, at least one member of L has a counterobjection.

> *Bargaining set $\mathcal{M}_2$*: $\mathcal{M}_2$ is the set of all proposals, such that if any

person i (that is, $K = \{i\}$) has an objection against L, L has a counterobjection.

We must answer two questions about $\mathcal{M}_1$ and $\mathcal{M}_2$. Do they exist and what kinds of predictions do they make? Here we confront the central dilemma of bargaining set theory. First, we simply note Peleg's result, that for superadditive, transferable-utility games, the bargaining set $\mathcal{M}_1$ necessarily exists. But we emphasize that existence here differs somewhat from the sort that we have formally invoked. Specifically, notice that since a person cannot object against himself, the proposal $(0, 0, \ldots, 0; \{1\}, \{2\}, \ldots, \{n\})$ necessarily is in all bargaining sets. In this sense, then, a bargaining set is never empty. But such "existence" is clearly uninteresting. Thus, those who study the bargaining set pose the question of existence differently. They ask: Are any of the several bargaining sets nonempty for "nontrivial" coalition structures? Peleg's result is that, for *any* coalition structure, there exists at least one proposal in $\mathcal{M}_1$.

To see the implication of this result, the reader can confirm that these seven proposals are the sole members of $\mathcal{M}_1$ for the simple three-person game, divide the dollar:

$(0, 0, 0; \{1\}, \{2\}, \{3\})$
$(\frac{1}{2}, \frac{1}{2}, 0; \{1, 2\}, \{3\})$
$(\frac{1}{2}, 0, \frac{1}{2}; \{1, 3\}, \{3\})$
$(0, \frac{1}{2}, \frac{1}{2}; \{2, 3\}, \{1\})$
$(\frac{1}{2}, \frac{1}{2}, 0; \{1, 2, 3\})$
$(\frac{1}{2}, 0, \frac{1}{2}; \{1, 2, 3\})$
$(0, \frac{1}{2}, \frac{1}{2}; \{1, 2, 3\})$

The advantage of $\mathcal{M}_1$, then, is that it eliminates discriminatory V-sets. Its disadvantage, aside from admitting $(0, 0, 0)$, is that it fails to exclude any coalition as a possible outcome. ($\mathcal{M}_2$ makes identical predictions in this case.)

We might regard the failure to exclude certain coalitions as a minor inconvenience, especially because a proof of general existence is a considerable accomplishment. But to understand the differences between $\mathcal{M}_1$ and $\mathcal{M}_2$ and the disadvantages of $\mathcal{M}_1$, consider this 13-person simple game with transferable utility:

$$v(C) = \begin{Bmatrix} 0 \\ 1 \end{Bmatrix} \text{ if } C \begin{Bmatrix} < \\ \geqslant \end{Bmatrix} 7$$

and the proposal

$$p = (\tfrac{1}{4}, \tfrac{1}{4}, \tfrac{1}{4}, \tfrac{1}{4}, 0, 0, \ldots, 0; \{1, 2, \ldots, 7\}, \{8, 9, \ldots, 13\}).$$

With p, then, a minimum-winning coalition forms, such that a minimum majority of that coalition (persons 1, 2, 3, and 4) proposes to reap all of the rewards. Clearly, it is difficult to believe that p actually would prevail as an outcome. The question is whether p is a member of any bargaining set.

Looking first at $\mathcal{M}_1$, suppose that some combination of persons 5, 6, and 7 object against 1, 2, 3, and 4. The worst possibility for $L = \{1, 2, 3, 4\}$ is that all three object by proposing to divide a small amount, e, among themselves [to satisfy condition (4a) in the definition of an objection] while giving four other persons, say, 8, 9, 10, and 11, $(1 - e)/4$ each [to satisfy condition (3a)]. That is, let $K = \{5, 6, 7\}$ object against L with the proposal

$$p' = (\mathbf{w}; \{5, 6, \ldots, 11\}, \{1, \ldots, 4, 12, 13\}),$$

in which

$$\mathbf{w} = (0, 0, 0, 0, e/3, e/3, e/3, (1 - e)/4, (1 - e)/4, (1 - e)/4, (1 - e)/4, 0, 0).$$

Recall that for p to be in $\mathcal{M}_1$, only one member of L must have a counterobjection. Furthermore, recall that that person can include some but not all of the original objectors in the counterobjection. Hence, person 1 can counter with

$$p'' = (\mathbf{y}; \{1, 7, 9, \ldots, 13\}, \{2, \ldots, 6, 8\}),$$

where

$$\mathbf{y} = (\tfrac{1}{4}, 0, 0, 0, 0, 0, e/2, 0, (1 - e)/4, (1 - e)/4, (1 - e)/4, e/8, e/8).$$

Readers should check as many objections to p as possible to convince themselves that a suitable counterobjection can meet every objection. Thus, p is in $\mathcal{M}_1$, which, given the intuitive implausibility of p, seems to render $\mathcal{M}_1$ an unreasonable definition of a solution, despite its guaranteed existence.

We can show that p is not in $\mathcal{M}_2$ by letting person 5 object against L with p'. Under $\mathcal{M}_2$, *all* members of L must counter with a single objection that gives each of them $\tfrac{1}{4}$. But this strategy would exhaust L's resources, so that if person 5 chooses an objection that gives everyone not in L some nonzero payoff, then L cannot attract three other persons into a winning coalition. Hence, there is an objection against p to which there is no counterobjection.

We can generalize this result about $\mathcal{M}_2$. First, notice that strong symmetric simple games with transferable utility exhibit *uniformly decreasing returns to scale*. That is, letting c denote the number of members of C,

$$v(C)/c < v(C')/c'$$

for all $C' \subset C \subset N$, with C' minimum winning. If $(\mathbf{u}, \mathbf{C})$ is in $\mathcal{M}_2$ for any such game and if $C \in \mathbf{C}$ is winning, then $u_i = u_j$ for all i and $j \in C$. Thus, the only proposals in $\mathcal{M}_2$ for the preceding game are of the form

$$(\tfrac{1}{7}, \tfrac{1}{7}, \ldots, \tfrac{1}{7}, 0, \ldots, 0; \{1, 2, \ldots, 7\}, \{8, \ldots, 13\}).$$

It appears, then, that $\mathcal{M}_2$ is the more promising solution hypothesis. Unfortunately, although the existence of $\mathcal{M}_1$ is guaranteed, that of $\mathcal{M}_2$ is not. Thus, either we have a solution that exists but that makes silly predictions, or we have one that makes plausible predictions but that may not exist. This is the frustrating problem with bargaining-set theory.

Despite these problems, bargaining-set theory is an attractive idea: Its assumptions about the bargaining process are intuitively reasonable, and many of its difficulties seem ultimately fixable. There are a great many permutations of assumptions in the definitions of objections and counterobjections that we do not explore and we might invent other variants. Some researchers suspect that one such variant, with perhaps a few modifications in definition, eventually will result in a solution that exists and that makes appropriate predictions.

9.6   The size principle

Neither the core, nor the V-set, nor bargaining-set theory makes coalition predictions. Political scientists judge this limitation to be a serious flaw. Often, coalitions are the only things we can observe. In parliamentary or legislative systems, for example, agreements on policy and legislation may be impossible to detect in advance of their actual implementation, and even then they may remain obscure. Although we can observe the allocation of ministries across parties in a parliament and although the immediate results of a legislative bargain, in terms of the passage of a particular bill, may be apparent, we may be unable to detect longer-term policy concessions and agreements or graft. Observing which coalition forms, however, may produce some insight into the policies that ultimately follow. It makes little sense to have a theory about things that we cannot observe while we ignore decisions that we can detect, and that may serve as an indicator of who has won, who has lost, and what they have won or lost.

As a partial response to the limitations of solution theory, William Riker formulates the size-principle hypothesis:

The size principle: In n-person, constant-sum cooperative games with transferable utility, only minimum-winning coalitions form.

Although the assumption of transferable utility is an unfortunate restriction and precludes application to vote-trading games and spatial committee games, the size principle nevertheless has received wide attention (and sometimes inappropriate application). To the extent that we can approximate the structure of political processes with the assumption of transferable utility, and because it has stimulated much research, the size principle warrants close inspection. But Riker's proof of the size principle does not rest on any solution hypothesis. Instead the proof derives from an examination of abstract cases. As such, it is difficult to follow and to relate to conventional developments, notation, and the like. But consider an argument that is not dissimilar from the one that we use to develop the core. First, if $p = (\mathbf{u}, \mathbf{C})$ is a proposal, then $\mathbf{u}$ must be Pareto-optimal for every coalition in $\mathbf{C}$, which, with transferable utility, requires

$$\sum_{i \in C} u_i \geq v(C) \quad \text{for all } C \in \mathbf{C}.$$

Recall that in our introductory treatment of the core, it seemed reasonable to suppose that if $\mathbf{u}$ is predicted, then *all* possible coalitions, C', should satisfy the condition

$$\sum_{i \in C} u_i \geq v(C'). \tag{9.1}$$

But imposing such a condition is equivalent to requiring that $\mathbf{u}$ be in the core, and the principal reason that we are contemplating solution concepts such as the V-set and the bargaining set is that we want to make predictions if the core is empty.

Suppose, however, that in examining a *specific* coalition structure we impose (9.1) not only for all coalitions in that structure but also for all subcoalitions of coalitions in $\mathbf{C}$. With the coalition structure $(\{1, 2, 3\}, \{4\}, \{5\})$, for example, suppose that (9.1) holds not only for $\{1, 2, 3\}$, $\{4\}$, and $\{5\}$, but also for $\{1\}$, $\{2\}$, $\{3\}$, $\{1, 2\}$, $\{1, 3\}$, and $\{2, 3\}$. Thus, instead of seeking predictions of the form "the coalition structure $\mathbf{C}$ will form and produce the outcome . . ." we opt instead for predictions such as "if the coalition structure $\mathbf{C}$ forms, then. . . ." In this instance we should reasonably consider once again the requirement that all admissible coalitions and defections satisfy (9.1), such that a defection means cases in which persons or subcoalitions break away from coalitions in C, to go their own way. We call any proposal, $p = (\mathbf{u}, \mathbf{C})$, that satisfies (9.1) for all $C' \subset C \in \mathbf{C}$, a *coalitionally rational proposal*. That is, the proposal p is coalitionally rational if, for no coalition in the coalition structure C, it is the case that any subset of persons has an incentive to defect unilaterally from the coalition of which it is a member.

Our interest in the concept of coalitionally rational proposal is that if we impose it as an assumption, then it implies the size principle for all symmetric zero-sum games with decreasing returns to scale. Recall that a game has uniformly decreasing returns to scale if

$$v(C')/c' > v(C)/c \quad \text{for all} \quad C' \subset C, \quad v(C) > 0, \qquad (9.2)$$

where c' and c are the respective sizes of C' and C. For example, if all coalitions are worth either 0 or 1, and if C' is a smaller winning coalition than is C, then we clearly have satisfied (9.2). Hence,

> 9.1. *If $p = (u, C)$ is a coalitionally rational proposal for any symmetric zero-sum game with uniformly decreasing returns to scale, if C is in $\mathbf{C}$ and if C is winning, then C is minimum winning.*

The logic of (9.1) is apparent in the following five-person game if we examine some typical payoff vectors for coalitions containing minimum-winning coalitions:

$$v(C) = \left\{\begin{matrix} 0 \\ -20 \\ -30 \\ 30 \\ 20 \\ 0 \end{matrix}\right\} \text{ if } C = \left\{\begin{matrix} 0 \\ 1 \\ 2 \\ 3 \\ 4 \\ 5 \end{matrix}\right\}.$$

Coalition structure	u_1	u_2	u_3	u_4	u_5
{1, 2, 3}, {4, 5}	10	10	10	−15	−15
{1, 2, 3, 4}, {5}	5	5	5	5	−20
{1, 2, 3, 4, 5}	0	0	0	0	0

Note first that this game exhibits uniformly decreasing returns to scale. For example,

$$0 = v(1, 2, 3, 4, 5)/5 < v(1, 2, 3, 4)/4 < v(1, 2, 3)/3.$$

Second, notice that only the first proposal, $(10, 10, 10, -15, -15; \{1, 2, 3\}, \{4, 5\})$, is coalitionally rational. In the second proposal with the four-person coalition $\{1, 2, 3, 4\}$, persons 1, 2, and 3 can split off, form a coalition, and increase their individual payoffs from 5 to 10. We observe the same instability for any distribution of 20 units among four persons, and clearly, for any distribution of 0 units among five persons (the third proposal above). But suppose that 4 tries to bribe his way into $\{1, 2, 3\}$ by agreeing to accept a payoff of -10 (compared with -15, if he is excluded). That is, he proposes $(10, 10, 10, -10, -10)$, with the coalition $\{1, 2, 3, 4\}$. This proposal is not coalitionally rational, and it is not likely to prevail.

The preceding argument does not prove the size principle, because coalitional rationality remains an assumption about outcomes, and is not a derived property. Nevertheless, result 9.1 serves to emphasize that if a minimum-winning coalition tries to add members, then it can do so under the assumption of uniformly decreasing returns to scale only by subtracting benefits from current members in such a way that these members will have an incentive to defect from the coalition. Stating the size principle in terms of coalitional rationality, then, reveals the underlying reason why this hypothesis is attractive in the context of political coalitions.

Although the size principle, when coupled with appropriate theoretical justifications, is an important hypothesis in political theory, we must emphasize some conjectures and hypotheses that this principle does not imply. Many researchers have sought to test Riker's idea in contexts of the formation of parliamentary government, in which parties that control different numbers of seats negotiate to form a governing coalition. If we narrow the analysis to those situations that we might reasonably classify as resembling zero sum or situations that create winning and losing coalitions, the question becomes: Do coalition governments of minimum-winning *weight* form? That is, suppose that a parliament consists of three parties with weights of 50, 40, and 30 seats, respectively. If a coalition of the first two parties forms, holding 90 seats, or 20 more than the winning coalition of parties 2 and 3, should we interpret its formation as a refutation of the size principle?

Coalitions larger than minimum-winning weight are common, and researchers often interpret them as counterexamples to Riker's size principle. But minimum-winning size and minimum-winning weight are distinct concepts. We have already seen why proto-coalitions, agreements to negotiate as a unit, might exist in politics to change a game's strategic character. For example, suppose that in a five-person, strong simple symmetric game, 1 and 2 form a bargaining unit, that 3 and 4 do the same, and that 5 chooses to negotiate alone. Letting B_{12} denote the bargaining unit of 1 and 2, and B_{34}, the unit of 3 and 4, if we assume transferable utility and that winning coalitions are worth 1 and that losing coalitions are worth nothing, then we have this three person game in characteristic-function form,

$$v(B_{12}) = v(B_{34}) = v(5) = 0$$
$$v(B_{12}, 5) = v(B_{34}, 5) = v(B_{12}, B_{34}) = 1$$
$$v(B_{12}, B_{13}, 5) = 1$$

which is a symmetric simple game, even though persons B_{12} and B_{34} outweigh 5 two to one. Thus, there is nothing in solution theory that implies that the minimum-winning but overweight coalition $\{B_{12}, B_{34}\}$ is more or less likely to form than either of the minimum weight coalitions.

Suppose, instead, that we model this parliamentary game as zero sum by using the characteristic function of the game that we used to illustrate Theorem 9.1. Hence,

$$v(B_{12}) = v(B_{34}) = -30; \; v(5) = -20$$
$$v(B_{12}, 5) = v(B_{34}, 5) = 30$$
$$v(B_{12}, B_{34}) = 20$$
$$v(B_{12}, B_{34}, 5) = 0.$$

In this instance, it appears that the minimum-weight coalitions, $\{B_{12}, 5\}$ and $\{B_{34}, 5\}$, have some advantage over $\{B_{12}, B_{34}\}$. This game, which exhibits decreasing returns to scale, in effect, assumes that the more weighty winning coalitions are worth less than the less weighty ones. At least from the perspective of bargaining set theory, however, this interpretation is deceptive, because each of the following proposals are in $\mathcal{M}_2$ (that is, nothing precludes $\{B_{12}, B_{34}\}$ from the predicted set of coalitions:

$$((10, -30, 20); \{B_{12}, 5\}, \{B_{34}\})$$
$$((-30, 10, 20); \{B_{34}, 5\}, \{B_{12}\})$$
$$((10, 10, -20); \{B_{12}, B_{34}\}, \{5\})$$

For an illuminating substantive example of the dangers of confusing minimum-winning and minimum-weight, consider the 1824 presidental election, which Riker himself interprets as a classic example of the size principle. In that election the 261 electoral votes were distributed as follows:

Andrew Jackson,	99 votes & a majority in	11 states
John Q. Adams,	84 votes & a majority in	7 states
William Crawford,	41 votes & a majority in	3 states
Henry Clay,	37 votes & a majority in	3 states

Because no candidate received a majority of the electoral votes, the election went to the House of Representatives, in which each state has one vote, rendering 13 states a majority necessary for election. But in the House, representatives need not vote in accordance with their respective state's electoral vote, and the lineup there became

Adams,   10 states
Jackson,   7 states
Crawford,   4 states
Clay,   3 states

The minimum-*winning* coalitions, then, were

{Adams, Jackson} with 17 votes
{Adams, Crawford} with 14 votes
{Adams, Clay} with 13 votes
{Jackson, Crawford, Clay} with 14 votes

and the minimum-*weight* coalition was {Adams, Clay}. The House, however, could consider only three candidates, and Clay, with the least support, was eliminated. The question became: To whom should Clay throw his support? Clay supported Adams, Adams became president with 13 votes, and despite objections to the "corrupt bargain," Clay became secretary of state. Thus, the unique minimum-weight coalition formed.

Certainly this historical event supports the size principle, because {Adams, Clay} is also minimum winning. But the size principle does not predict that {Adams, Clay} is the unique possibility, since there are three other minimum-winning coalitions. Indeed, if we make some assumptions about payoffs, then there are stable coalition structures and proposals that involve each of these coalitions. Suppose that the "unions of support" of Adams, Jackson, and so forth are inviolate with weights 10, 7, 4, and 3, respectively, and that the situation corresponds to a zero-sum game. In particular, suppose that the value of a winning coalition decreases as its size increases, so that $v(C) = -c$ if C is losing and $v(C) = 24 - c$ if C is winning. Then,

$$v(\text{Adams}) = -10; \ v(\text{Jackson}) = -7; \ v(\text{Crawford}) = -4; \ v(\text{Clay}) = -3$$

and for the minimum-winning coalitions,

$v(\text{Adams, Jackson}) = 7$
$v(\text{Adams, Crawford}) = 10$
$v(\text{Adams, Clay}) = 11$
$v(\text{Jackson, Crawford, Clay}) = 10$

Notice now that

(6, −7, −4, 5; {Adams, Clay}, {Jackson, Crawford})

is in $\mathcal{M}_2$, and so is

(6, 1, −4, −3; {Adams, Jackson}, {Crawford, Clay}).

That is, $\mathcal{M}_2$ does not exclude the maximum-weight, minimum-winning coalition of Adams and Jackson. If Jackson objects with

(−10, 2, 3, 5; {Adams}, {Jackson, Crawford, Clay}),

for example, then Adams can counter with (6, −7, −4, 5) by coalescing with Clay, or with (6, −7, 4, −3) by coalescing with Crawford.

9.7   Problems with nontransferable utility games

We suspect that Clay's coalition with Adams, while consistent with the hypothesis of minimum-winning coalitions, also reflected the fact that no other stable coalition could offer Clay more, and reflected also the ideological conflict among the candidates. That is, the issues in 1824 and the preferences of each candidate's supporters may have prevented any effective coalition between Adams or Clay with Jackson. The coalition of Adams and Clay may have required far less compromise on the issues than any other alternative. Because they would involve fundamental realignments of power, it is perhaps too great an abstraction to suppose that coalitions incorporating Adams, Jackson, and their supporters could simply divide some easily transferable resource that they could use to compensate either party for any issue compromises.

This argument suggests that we ought to conceptualize the election of 1824 spatially, that an understanding of why {Adams, Clay} formed requires a specification of the issues and the candidates' positions on those issues. But we already know that with this conceptualization, transferable utility, as well as the size principle as Riker formulates it, cannot be used to model the corresponding cooperative game. Curiously, though, many applications and tests of the size principle in legislatures and parliaments adopt a spatial conceptualization, and the hypotheses that scholars offer as alternatives to, or modifications of, the size principle rely heavily on this conceptualization. These hypotheses include the predictions that the "favored" coalitions are those that are either *connected* or *compact*. Connected means that, for single-issue contests, if 1 and 2 can form a winning coalition but if 3 is between 1 and 2 on the issue, then the coalitions {1, 3} and {2, 3} are more likely to form than is the coalition {1, 2}. Compact means that if 1 and 2 are a winning coalition, but if 3 is closer to 1 than he is to 2, then {1, 3} (assuming that it is also winning) is more likely to form than is $\{1, 2\}$.

Although such hypotheses seem ad hoc and remain unrelated to solution theory in a formal way, the attention they receive by observers familiar with parliamentary systems, as well as their empirical support, indicate the relevance of a spatial conceptualization. For example, we might argue that not only was the coalition $\{$Adams, Clay$\}$ minimum winning, but also, because Clay was closer to Adams than to Jackson on most issues, and since Adams and Jackson probably defined the extremes on those issues, $\{$Adams, Clay$\}$ was also connected and compact. Hence, given the intuitive appeal of such hypotheses, it is important to ascertain whether we can extend the developments of the preceding sections to spatial committee games.

The analysis in Figure 9.2, which shows the V-set for a three-person, two-dimensional spatial game, suggests that such extensions are possible. Unfortunately, three-person spatial games are a *very* special case, and an analysis of more general possibilities suggests that we require different solution concepts for most cooperative political games. We begin by turning to a modified version of the $\mathcal{M}_1$ bargaining set that Robert Wilson first proposed, called the *strong bargaining set*. Wilson criticizes bargaining-set theory for not requiring that counterobjections have stability properties of their own. If counters are not credible, then there is no reason to suppose that they can thwart objections. Notice that this criticism finds its rationale in the arguments that Chapters 5 and 6 use to argue that beliefs and actions ought to be consistent: If a counterobjection is to be credible, then it ought to be a predicted outcome. Hence, as a partial resolution of his criticism, Wilson proposes this modification in the definition of the bargaining set:

> $\mathcal{M}_1$ is a *strong bargaining set*, if for every objection against a proposal in $\mathcal{M}_1$, there is a counterobjection in $\mathcal{M}_1$.

This definition yields Theorem 9.2:

> 9.2. (Wilson): *For simple games in characteristic function form, the set of payoff n-tuples of the main-simple V-set, along with the appropriate minimum-winning coalitions, is the unique strong bargaining set.*

The reader should verify that the three outcomes corresponding to a main-simple V-set and to a bargaining set for the vote-trading game in Figure 9.1, $(1, 1, 0)$, $(1, 0, 1)$, and $(0, 1, 1)$, constitute a strong bargaining set. Similarly, the three outcomes o_1, o_2, and o_3 for the spatial game in Figure 9.4b also form a strong bargaining set. If person 3 objects with o''' and the coalition $\{2, 3\}$ against 1 with respect to the

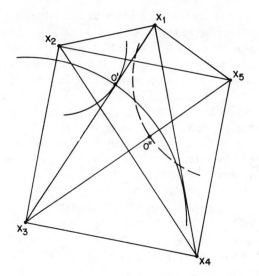

9.7 A spatial game with an incomplete bargaining set.

outcome o_2 and if 1 can counter with $(o_1, \{1, 2\})$, then 1 recovers to the utility that he is getting at o_2, and his new coalition partner, 2, prefers o_1 to o'''. A similar argument holds for any other possible objection.

The strong bargaining set, then, blends bargaining-set theory's behavioral motivations with the concept of the main-simple V-set, and it thereby provides a justification for supposing that minimum-winning coalitions form in spatial committees. Specifically, although earlier we criticized the main-simple V-set as an ad hoc way to exclude discriminatory V-sets as predictions, the definition of a strong bargaining set provides a rationale for supposing that outcomes prevail only if we can identify them with specific minimum-winning coalitions as proscribed by the definition of main-simple.

But this blending of motivations is not our principal reason for considering the strong bargaining set. Our intent is more sinister, because with this blend we can show that spatial committee games in general do not have main-simple V-sets, and that their bargaining sets yield implausible predictions. Consider the simple majority-rule, five-person spatial game in Figure 9.7 and the proposal,

$$p' = (o'; \{1, 2, 3\}).$$

Suppose that 3 objects against 2 with the proposal,

$$p'' = (o''; \{3, 4, 5\}).$$

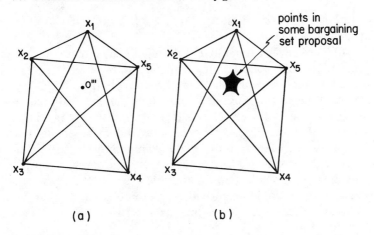

9.8 Bargaining set points.

This is a legitimate objection, because

o'' is in $v^*(3, 4, 5)$
$o'' P_3 o'$
$o'' R_4 o'$ and $o'' R_5 o'$

To see that 2 does not have a counter, recall that 2 cannot use the objector, 3, in a counter, nor can 2 find a point that makes him as well off as he is with o' while making either 4 or 5 as well off as they are with o''. That is,

$\{o \in O \mid o R_2 o'\} \cap \{o \in O \mid o R_4 o''\}$ is empty
$\{o \in O \mid o R_2 o'\} \cap \{o \in O \mid o R_5 o''\}$ is empty

Hence, 2 cannot lure 4 or 5 into a counter that "defends" what 2 is getting from o', in which case the proposal, p', cannot be in any bargaining set.

Readers should convince themselves that for every point that is Pareto-optimal for the coalition $\{1, 2, 3\}$, that is, for all outcomes in $v^*(1, 2, 3)$, either 1, or 2, or 3 has an objection that no one can counter. This argument, moreover, applies equally well to the coalitions $\{2, 3, 4\}$, $\{3, 4, 5\}$, $\{4, 5, 1\}$, and $\{5, 1, 2\}$. Hence, we can interpret bargaining-set theory as predicting that these coalitions will not form. But bargaining-set theory does not exclude all minimum-winning coalitions. We can also verify that the proposal $p = (o'''; \{1, 3, 4\})$ in Figure 9.8a can find a defense against every objection, and that the set of points in the shaded region of Figure 9.8b corresponds to proposals by the coalitions $\{1, 3, 4\}$, $\{2, 4, 5\}$, $\{3, 5, 1\}$, $\{4, 1, 2\}$, and $\{5, 2, 3\}$, which are in $\mathscr{M}$.

Later, we review some experimental evidence that contradicts these predictions. But consider the example's implications for V-sets. Our preceding analysis establishes that the strong bargaining set is empty, at least with respect to 5 of the 10 minimum-winning coalitions. From the relation of the main-simple V-set to the strong bargaining set that Theorem 9.2 establishes, and because the main-simple V-set is an admissible set of proposals by the minimum-winning coalitions, the main-simple V-set must be empty for this game as well. Thus, instead of providing a rationalization for the main-simple V-set, the strong bargaining set, in conjunction with our example, shows that we require additional theoretical developments for the analysis of cooperative committee games.

At this point, it is useful to review bargaining-set theory to see what occasions its problems with respect to games without transferable utility. Briefly, notice that the motivation of bargaining-set theory derives from the view that if a person is getting or demanding "too much," then he should give up some of his payoff to persons with "legitimate" objections. But in spatial games people cannot freely transfer utility, and proposals necessarily exhibit a form of externality that renders them vulnerable to objections. In our example, 2 cannot defend the utility that he gets from o', but he also cannot be excluded from this payoff if 1 and 3 seek to form $\{1, 2, 3\}$, with an outcome that is Pareto-optimal for the coalition. At o', it is impossible to increase 1 or 3's payoffs at the expense only of 2. If 3's payoff grows larger because 2's declines, then 1 is hurt as well, in which case 1 also can object against 3. Hence, in forming a winning coalition, 1 and 3 must necessarily give 2 more than he is "worth," more than he can defend. This condition violates the requirement of bargaining-set theory, that *every* person must defend his payoff against *all* possible objections. Because these externalities are inescapable, and because 2 cannot give 1 and 3 any "excess" payoff, no stable proposal exists for the coalition $\{1, 2, 3\}$.

9.8   The competitive solution

Until recently, game theory's principal applications occurred in economics, a discipline that studies situations in which it is reasonable to assume that the primary measure of value, money, is a transferable commodity, and that utility is linear and separable in this commodity. Thus, transferable utility seems to be an acceptable assumption for modeling those situations, and the various problems that we encounter in applying solution concepts such as the V-set and $\mathcal{M}_1$ appear tolerable. But we cannot represent many games, such as those in which preferences have a spatial representation, with transferable utility without abandoning the essential character of the situation under study. This problem is especially bothersome in politics, because without an appropriate solution hypothesis, we cannot proceed with the major purposes of our research. For example, we cannot compare the institution of unrestricted majority rule with, say, two-candidate elections; we cannot evaluate the implications of removing procedural restrictions such as issue-by-issue voting; and we cannot deduce the consequences of people short-circuiting procedural constraints if transaction costs are low or if the issues under consideration are sufficiently important.

A third solution hypothesis, the competitive solution, tries with some success to resolve such problems. This hypothesis begins with the anthropomorphic view that coalitions, not people, are the active participants in a game, and that coalitions must bid for members in a competitive environment of other coalitions that simultaneously are attempting to form. Thus, to be successful, to have a chance of forming, coalitions must bid effectively by appropriately rewarding their "critical" members. If any one person or set of persons is pivotal between two coalitions, and if each coalition is to have a chance of forming, then those who are *pivotal* should be indifferent *as a set* between the offers of each coalition, lest their preferences ensure that one of the two coalitions cannot form. This competition among coalitions can result, of course, in certain ones being unable to attract members, and these coalitions will be unable to form.

To formalize this view we restrict discussion to strong simple games. To represent the ideas of pivots and preference, let $p = (\mathbf{u}, C)$ and $p' = (\mathbf{u}', C')$ be two proposals, and define,

> p is *viable* against p' if at least one person common to C and C' (the "pivots" in $C \cap C'$) prefers $\mathbf{u}$ to $\mathbf{u}'$; the proposal p is *strictly viable* against p' if the pivots between C and C' all prefer $\mathbf{u}$ to $\mathbf{u}'$.

Let $\mathcal{K}$ be a set of proposals, such that no coalition is associated with more than one proposal in $\mathcal{K}$. This constraint on $\mathcal{K}$ mirrors the ideal that proposals are strategies that coalitions use to attract members: Just as people cannot adopt more than one strategy in a game, so coalitions are similarly restricted.

> $\mathcal{K}$ is *balanced* if every proposal in $\mathcal{K}$ is viable against every other proposal in $\mathcal{K}$.

Hence, if $\mathcal{K}$ is balanced and if $(\mathbf{u}, C)$ and $(\mathbf{u}', C')$ are two proposals in $\mathcal{K}$, then the persons common to C and C' do not all prefer $\mathbf{u}$ to $\mathbf{u}'$ or $\mathbf{u}'$ to $\mathbf{u}$. Next,

> p *upsets* $\mathcal{K}$ if p is viable against every proposal in $\mathcal{K}$ and if p is strictly viable against some proposal in $\mathcal{K}$.

Finally

> *Competitive solution*: $\mathcal{K}$ is a competitive solution if it is balanced and if there is no proposal that upsets it.

Although $\mathcal{K}$ might appear unrelated to the core, the V-set and the bargaining set, there are important relationships between $\mathcal{K}$ and each of these concepts. One way to see these relationships, as well as the distinctiveness of $\mathcal{K}$, is through this equivalent restatement of its definition:

> The set of proposals, $\mathcal{K}$, is a competitive solution if the following three conditions are satisfied:
>
> 1. *Internal stability*: No proposal in $\mathcal{K}$ is strictly viable against any other proposal in $\mathcal{K}$,
> 2. *External stability*: If any proposal, q, not in $\mathcal{K}$ is strictly viable against a proposal in $\mathcal{K}$, then there is some other proposal in $\mathcal{K}$ that is strictly viable against q,
> 3. No coalition is associated with more than one proposal in $\mathcal{K}$.

This definition almost directly parallels that of the V-set, but there are important differences. First, by considering proposals, and not utility vectors, $\mathcal{K}$ directly links payoffs to coalition predictions. Second, $\mathcal{K}$ uses the notion of viability, and not that of domination. This is perhaps the most crucial departure. Recall that $(\mathbf{u}, C)$ dominates $(\mathbf{u}', C')$ only if C can get $\mathbf{u}$ (that is, $\mathbf{u} \in v(C)$) and if *all* members of C *strictly* prefer $\mathbf{u}$ to $\mathbf{u}'$. But strict viability does not constrain preferences over $\mathbf{u}$ and $\mathbf{u}'$ of the nonpivotal members between C and C'. The concept of strict viability focuses instead on pivotal players as the key actors in deciding whether one proposal has an advantage over another. Third, although something in V must dominate each outcome outside of V, the definition of $\mathcal{K}$ requires that if p' is not in $\mathcal{K}$, then we must find a proposal in $\mathcal{K}$ that is strictly viable against p' only if p' is strictly viable against something in $\mathcal{K}$. Thus, the elements of $\mathcal{K}$ need not dominate everything else, but only those things that might threaten $\mathcal{K}$'s stability. Finally, in conformity with the perspective that the active players are the coalitions, we can associate no coalition with more than one proposal.

Despite these important differences, we can establish the following relationship between the core, V, and $\mathscr{K}$:

9.3. *If* $\mathbf{u}$ *is in the core, then* $p = (\mathbf{u}, N)$ *is a competitive solution; and if* V *is a main-simple V-set, then there is a competitive solution,* $\mathscr{K}$*, such that* $(\mathbf{u}, C) \in \mathscr{K}$ *implies that* $\mathbf{u} \in V$ *and* C *is minimum winning.*

This result, in conjunction with the correspondence that Theorem 9.2 establishes between the strong bargaining set and the main-simple V-set, shows that *the strong bargaining set, if it exists, is a competitive solution.*

The proof of 9.3 illustrates the definition of the competitive solution. To establish the correspondence between $\mathscr{K}$ and the core, suppose that $\mathbf{u}$ is an element of the core and that $(\mathbf{u}, N)$ is the sole proposal in K. If $(\mathbf{w}, C)$ is any other proposal, then $(\mathbf{w}, C)$ upsets $\mathscr{K}$ if and only if each member of C strictly prefers $\mathbf{w}$ to $\mathbf{u}$, since all members of C pivot between C and N. But if $\mathbf{u}$ is in the core, then there cannot exist another n-tuple, $\mathbf{w}$, such that $w_i > u_i$ for all $i \in C$ and $\mathbf{w} \in v(C)$. In this event, no $(\mathbf{w}, C)$ is strictly viable against $(\mathbf{u}, N)$, so $(\mathbf{u}, N)$ is a competitive solution.

For the part of Theorem 9.3 that establishes the relationship between $\mathscr{K}$ and the main-simple V-set, let $\mathscr{K} = \{p', p'', \ldots, p'''\}$ be the set of proposals whose associated payoffs, $\{\mathbf{u}', \mathbf{u}'', \ldots, \mathbf{u}'''\}$, correspond to such a V-set, and let C' be the coalition associated with proposal p', and so on. From the definition of main-simple, $\mathbf{u}' \geq \mathbf{u}''$, for all members of C', which implies that p' is viable against p'' (that is, $u_i' \geq u_i''$, for all $i \in C' \cap C''$). By a similar argument, p'' is viable against p', so $u_i'' \geq u_i'$ for all $i \in C' \cap C''$. But then $u_i'' = u_i'$ for all $i \in C' \cap C''$, so $\mathscr{K}$ is balanced. Now suppose that $p^* = (\mathbf{u}^*, C^*)$ is any feasible proposal not in $\mathscr{K}$. To simplify the discussion suppose that C^* is also minimum-winning (the reader should try to reconstruct this proof by supposing that C^* is any winning coalition). From the external-stability property of V-sets, there must exist a proposal in $\mathscr{K}$, say $p''' = (\mathbf{u}''', C''')$ such that $\mathbf{u}'''$ dominates $\mathbf{u}^*$. Because C^* and C''' must both be winning, the intersection of C^* and C''' cannot be empty. But since $\mathbf{u}'''$ dominates $\mathbf{u}^*$, the pivots between C''' and C^* must strictly prefer p''' to p^*, in which case p''' is strictly viable against p^*. Hence, p^* cannot upset $\mathscr{K}$. Since p^* is an arbitrary proposal, this argument holds for any p not in $\mathscr{K}$, so $\mathscr{K}$ is a competitive solution. This completes the proof of Theorem 9.3.

The links that Theorem 9.3 establishes between $\mathscr{K}$ and other solution concepts are important. We already know much about the core and the main-simple V-set. Work reported in a vast empirical and experimental literature tests the core, V, and the bargaining set in various

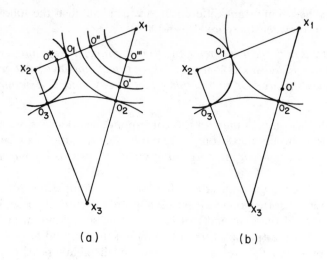

(a)   (b)

9.9 The competitive solution.

transferable-utility settings. This theorem makes that literature relevant to $\mathcal{K}$. Furthermore, the definition of $\mathcal{K}$ provides a useful behavioral justification for supposing that outcomes in the core and outcomes in main-simple V-sets prevail, although since $\mathcal{K}$ is necessarily a finite set (owing to the assumption that no coalition makes more than one proposal in $\mathcal{K}$), $\mathcal{K}$ also eliminates the nonsensical discriminatory V-sets from consideration.

To illustrate these links, consider again a spatial three-person, simple committee game with circular indifference contours. Referring to Figure 9.9a, suppose that when bargaining begins, 1 and 2 initially agree on the outcome o_1. Imagine this sequence of events: person 3, not wanting to be excluded from an agreement, offers 1 the outcome o', which 1 prefers to o_1. Hence, 1 switches his "loyalty" to 3 and to the outcome o'. Person 2, however, counters with an even better offer to 1, o'', whereupon 3 "raises the stakes" by proposing o'''. Thus, in their desire to be included in an agreement, and to receive a favorable payoff, 2 and 3 offer proposals that are ever closer to the pivotal player's ideal. Before this process converges to 1's ideal point, however, 2 sees the direction of events and instead approaches 3 with the proposal o_3. Clearly, 3 prefers o_3 to o''', his last offer to 1. With the coalition $\{2, 3\}$ now a real possibility, person 1 realizes that 2 and 3 have offered him too much. He might then return to 2 and assert his willingness to abide by their original agreement, o_1. If he overreacts by giving up too much, say, by offering 2 the outcome o^*, then 3 can respond with o_2, which both he and 1 prefer to 1's excessive compromise, o^*. Finally, consider the three outcomes o_1, o_2, and o_3. Referring to Figure 9.9b, if $\{1, 2\}$ offers o_1, if $\{2, 3\}$ offers o_3, and if $\{1, 3\}$ offers o_2, then the pivot between any pair of proposals is indifferent. Can a bidding war begin now? That is, can a new proposal upset these three proposals taken as a group? To see that it cannot, suppose that 3, in trying to attract 1 away from any agreement with 2, offers o' to 1. Although such an offer might seduce person 1, 2 could simply respond by bringing o_3 to 3's attention, which 3 prefers to o'. Person 3 is the pivot, and he logically should choose o_3. Person 1 may foresee this possibility and recognize that by allowing himself to be seduced by o', in effect he is making o_3 more probable. Hence, he refuses o' and expresses satisfaction with all current offers.

Bargaining need not proceed in this fashion, but this scenario illustrates the bidding up to pivots that $\mathscr{K}$ models, as well as the ability of coalitions to upset certain arrangements if pivots are getting too much or too little from the active coalitions. The intent of the competitive solution, then, is to define proposals that are balanced with respect to pivots and that the introduction of new proposals cannot upset. The preceding analysis of a three-person spatial game, though, should not convince us that $\mathscr{K}$ has any advantage over V-sets and bargaining sets, because those concepts also handle this situation adequately. What we are especially interested in is whether $\mathscr{K}$ makes predictions for other spatial committee games that correspond to the intuitions that observers of parliamentary coalitions have developed, and whether $\mathscr{K}$ treats spatial games, such as the one that Figure 9.7 depicts, more plausibly than does V or $\mathscr{M}_1$.

First, as we note earlier, students of parliamentary government have offered a variety of hypotheses about coalition formation if Euclidean distance represents party preferences. Hypotheses such as the prediction that minimum-variance or connected coalitions form, while unrelated to any specific solution hypothesis, summarize the intuition that coalitions should be sensitive to the negotiating parties' spatial locations.

To illustrate $\mathscr{K}$'s sensitivity to spatial location, consider Figures 9.10a–b. Each figure assumes that person 1 has two votes, and persons 2, 3, and 4 have one vote each. Thus, any coalition with person 1 and one other member is minimum winning, and the only minimum-winning coalition that excludes 1 is $\{2, 3, 4\}$. Figure 9.10a shows $\mathscr{K}$ and indicates that in this instance, a proposal represents each minimum-winning coalition. Turning to Figure 9.10b, notice that the locations of ideal points here are identical to those that Figure 9.10a depicts, except that

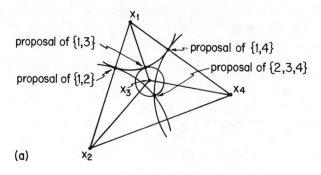

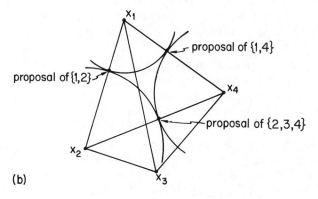

9.10(a) $\mathcal{K}$ with a proposal by $\{1, 3\}$. (b) $\mathcal{K}$ without a proposal by $\{1, 3\}$.

3's ideal point now falls outside the triangle formed by the ideal points of 1, 2, and 4. Thus, $\{1, 3\}$ no longer has a proposal in $\mathcal{K}$. The reader should check that if $\{1, 3\}$ tries to upset the competitive solution shown in Figure 9.10b, then any proposal it formulates is either not viable against the proposals of $\{1, 2\}$ and $\{1, 4\}$ or against the proposal of $\{2, 3, 4\}$: Either 1 is a pivot or 3 is a pivot, and no proposal is Pareto-optimal for $\{1, 3\}$ and gives 1 as much as he gets from $\{1, 2\}$ and $\{1, 4\}$ or gives 3 as much as he gets from $\{2, 3, 4\}$.

This sensitivity of $\mathcal{K}$ to the spatial location of ideal points matches our intuition about parliamentary coalitions and we can only wonder whether the preceding argument might not help us understand better why a coalition of Adams and Jackson in 1824 was unlikely. But before we explore other ideas about such coalitions, including the size principle and the connected-coalition hypothesis, let us reexamine the game in

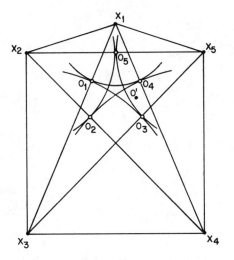

9.11 $\mathcal{K}$ for a five-person spatial game.

Figure 9.7. Recall that with simple majority rule, this configuration of preferences does not yield a main-simple V-set, and at the same time, the bargaining set seems inappropriate, owing to externalities. Figure 9.11 reproduces this game and the five specific outcomes, o_1 through o_5, that constitute a competitive solution.

$$\mathcal{K} = \begin{bmatrix} p_1 = (o_1, \{1, 2, 3\}) \\ p_2 = (o_2, \{2, 3, 4\}) \\ p_3 = (o_3, \{3, 4, 5\}) \\ p_4 = (o_4, \{4, 5, 1\}) \\ p_5 = (o_5, \{5, 1, 2\}) \end{bmatrix}.$$

To check whether $\mathcal{K}$ is a competitive solution, consider any two proposals in $\mathcal{K}$, say p_1 and p_2. In this instance, the pivots are persons 2 and 3, and while 2 prefers o_1 to o_2, 3 prefers o_2 to o_1. Hence, neither proposal is strictly viable against the other. Similarly, 5 is the pivot between p_3 and p_5, and as required, he is indifferent between them. An exhaustive search of this sort shows that $\mathcal{K}$ is balanced. Now consider any other outcome, say o' and a coalition such as $\{2, 4, 5\}$. Clearly, p_1 is strictly viable against the proposal $p' = (o', \{2, 4, 5\})$, because the pivot, person 2, strictly prefers o_1 to o'. Hence, p' cannot upset $\mathcal{K}$. Expanding this search reveals that no new proposal is viable against everything in $\mathcal{K}$ and strictly viable against something in $\mathcal{K}$. Hence, $\mathcal{K}$ is a competitive solution.

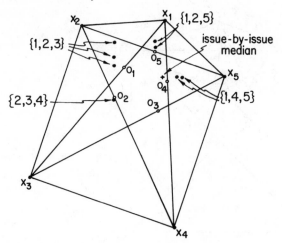

9.12   Experimental outcomes under majority rule.

9.9   Some experimental evidence

The preceding discussion gives us some hope that $\mathscr{K}$ provides a useful analysis of spatial-committee games, and Theorem 9.3 shows that $\mathscr{K}$ also treats other games, including transferable utility games, that the core and the main-simple V-set handle. We now review the results of some experimental studies, because they illustrate the application of solution theory to political voting games.

Spatial committees

Chapter 6 reviews a series of experiments that use the preference configuration in Figure 9.11 (see Figure 6.19). But in those experiments, subjects must propose and formally vote on motions that differ from a status quo, one issue at a time. The general conclusion is that if participants can costlessly communicate and coordinate, and thus can short-circuit procedures, the issue-by-issue median preference is less attractive as a prediction about eventual outcomes. But as the severity of procedural constraints diminishes, to what alternative points are outcomes drawn? Presumably, if we remove all restrictions on cooperation and communication, then the relevant predictions should be those that solution theory supplies. First, the bargaining set predicts points in the interior pentagon formed by contract lines in Figure 9.12 (see Figure 9.8), while the competitive solution predicts the five points shown on the sides of this pentagon. A series of experiments can provide a critical test of these two solution hypotheses. At the same time, the experiments form a logical extension of the experiments that Chapter 6 reports.

Figure 9.12 reports the results of eight experiments using the unconstrained bargaining format that solution theory assumes (we denote the coalitions enforcing these agreements alongside the outcomes). As before, we induce preferences using monetary payoffs that decrease monotonically from the respective ideal points, and to maintain the nonsidepayment character of the game, subjects cannot reveal their payoff schedules to each other, nor can they discuss schemes to divide payoffs at the termination of the experiment. Notice first, that although the predictions of bargaining-set theory fare poorly (no outcome is in the interior pentagon), $\mathcal{K}$ predicts outcomes remarkably well. The chosen outcomes are close to those that $\mathcal{K}$ predicts, and the appropriate coalitions enforce those outcomes. This figure also provides a sharp contrast to the experiments reported in Figures 6.19a–b. Specifically, the attractiveness of the issue-by-issue median is nowhere in evidence in Figure 9.12.

These experiments, then, tell us what happens in the next step of the progression, as we eliminate any institutional constraints entirely, aside from the majority-rule requirement. With strong constraints, the relevant predictive hypothesis is a noncooperative equilibrium (the issue-by-issue median), but with no restrictions the relevant hypothesis is a cooperative solution concept, $\mathcal{K}$ in this instance. And the results in Figures 6.20a–b reveal that if the severity of restrictions falls between these two extremes, then outcomes also lie between the predictions of both models. We can use this observation to infer other hypotheses about committees. For example, we can infer that as the importance of issues increases relative to the costs of circumventing institutional constraints, the results will tend toward cooperative outcomes. But as the costs of bypassing or changing procedures increase or as the importance of issues declines, procedural constraints will impose noncooperative equilibria as outcomes.

Weighted voting systems

Perhaps the most extensive application of cooperative game theory to politics concerns the formation of parliamentary coalitions, which are the governments that the political parties establish. Research in this area proceeds in two general directions. The first assumes that ministries or cabinet posts are a transferable resource that serves the same theoretical function as money for economists. The second direction, and the one that we focus on here, assumes a spatial conceptualization of party positions and preferences. This direction, then, supposes that ministries, at best, are indicators of policy agreements on issues, and accordingly, it requires a nontransferable-utility formulation. Thus, although researchers in this area remain sympathetic to the focus of the size principle on coalition predictions, the assumption of transferable utility seems unduly restrictive. At the same time, solutions such as the core and V-set stand mute on the question of which coalitions are most likely to form. This absence of prediction seems especially frustrating to political scientists, because coalitions often are the only observable variable. As a result, several scholars have proposed a variety of alternative hypotheses that seem reasonably connected to empirical evidence and intuition, even though they do not derive formally from solution theory.

One of the principal hypotheses of this sort is that "connected" coalitions have an advantage over others. "Connected" means that the outcomes that are Pareto-optimal for the coalition should not contain the ideal points of excluded parties. Because $\mathcal{K}$ is specifically concerned with spatial games and with coalition predictions, it should say something about this hypothesis. To this end, recall that $v^*(C)$ denotes the Pareto-optimal outcomes for the coalition C. Using this notation, we define,

> *External coalition*: A winning coalition C is external if there is another winning coalition C' with $C \cup C' = N$ and $C \cap C' = \{i\}$, for some $i \in N$ and $v^*(C) \cap v^*(C') = \{x_i\}$, the ideal point of the pivot between C and C'. Any winning coalition that is not external is internal.

Hence, a minimum-winning coalition, C, is external if the Pareto-optimals of the complementary (losing) coalition, $N - C$, do not include any part of $v(C)$, and if the Pareto-optimals of C do not contain the ideal points of $N - C$ [except perhaps on the boundary of $v^*(C)$]. For example, the coalition $\{1, 2, 3\}$ in Figure 9.11 is external, because we can find another minimum–winning coalition, $\{1, 4, 5\}$, such that Pareto–optimals of both coalitions intersect only at $\{x_1\}$. And notice that v^* of the complement of $\{1, 2, 3\}$, the coalition $\{4, 5\}$, does not intersect $v^*(1, 2, 3)$. Thus, an external coalition is connected; but additionally, its complement is connected as well.

The relevance of this definition is a result that we can establish for $n \leqslant$ 5-person games, but that remains a conjecture for larger games: *In any spatial game of the sort considered in this chapter (that is, majority-rule, Euclidean preferences, and so forth) if the competitive solution is not empty, then there exists one solution such that if the proposal $p = (\mathbf{u}, C)$ is in $\mathcal{K}$, the coalition, C, is external.*

We word this conjecture cautiously for two reasons. First, we have not established a general-existence theorem for $\mathcal{K}$. Second, if a core exists, such as occurs if the spatial game concerns only a single issue, then there are multiple competitive solutions; any proposal consisting of a core outcome in conjunction with *any* winning coalition. Of course, this qualification is only reasonable because if a (unique) core exists, then the eventual policy outcome is insensitive to the specific coalition that forms. Those who study coalitions in parliaments have largely ignored this insensitivity. (Implicitly, then, those studies of parliamentary-coalition formation that posit the hypothesis of connected coalitions assume that a core does not exist.) Aside from these qualifications, the competitive solution formalizes the intuition leading to the hypothesis that connected coalitions are favored over others. Indeed, it strengthens this hypothesis by highlighting its dependence on the existence of the core. Notice, also, that although $\mathcal{K}$ predicts connected and external winning coalitions in Figure 9.12, the bargaining set predicts internal winning coalitions and, thus, disconnected losing coalitions (see Figure 9.6a).

Turning to actual parliamentary systems, research either can use the "laboratory" of parliaments to test hypotheses or it can use a specific hypothesis to discover those features of a situation that are common to parliaments in general. Ideally, we should combine these procedures, but a difficulty in testing $\mathcal{K}$ (or any other hypothesis) with parliamentary data (compared with applying a hypothesis) is that outcomes vary greatly with specific assumptions about party positions, the costs of negotiation, and historical features peculiar to one country or another. Parliaments, then, provide a poor laboratory for *initial* evaluations of abstract hypotheses. Thus, although the experimental laboratory has its own disadvantages, we turn again to this alternative.

A useful procedure is to take a spatial representation of party positions, along with the voting weights of parties, as reflected in the seats they control. We then derive the various predictions of applicable solution hypotheses and test these predictions against those observed in an appropriate experimental setting. For example, Figure 9.13 shows the spatial preferences that Converse and Valen ascribe to the Norwegian parliament in 1969. Owing to the parties' voting weights, the minimum-winning coalitions are (1) the Social Democrats (SD) with any other party, and (2) all parties against the Social Democrats. Figure 9.13 also shows the four (unique) competitive solution points associated with these parties. The bargaining set, on the other hand, predicts the coalitions {L, SD} and {CP, SD}, which we associate with proposals in $\mathcal{K}$, as well as {Ctr, SD} and {C, SD}, which the competitive solution predicts will not form. Hence, we should take the occurrence of {Ctr,

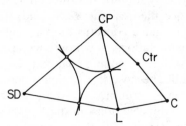

Party	Voting Weight
Center (Ctr)	18
Christian Peoples (CP)	15
Conservative (C)	31
Liberal (L)	18
Social Democrat (SD)	68

9.13 Competitive solution.

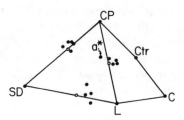

9.14 Experimental outcomes.

SD} and {C, SD} as support for bargaining-set theory, and the formation of {L, CP, Ctr, C} as support for $\mathcal{H}$.

Figure 9.14 summarizes fourteen experiments that used the same procedures as those of the experiments that Figure 9.12 reports. Outcomes essentially are evenly divided between the three points that $\mathcal{H}$ predicts. A coalition of the whole supported the outcome denoted a^*, and fairness emerged as an explicit criterion in the subjects' discussion there; otherwise, all observed outcomes reflect the minimum-winning coalitions that $\mathcal{H}$ predicts. Hence, the five outcomes that the coalition {L, CP, Ctr, C} adopts dispute bargaining-set theory (as well as those hypotheses that predict winning coalitions with a minimal number of parties or minimum variance), while two of the four coalitions with proposals in the bargaining set, {SD, Ctr} and {SD, C}, never form.

These experimental outcomes give us some overall confidence in $\mathcal{H}$ as a hypothesis about the formation of parliamentary coalitions. But we should not become overconfident. These experiments take place in the rarefied atmosphere of the laboratory, which may or may not be relevant to coalition processes in parliaments. Although any adequate theory must explain what occurs in the laboratory as well as in a parliament or legislature, the laboratory is a useful device for eliminating hypotheses from consideration. But it is not necessarily a good device for assessing an idea's general validity.

Legislator	Bill 1	Bill 2	Bill 3	Bill 4	Bill 5
1	10	−2	−4	−4	−2
2	−2	10	−2	−4	−4
3	−4	−2	10	−2	−4
4	−4	−4	−2	10	−2
5	−2	−4	−4	−2	10

9.15 Vote-trading experiment with empty core.

Vote trading

Earlier, we discussed a series of vote-trading experiments with a core but in which the core fares poorly (see Section 8.7). Since $\mathcal{K}$ corresponds to the core if the core exists, we can interpret those experiments as evidence against $\mathcal{K}$ (as well as against all other game-theoretic solutions). Paradoxically, though, $\mathcal{K}$ performs better if the core is empty. Specifically, consider the vote-trading game in Figure 9.15.

Here, the sincere-voting outcome fails all bills, and yields the payoff 5-tuple (0, 0, 0, 0, 0). Passing bills 1, 2, and 3, though, dominates this outcome with respect to the coalition {1, 2, 3}, so that by Theorem 2.4, no Condorcet winner exists and, by Theorem 8.2, no core exists. This game also appears to have significant symmetries among all minimum-winning coalitions. Nevertheless, the unique competitive solution involving minimum-winning coalitions is limited to these five outcomes and coalitions:

Coalition	Outcome	Payoffs
{1, 2, 3}	(1, 1, 1, 0, 0)	(4, 6, 4, −10, −10)
{2, 3, 4}	(0, 1, 1, 1, 0)	(−10, 4, 6, 4, −10)
{3, 4, 5}	(0, 0, 1, 1, 1)	(−10, −10, 4, 6, 4)
{4, 5, 1}	(1, 0, 0, 1, 1)	(4, −10, −10, 4, 6)
{5, 1, 2}	(1, 1, 0, 0, 1)	(6, 4, −10, −10, 4)

It might seem strange that $\mathcal{K}$ predicts only these minimum-winning coalitions. But it is straightforward to see that this set is balanced. To show that some other proposal cannot upset this set, consider, for example, the proposal by the coalition {1, 3, 5} to pass bills 1, 3, and 5, which seems to be a logical agreement for them. The payoff vector from this outcome is (4, −8, 2, −8, 4). But person 3 pivots between {1, 3, 5} and {2, 3, 4}, and he strictly prefers the proposal to pass bills 2, 3, and 4. Thus, {1, 3, 5} cannot upset $\mathcal{K}$. An equivalent argument shows that no excluded coalition can upset the preceding configuration of proposals, and that it constitutes a competitive solution.

Preference Orders

Person 1	Person 2	Person 3	Person 4	Person 5
n	j	b	l	b
j	o	h	e	a
f	m	a, f	←	←
→ i	e, f	i	d ⌉	e, h, ⓖ
k	i	k, m, d, j	a, o, ⓖ	k
ⓖ, o, h	k, d	g	m	d
d, m	g, b	e, c	i	m
b	h, c	l, n	k, b	o, c
a, c	l, n	o	f, c	l, n, j
l	a		n, j	f, i
e			h	

V-set = {d, j, k, m}

$\mathcal{M}_1$ = {(d, {1, 3, 4}), (g, {1, 4, 5}), (g, {1, 2, 5}), (g, {1, 2, 4, 5}), (m, {2, 3, 5})}

$\mathcal{K}$ = {(a, {3, 4, 5}), (e, {2, 4, 5}), (f, {1, 2, 3}), (h, {1, 3, 5}), (o, {1, 2, 4})}

9.16 Testing $\mathcal{K}$ with finite alternatives.

In 10 experimental trials using this game and an essentially equivalent version, outcomes in $\mathcal{K}$ prevail in every instance. Coalitions such as {1, 3, 5} never occur. These results are suprising in the light of the core's "failure" to predict outcomes, with anything approximating this success rate, in a similar vote-trading game. Of course, any explanation for this difference is at best pure conjecture. There is much about the *process* whereby people form coalitions and formulate viable proposals that we do not understand. All we can say at this point is that the competitiveness and instability of games such as the one in Figure 9.15 appear to draw out the appropriate proposals, as players seek viable alternatives.

Experimental anomalies

We should not interpret the preceding survey of experimental findings to mean that all experimental research supports $\mathcal{K}$ or some other solution hypothesis. For example, consider the preference orders over 16 alternatives, denoted a through p, in Figure 9.16. Excluding lotteries, the competitive solution consists of the five proposals (a, {3, 4, 5}), (f, {1, 2, 3}), (e, {2, 4, 5}), (o, {1, 2, 4}), and (h, {1, 3, 5}). But the V-set consists of the alternatives {d, j, k, m}, while the bargaining set involves only {d, g, m}.

In a series of 58 experiments using these and similar preferences, the players chose $\mathcal{K}$ 47 times while they chose outcomes in V or in the bargaining set only 9 times. Hence, the results of these experiments, together with the successes reported for $\mathcal{K}$ in the vote-trading and spatial-committee games, suggest that $\mathcal{K}$ is indeed the appropriate solution concept for majority-rule games. Certainly, to assert the viability of V or of any bargaining set we would have to explain away a great deal. But Figure 9.16 also indicates another position for alternative g, by raising it in the preference orders of persons 1, 4, and 5. Certainly, g now seems to be more attractive, and, indeed, in a new series of 15 bargaining experiments the players selected g 9 times. Interestingly, though, g is not in $\mathcal{K}$, and raising g does not change the proposals in $\mathcal{K}$. This experimental series, then, poses a problem for the competitive solution.

As a final variation on these experiments, Miller and Oppenheimer report results in which they manipulate the success rate of $\mathcal{K}$ by varying the cardinality of the payoffs to the subjects. Briefly, these experiments use the following simplified preference profiles, which involve only the five competitive solution points plus a sixth alternative, d, that falls exactly in the middle of each person's scale:

Person 1	Person 2	Person 3	Person 4	Person 5
f	o	h	e	a
o, h	e, f	a, f	a, o	e, h
d	d	d	d	d
a	h	e	f	o
e	a	o	h	f

The competitive solution for this configuration remains "everything but d." Nevertheless, by varying the relative payoffs of d, $\mathcal{K}$'s success rate in 35 trials varies between 0% and 80% with an overall success rate of only 20%. The interpretation that the experimenters give to these results is that if the payoff from d is sufficiently high (approximately one-half the expected value associated with an equiprobable lottery over e, a, o, h, and f), then subjects avoid the uncertainty of competition (the lottery between a high payoff if one is included in a winning coalition and a low payoff if one loses) in favor of the safe and fair outcome d.

The procedure that Miller and Oppenheimer use to induce preferences in these experiments differs from the procedures used with the preferences in Figure 9.16, which may account for some differences in the success rates of $\mathcal{K}$. But it is evident that subjects respond to the cardinal values of their payoffs, and that changes in these values affect the frequency with which specific outcomes prevail. The cardinality of preferences appears to affect the extent to which subjects play the cooperative game competitively. We should anticipate that similar phenomena will occur in politics. Thus, if the issues are of sufficient importance, then coalitions and outcomes probably will accord with $\mathcal{H}$'s predictions; but if "fair" or risk-avoiding outcomes exist, if the potential benefits of being included in some minimum-winning coalition do not greatly exceed the payoffs associated with such fair outcomes, and if the costs of being excluded are great, then players might coordinate their strategies beforehand to avoid the instabilities associated with competitive negotiations. Indeed, this provides one hypothesis about the emergence of universalism in the U.S. Congress with respect to public works legislation.

One objection to experimental methods in general, however, is that they do not mirror politics, that they are artificial environments, and that we can learn little from undergraduates whose payoffs take the form of "hard cash." But this unscientific view, which is supported if a discipline depends exclusively on historical observation and journalistic interpretation as against the development and testing of general theories, ignores the fact that an adequate theory should be viable in every environment. Thus, we must take seriously negative as well as positive results. They also offer a general lesson. Nearly everyone involved in experiments is surprised by the strategic complexity that subjects in even the simplest experimental situations display. Deception, taking advantage of the willingness of other subjects to contemplate considerations of equity, threats, and counterthreats, and the attempt to "mold" one's personality to fit the sympathies of other subjects, all emerge in most experimental sequences. Yet outcomes appear to fit certain patterns. And although our theories cannot accommodate all such patterns, they can account for a considerable amount of experimental data. It certainly will always be possible to devise experiments that yield outcomes at odds with some theoretical proposition. Experimental anomalies open the door to future research, while patterns that we do understand suggest that we can study the presumed complexity of political processes in an orderly way.

9.10   A noncooperative view and alternative ideas

We commented earlier that one problem with the V-set is that it lacks a convincing "behavioral" foundation. But although the bargaining set and the competitive solution both enjoy reasonable behavioral motivations, at the same time they seem to be disjointed from noncooperative game theory. Certainly, they would be more convincing hypotheses if we could justify them in terms of Nash equilibrium notions, even if the complexities of modeling a cooperative game in extensive form necessitate some simplifications, such as the one that the characteristic function introduces.

While admitting that the weak connection between cooperative and noncooperative game theory remains a problem for the competitive solution as well, consider the following possibility: Suppose, again, that we conceptualize bargaining in a strong simple game as a noncooperative game between the $2^n - 1$ nonempty coalitions. Next, let the proposal, $p_c = (\mathbf{u}, C)$, be the "strategy" of the winning coalition, C, let P be a set of proposals such that no more than one proposal in P represents each coalition, and let $\mathbf{C}$ be the set of winning coalitions with proposals in P. Now define the "payoff function to the coalition" thus:

$$U_c(P) = \min_{C' \in \mathbf{C} - C} \max_{i \in C \cap C'} [u_i(p_c) - u_i(p_{c'})], \; p_c \in P. \tag{9.3}$$

This expression defines the payoff to the coalition C thus: For every other coalition $C' \in \mathbf{C}$, look at the pivots between C and C' and compute the maximum differential that a pivot associates with p_c compared with $p_{c'}$. $U_c(P)$ is the minimum of these maxima. If $U_c(P) \geq 0$, then p_c must be viable against every proposal in P; otherwise, if it is not viable against $p_{c'}$, say, then all players that pivot between C and C' strictly prefer $p_{c'}$, in which cases

$$\max_{i \in C \cap C'} [u_i(p_c) - u_i(p_{c'})] < 0,$$

so that, no matter what value this term assumes for other coalitions, the minimum of the maxima is negative. This reasoning directly establishes that if $\mathcal{K}$ is a competitive solution (if no proposal in $\mathcal{K}$ is strictly viable against any other proposal in $\mathcal{K}$), $U_c(\mathcal{K}) \geq 0$ for all coalitions with proposals in $\mathcal{K}$.

Now consider the noncooperative game among all coalitions, in which each coalition's strategy is a proposal, and in which expression (9.3) gives its payoff function. It is possible to show (we spare the reader a proof) that if P^* is a (strict) Nash equilibrium to the corresponding noncooperative game, such that $U_c(P^*) \leq 0$ for all C, and if we let $\mathcal{K}$ be the set of proposals, such that $U_c(P^*) = 0$ if p_c is in $\mathcal{K}$, then $\mathcal{K}$ is a competitive solution. Thus, certain kinds of Nash equilibria for the noncooperative game among coalitions correspond to competitive solutions; a link is thereby established between $\mathcal{K}$ and noncooperative game theory.

The principal problem with the competitive solution, however, is that there does not yet exist any general existence theorems for specific categories of games. Of course, Theorem 9.3 explains that $\mathcal{K}$ exists whenever the core or a main-simple V-set exists. But the purpose of $\mathcal{K}$ is to make predictions for games that the core and V cannot handle. Laing and Olmstead report a five-person spatial game that appears not to have a competitive solution. For this reason, Gretlein develops the *defensible set*, which resembles but is not identical to $\mathcal{K}$. Briefly, the idea behind the defensible set is that each person's strategy is a utility level, l_i, such that i agrees to settle for any payoff at or above this level. Each player must defend this level thus: If $p = (\mathbf{u}, C)$ gives each member, i, of C at least l_i, and if $p' = (\mathbf{u}, C')$ drops i below his level, then there must exist a $p'' = (\mathbf{u}'', \mathbf{C}'')$, such that p'' is at least as good as p' for the members of C'' and $u_i'' \geq l_i$. Letting $\mathbf{l} = (l_1, l_2, \ldots, l_n)$ and $p_i(\mathbf{l})$ be all proposals that simultaneously are above i's level and above the levels for winning coalitions, we define

$$u_i(\mathbf{l}) = \begin{cases} l_i \text{ if } i \text{ can defend all proposals in } p_i(\mathbf{l}) \\ \text{some minimal payoff otherwise.} \end{cases}$$

The corresponding noncooperative game, then, finds all persons maximizing their levels, and the (possibly nonunique) Nash equilibrium levels define the cooperative solution. That is, given a Nash equilibrium of levels, all proposals satisfying the levels for winning coalitions constitute the defensible set. With the defensible set, then, instead of demanding specific payoffs, people act on some notion of an acceptable payoff. But the level of utility that people are willing to accept depends on strategic possibilities and the levels of others.

Although the details of this solution hypothesis are incomplete, the defensible set or some similar idea might generalize and synthesize both bargaining-set theory and $\mathcal{K}$. Bargaining-set theory requires that people be able to defend every payoff that they receive, including those payoffs that result from fortuitous externalities. But with a defensible-set formulation, players need only defend up to some minimal level. Furthermore, the defensible set places no restrictions on what a person's coalition partners receive when he formulates either an objection or a counterobjection. This hypothesis, then, solves games that bargaining sets treat unsatisfactorily. For example, the reader should check that the indifference contours describing the competitive solution in Figure 9.11 correspond to a Nash equilibrium of levels, and that the defensible set is identical there to $\mathcal{K}$ (although Gretlein presents examples in which this equivalence does not hold).

One problem with the defensible set is that it remains unclear how we should extend it if a Nash equilibrium of levels does not exist. Mixed strategies seem farfetched in this context. Elaine Bennett offers a variant of the defensible set, called the *bargaining equilibrium*, which she defines thus: Let l be a vector of levels; let $A(p)$ be the outcomes in O that satisfy the levels for a majority of players; let $W(p)$ be the winning coalitions associated with outcomes in $A(p)$; and let $W_i(p)$ be the winning coalitions in $W(p)$ of which i is a member. Then, player i *is vulnerable to player j at p* if i needs j's cooperation, but j does not need i's cooperation to be included in a possible winning coalition, given everyone's levels. The n-tuple of levels, l, corresponds to a bargaining equilibrium if for all i, $W_i(p)$ is not empty, if no player is vulnerable, and if l is a strong Nash equilibrium [that is, neither a single player nor group of players can raise his or their levels without becoming vulnerable or without $W_i(p)$ becoming empty]. Again, though, as yet no general existence theorem supports this idea.

Although the preceding discussion might reenforce the view of a multitude of solution concepts, this research promises ultimately to yield a different result. That defensible-set notions make the competitive solution appear as a generalization of the bargaining set for majority-rule spatial games suggests that a synthesis is possible, or if not a synthesis, at least a generalization that remains consistent with the strengths of each hypothesis.

9.11 Summary

This chapter reviews three solution hypotheses that treat cooperative games without cores: the V-set, the bargaining set, and the competitive solution. Familiarity with the following concepts is required before the strengths, weaknesses, and applicability of any of these hypotheses can be understood:

stable set, V	internal stability	external stability
discriminatory V-set	main-simple V-set	proposal
objection	counterobjection	$\mathcal{M}_1$ and $\mathcal{M}_2$
coalition structure	the size principle	coalitionally rational
minimum winning	minimum weight	proposal
compact coalition	strong bargaining set	connected coalition
strictly viable	pivotal	viable
upset	competitive solution	balanced
defensible set	bargaining equilibrium	external coalition
uniformly decreasing		vulnerable
returns to scale		

Once again, though, we should remind the reader that the solution hypotheses reviewed here can be applied legitimately to a situation only if an appropriate characteristic-function representation of the situation is at hand. This chapter's examples are concerned mainly with committees that use simple majority rule in an otherwise procedure-free context. But if specific procedures otherwise constrain what people can do – what constitutes a winning, blocking, and losing coalition – then $v(C)$ and the dominance and viability relations must reflect these constraints before the various solution hypotheses are applied.

Admittedly, the focus on simple majority rule can be justified by the argument that if there are no costs to negotiation, coalition formation, and coalition maintenance, then procedures probably matter little since some coalition, generally a majority, can short-circuit any procedural constraint. Reality, then, lies somewhere between the analysis offered in this and the previous chapters and the analysis in Chapter 6. Procedural details are rarely if ever wholly decisive, but transaction costs are probably never zero.

With respect to the advantages and disadvantages of each of the three solution hypotheses that this chapter reviews, Figure 9.17 provides a general summary and conveys the opinion that, of the three concepts, the competitive solution seems to be the most promising. But because we cannot guarantee the existence of $\mathscr{K}$, its definition and ideas such as the bargaining equilibrium and the defensible set will be subject to revision as research proceeds.

Before we conclude this discussion, though, we should note that in spite of the considerable effort devoted to solution theory, there nevertheless remains a disquieting aspect to this research. The concepts reviewed here rely on the characteristic function, and to the extent that $v(C)$ abstracts away some of the strategic character of a situation, and in particular, eliminates consideration of specific individual choices, the definitions of these concepts share an uncomfortable ad hoc flavor. Certainly, anyone who has studied these concepts understands that at times the choice of a formal assumption seems arbitrary, and we often choose a particular one not to model some substantive behavioral idea, but instead to facilitate the proof of a result. Of course, one alternative to the characteristic function is to maintain a noncooperative representation and to suppose that the alternatives that coordination and communication admit augment each person's strategies. But this approach must contend with the complexity of a cooperative game's extensive form.

Aside from complexity, viewing cooperative games as noncooperative has a logical appeal. Suppose that two people agree to some coordinated action, and that we label them a coalition. But suppose, further, that the coalition has no strictly legal means or "third parties" available for

	Advantages	Disadvantages
V-set solution	Concise definition applicable to games with and without side-payments and transferable utility. Highlights philosophy of solution theory.	Multiple (discriminatory) solutions; no behavioral rationale and hence, no coalition predictions; for spatial games only discriminatory solutions exist in general.
Bargaining set theory	Imaginative bargaining rationale; predictions similar to those of V-sets but avoids discriminatory solutions.	Defined with transferable utility games in mind but takes account of externalities inadequately; definitions seem arbitrary at times.
$\mathcal{M}_1$	General existence theorem for transferable utility games.	Implausible predictions for some simple games with transferable utility; no coalition predictions.
$\mathcal{M}_2$	Plausible predictions for simple games with transferable utility.	No general existence theorem; coalition predictions for spatial games seem unreasonable
$\mathcal{K}$	Behavioral rationale with non-cooperative interpretations; related to the core, main-simple V-set, and strong bargaining set. Eliminates discriminatory solutions, make coalition predictions, good experimental support.	No general existence theorem, and may not exist for majority-rule spatial games; unknown properties for games with transferable utility; difficult to compute, and may require modification for general existence.

9.17 Summary of solution concepts.

enforcing this agreement. If the coalition is to survive, then it must do so because it is in the interests of both persons, acting individually, to maintain it. This noncooperative view of coalitions, then, asserts that coalitional agreements arise and persist because it is in the interests of the persons involved to avoid choosing unilaterally some other action. Organizations such as the United Nations enjoy ephemeral support when compared with the support for national interests. One set of national leaders maintains agreements with another because it is in their interest to do so, and not because some international court reviews their actions. Similarly, people do not maintain agreements to trade votes because one member gives up his seat or voting button to another or because one member can sue another for violating a verbal contract. Instead, they maintain implicit contracts because, once they agree to it, it remains in each member's self-interest to abide by the contract.

Consistency with the theoretical foundations of individual preference and choice requires this self-interest perspective. One hypothesis about vote trades and international treaties, then, is that their durability stems from properties of the larger game of which they form a part. If a legislator breaks a vote-trading agreement, then his colleagues are unlikely to trust his word in the future. The self-interest or self-enforcing character of a congressional coalition derives from each legislator's desire to keep open the possibility of future trades and to be an effective member of Congress. Similarly, abiding by an international treaty, even if it means less that optimal choices in the short run, maintains for a nation the possibility of profitable future coordination. Thus, coalitions emerge and survive because of an implicit *supergame* of which they are but a small and transitory part. By supergame, we mean a repetition of similar or dissimilar games over time either with a finite or infinite horizon. A supergame, then, consists of stringing together a series of shorter extensive-form games. Because it is a repetition of noncooperative games, a supergame is itself noncooperative. But a strategy that a player adopts within any subgame of the supergame takes on new meaning, because it signals to others how one intends to play future games, and it reflects what one has learned from games played in the past.

Much of this discussion seems unexceptional. But earlier we argue that we should not try to model too much of reality at one time. Short of trying to accomplish the impossible task of modeling all of the future, we necessarily confined any particular game to a "snapshot" of real time. Thus, classical cooperative game theory simply assumes the mechanisms with which people enforce coalition agreements. It then proceeds with a characteristic-function representation of the main phenomena under investigation. And despite solution theory's obvious abstractions and simplification, few ideas give us the equivalent predictive power in the experimental laboratory, while simultaneously promising consistency with empirical regularities such as connectedness and minimum size. Nevertheless, we cannot reject the hypothesis that a universally adequate solution theory requires some model of the dynamic and temporal features of a situation, a subject that Chapter 10 explores.

CHAPTER 10

Repeated games and information: some research frontiers

No matter what theorems we prove and what empirical support we muster, Chapter 9 suggests that solution theory will retain some ad hoc quality unless we can relate it to an extensive-form model of cooperation. Ultimately, we want to rationalize solution theory by appealing directly to assumptions about individual rationality and utility maximization, and not to coalition formation hypotheses. Although this is a goal that solution theory does not meet today, researchers have taken some cautious first steps toward that goal, and this chapter reviews some of them.

Since these steps are research on the frontier, the parts of this chapter are not necessarily well integrated. First, we look at the prisoners' dilemma when the same persons play it successively. Because there is a significant experimental literature that reports the emergence of cooperative action in repeated prisoners' dilemma games, the analysis of this game might yield some insight into how cooperation can arise in general and how people can sustain it in the absence of formal enforcement mechanisms.

Although the analysis of repeated games helps us to understand cooperation in politics, we must consider another complication. Chapter 4 analyzed elections with incomplete information and showed how a succession of polls can yield outcomes that are identical to those that we would predict if information is complete. Chapter 5 considered the voting calculus in a simple election if voters do not know the costs of voting for other citizens. And Chapter 6 examined voting under binary agendas with incomplete information. The treatment of incomplete information in sequential games is even more complex than the preceding "simple" formulations. Strategies not only dictate outcomes directly, but also communicate something about one's preferences to others, and the information that strategies communicate ought to be a factor in each person's strategic considerations. Focusing on a particular example, the chain store paradox, we consider how people might select strategies in a game's early stages to reveal information that affects subsequent choices.

The third part of this chapter looks at two-person bargaining. Here

441

we review some classical approaches and normative applications to the evaluation of voting systems. But two-person bargaining models may be the most problematical area of inquiry because scholars remain unsure even of how best to represent such situations as games. The final part of this chapter reexamines repeated prisoners' dilemmas and offers a model that might explain the emergence of reciprocity norms in a legislature. It also shows how we might sustain at least one hypothesis about two-person bargaining as a noncooperative equilibrium.

10.1   The repeated prisoners' dilemma

A useful place to begin a reanalysis of cooperation is the prisoners' dilemma, because it, as much as any other game, highlights the problem of maintaining cooperative agreements. Recall that in a single play of this game, a Pareto-inefficient equilibrium seems inescapable. If no enforcement mechanisms exist, then even if the players can communicate beforehand, each has an incentive to defect from a cooperative agreement. Nevertheless, to repeat Chapter 7's discussion of $v(C)$ and enforceable contracts, the characteristic function for a prisoners' dilemma with transferable utility, such as the one that Figure 7.15a describes, is

$$v(1) = v(2) = -2; v(1, 2) = 20.$$

Thus, a core exists and corresponds to the set $\{u \mid u_1 + u_2 = 20, u_i \geq -2\}$. Notice that attaining any outcome in the core requires that both persons choose their first (cooperative) strategy. But each person has an incentive to defect, so without enforcement, the core will not prevail. Asserting that the core predicts outcomes for this game presupposes enforcement and illustrates that $v(C)$ ignores the question of why maintaining coalitional agreements is in a coalition member's self-interest.

Instead of assuming the existence of enforcement mechanisms, we note that a considerable experimental literature reports the emergence of implicit cooperation in repetitive prisoners' dilemmas. This literature interests social scientists because of its potential substantive significance. Chapter 5 shows that many justifications for government intervention in private markets concern the resolution of prisoners' dilemmas occasioned by public goods. But if people can coordinate and cooperate even if it is not in their short-term interest to do so, then government intervention may remain unnecessary to resolve all such dilemmas. This reasoning ultimately may alter our views about the proper functions of government and about the nature of the dilemmas that governments themselves induce.

This literature also interests game theorists, who model repeated versions of the prisoners' dilemma to learn something about how they might represent cooperative agreements as Nash equilibria. But we cannot solve the problem of rationalizing cooperation merely by supposing that a game is played repeatedly. Consider the situation in which the prisoners' dilemma is played several times by the same two persons who both know beforehand the number of repetitions. We might speculate that a person should cooperate early, hoping to induce the other player's cooperation in future trials. Suppose, however, that we analyze this situation the same way that we analyzed roll call voting in Chapter 3. That is, we compute the subgame perfect equilibrium by first finding the strategies that the participants ought to select in the last play of the game; second, we use this information to infer a choice on the next to the last play; and so on. With respect to the prisoners' dilemma, notice that there is no reason to cooperate on the last trial, because cooperation cannot yield any subsequent rewards. On the last play, each person should simply choose his dominant strategy, which is to defect and not cooperate. But this conclusion means that cooperation on the next-to-the-last trial is fruitless as well because the participants know that it cannot induce cooperation on the last trial. This argument yields the inevitable conclusion that both persons should defect on the very first play of the game and in every play thereafter. Put differently, the Nash equilibrium resulting in a Pareto-inefficient outcome is the unique subgame perfect equilibrium to the finite supergame.

Although this argument shows that simply repeating a game does not always induce cooperation, there is a difference, however, between this case and many of the dilemmas that we anticipate in politics. In many instances, people may not know when a game will end. Perhaps only lame ducks in a legislature can identify their last vote, but even they may harbor future political ambitions. For example, vote trading occurs in an environment in which participants cannot measure the precise number of profitable future trades. And in international relations, national leaders would prefer to believe that the time horizons of the nations that they represent are infinite. Modeling such situations, then, requires an extensive form of unknown or infinite duration. But we cannot analyze such extensive forms by backward reduction and the concept of subgame perfectness, because the terminal node is unknown, or, in the case of an infinite repetition, because such a node does not exist.

Hence, we have two possibilities with respect to the repeated prisoners' dilemma: There can be an infinite or unknown number of repetitions, or the number of repetitions can be finite and known. In the second case, if the information of all participants about the game's structure is complete, then the unique subgame perfect Nash equilib-

Person 2

		c	d
Person 1:	c	x, x	z, y
	d	y, z	v, v

10.1 Prisoners' dilemma.

rium entails noncooperative action at every point in the sequence. But we must analyze the first case differently. To see how the analysis might proceed, let us consider in particular the game in Figure 10.1, where, to be a prisoners' dilemma, it must be the case that $y > x > v > z$.

Assuming that the players repeat this game an infinite number of times, notice that unless each person discounts future payoffs, the expected payoff from the supergame need not be finite for any strategy sequence. In this event we cannot deduce optimal strategies. Since it is unreasonable to suppose, however, that anyone would regard, say, receiving \$50 today as equivalent to receiving the same amount a year from now, suppose that each player discounts future payoffs. Specifically, let the relative weight given to a payoff received in period t be w^{t-1}, where w is some number between zero and one. Hence, if u_{it} is player i's payoff in period t, then he evaluates that payoff *now* to be worth

$$u_{it}w^{t-1}.$$

Put simply, the higher the value of w, the more weight i places on late moves, as compared with the value that he places on the outcome of the current period. Hence, the *present value* of the stream of payoffs, u_{i1}, u_{i2}, ..., is

$$u_{i1} + u_{i2}w + u_{i3}w^2 + \cdots + u_{in}w^{n-1} + \cdots \qquad (10.1)$$

Having thus specified a general form for the present value of a stream of payoffs from an infinite repetition of a game, we must now specify strategies for that game. Formulating the normal form of an infinitely repeated game, though, requires that we make some simplifying assumptions about available strategies. In an infinite repetition of any game, there are always more strategies than we can describe, because, with an infinite number of plays, there are an infinite number of conditional responses. But it seems reasonable to suppose that people simplify their decisions and adopt certain heuristics as strategies in the repeated game when confronting such complexity. That is, they limit

Person 2:	$a+$	T	$c+$	$d+$
$a+$	x, x	x, x	x, x	$z(1-w), y(1-w)$
T	x, x	x, x	x, x	$z(1-w), y(1-w)$
$c+$	x, x	x, x	x, x	z, y
$d+$	$y(1-w), z(1-w)$	$y(1-w), z(1-w)$	y, z	$0, 0$

(Person 1: labels the rows)

10.2 Prisoners' dilemma supergame.

their conditional responses to a few predetermined patterns. Although we can imagine many interesting possibilities, the most commonly considered heuristics in the context of the prisoners' dilemma are these:

$c+$. Choose c on every occasion and hope that the other player eventually reciprocates.

$d+$. Choose d on every occasion without regard to the other player's choices.

$a+$. Choose c for the first trial; choose c if the other person chooses c in the previous trial. Once the other person chooses d, choose d thereafter.

T. (Tit for tat): Choose c on the first trial; thereafter, match the choice of the other person in the preceding trial.

The normal-form game in Figure 10.2 corresponds to the supergame limited to these four strategies after we set the noncooperative outcome, v, equal to zero, and after we simplify expected payoffs by using the following mathematical identity for values of w between 0 and 1:

$$1 + w + w^2 + w^3 + \cdots + w^n + \cdots = \frac{1}{1-w}. \qquad (10.2)$$

Some payoffs in the cells of Figure 10.2 are self-explanatory. For example, if both persons choose $d+$, then each receives a payoff of $v = 0$ every time, so that with or without discounting, $(0, 0)$ is entered in the lower right cell. For another example, if both persons choose $a+$, then they both choose c on the first trial as well as on all subsequent trials, in which case their payoff in each trial is x. The discounted value of this stream of x's, calculated from expressions (10.1) and (10.2), is $x/(1-w)$. Without affecting matters, however, and provided that we do the same in all other cells, we can multiply both persons' payoff by the constant $(1-w)$, and thus we enter (x, x) in the upper left-hand cell of the figure. Alternatively, if 1 chooses $a+$ and 2 chooses $d+$, then 1 receives a payment of z in the first period, and 0 thereafter, and 2 receives y in the first period and 0 thereafter. The present value of their payoffs, then, are z and y, respectively, which we must multiply by the constant, $(1 - w)$. Thus, the payoff 2-tuple $(z(1 - w), y(1 - w))$ is entered in the upper right-hand cell of Figure 10.2.

Having thus specified the normal form to the infinitely repeated prisoners' dilemma, we can now identify a variety of equilibria. Specifically, notice that if $x \geq y(1 - w)$, then $(a+, T), (a+, a+), (T, a+)$, and (T, T) are all Nash-equilibrium strategy pairs, and they are all pairs in which both persons choose the cooperative alternative, c, in every trial. That is, if people do not discount the future too much, then the cooperative outcome can prevail as an equilibrium. Of course, $(d+, d+)$ is also an equilibrium [because $z < v = 0$, and $z(1 - w)$ is negative], but there is no longer any certainty that this Pareto-dominated outcome will occur. Thus, although equilibria are neither equivalent nor interchangeable, the infinitely repeated prisoners' dilemma makes cooperation an equilibrium and therefore a possible outcome.

This example is important because the kind of cooperation that emerges is not the sort that cooperative game theory necessarily envisions. We add no new strategies of communication; indeed, communication may be prohibited. Nevertheless, except for $(d+, d+)$, the choices of both persons in all other equilibria are identical to their choices if they explicitly cooperate. And if they can communicate, they can avoid the mutually distasteful equilibrium, $(d+, d+)$. Thus, it may be impossible to detect any difference between a repeated noncooperative prisoners' dilemma and dilemmas whose outcomes are negotiated separately with provisions for some third-party enforcement of agreements. For example, then, if new members of Congress early in their careers learn norms that avoid disruptive conflict, then subsequent *explicit* enforcement of those norms, or enforcement by some exogenous agent, is unnecessary.

There remains in our analysis of Figure 10.2 the problem that equilibrium strategies are not interchangeable, so that people might not reach an equilibrium without coordination. This is a common problem in the analysis of supergames and we could ignore it when talking about interactions between the same two persons or we could hypothesize that if the game is sufficiently important, then some initial communication should readily resolve the issue. But how, in general, might mutually beneficial cooperation emerge if different persons interact over time?

Robert Axelrod suggests an imaginative answer to this question. Suppose that the members of a collectivity such as a legislature, denoted

S, are all of one type, people who all abide by a specific strategy. Suppose that other people, denoted I, of another type invade this society. Axelrod asks and answers several questions about this situation: With what strategies can the members of I survive in the sense that it does not pay for them to shift to the strategy that the larger group, S, uses? Alternatively, what strategies resist invasion in the sense that the members of I are better off by abandoning their initial strategy and conforming to S?

Suppose, first, that I is a single player i, that i interacts with the members of S randomly, and that each member of S knows how i previously played the game with other persons. Hence, if i abides by the strategy $d+$, then all interactions after the first provoke d from members of S if they use T. Thus, i's discounted payoffs become $y + vw + vw^2 + \cdots = y$, for $v = 0$. But this calculation simply traces the development of Figure 10.2 and implies that i is better off adopting the tit-for-tat strategy T if $x > y(1 - w)$. Thus, if the members of S recall how i played previously, if w is sufficiently large, and if i does not discount the future at too great a rate, then a single uncooperative person cannot invade S.

Now let I be a cluster of persons, but suppose that I is small in number, and that the members of S interact predominately with their own kind, and that in any given trial, a member of I interacts with his own kind with probability p and with a member of S with probability $(1 - p)$. Let $V_I(I)$ denote the discounted value of interactions by members of I with each other, let $V_I(S)$ denote the discounted value of interaction by members of I with members of S, and let $V_S(S)$ denote the discounted value of interactions of persons using S's strategy against similar types of persons. Then, $pV_I(I) + (1 - p)V_I(S)$ is the average score of someone using the strategy of group I. But a member of S almost always interacts with people of the same type, and the result is an average payoff approximately equal to $V_S(S)$. Thus, a cluster, I, can invade S and survive in the sense that no member of I has an incentive to defect and become a member of S if

$$pV_I(I) + (1-p)V_I(S) > V_S(S). \tag{10.3}$$

Notice that this formulation supposes that interactions are not wholly random. Although I is small relative to S, the members of I interact in proportion to their numbers: that is, they remain a community within S.

To see the implications of this perspective, suppose that S consists of people who all choose $d+$, and that I consists of "tit-for-tatters." In this instance $V_I(I) = x/(1 - w)$, $V_S(S) = v = 0$, and $V_I(S) = z$. That is, members of I always cooperate with each other, and except in their first interaction, they always choose d against a member of S. In this case I

can invade S (I can maintain its discriminatory strategy) if, from equation (10.3), $px/(1 - w) + (1 - p)z > 0$. For example, suppose that $x = 2$, $z = -1$, and let the discount rate, w, be .9. The preceding expression solves to $p > 1/21$. Thus, even if "tit-for-tat invaders" interact with each other less than 5% of the time and with persons who play d otherwise, it pays for them to maintain their identity and strategy.

Alternatively, if one person cannot invade a group whose members use *nice rules* (who never defect first but use a strategy such as tit for tat), then a cluster also cannot invade such a group. For such a group, S, to be invaded, expression (10.3) must be satisfied. But if the strategy that S uses is nice, then $V_I(I) \leq V_S(S) = x/(1 - w)$, which is the highest payoff possible when two persons match strategies. Thus, $V_I(I) < V_S(S)$, in which case we can satisfy expression (10.3) only if $V_I(S) > V_S(S)$. But this inequality is equivalent to a single individual being able to invade S. And we already know that on an individual basis a user of $d+$ cannot invade a group that uses T, tit for tat. Hence, the strategy does not have the weakness that $d+$ shows, the susceptibility to invasion by those who use rules of an opposite type.

This mode of analysis is new, but its applications to politics are evident. For example, we could model a legislature in which new members constitute I and reelected incumbents correspond to S, and then ascertain whether new norms can survive in the legislature and the extent to which survival depends on legislative turnover (p) and career objectives (that is, a low w for those who view election to the legislature as a stepping stone to some other office, and a high w for those who hope to make the legislature their home). What makes the single-trial prisoners' dilemma problematical for the participants is that no amount of preplay communication can resolve the dilemma if both persons know that the game will be played only once. But infinite repetition, preplay communication, tradition, and evolutionary selection may yield a cooperative equilibrium that requires the enforcement activities of no external agent.

10.2   The balance of power

Like prisoners' dilemmas, most other games change their forms when viewed as a sequential process. Consider the concept of balance of power in international relations theory. Using the tools of cooperative game theory, many scholars have sought to formalize the idea that nations act to ensure that a coalition or potential coalition of other nations does not become dominant. But one difficulty with establishing a theoretical basis for balance-of-power notions is that if we model international conflict as a simple zero-sum, transferable utility game, then solution concepts predict that some coalition (presumably minimum winning) will predominate at the expense of the rest of the players. That is, something other than the equilibrium of nation states will prevail.

Imagine, though, the following scenario: Each of three nations, $i = 1$, 2, 3, is endowed with some resource, r_i. Let $r_1 + r_2 + r_3 = R$. Suppose that if any nation secures over one-half of R, then it can eliminate either or both of the other nations and absorb their resources. Also, suppose that any coalition of two nations, if the resources that it controls exceed $R/2$, can absorb and eliminate the third. If (140, 100, 60) is the initial resource distribution, then the rules just stated appear to imply the following characteristic function:

$$v(i) = 0, i = 1, 2, 3, \quad \text{since no state can ensure against its own destruction,}$$

$$v(1, 2) = v(2, 3) = v(1, 3) = R, \quad \text{since any coalition of two nations can eliminate the third nation,}$$

$$v(1, 2, 3) = R, \text{ since total resources are constant.}$$

The main-simple V-set, the various bargaining sets, and the competitive solution, when applied to this characteristic function, all predict that one of these three proposals will prevail:

$$((150, 150, 0), \{1, 2\}, \{3\})$$
$$((150, 0, 150), \{1, 3\}, \{2\})$$
$$((0, 150, 150), \{2, 3\}, \{1\})$$

But consider the response of nation 3, say, to a coalition between 1 and 2. Nation 3 knows that if it gives 10 units of resources to 1, and if all nations can look ahead to the consequences of their actions, then none will have an incentive to subsequently form a coalition with 1. To permit 1 to have more than 150 units means, ultimately, that 1 will dominate and absorb both 2 and 3. In partial recognition of the costs of war, if we further assume that every nation prefers to increase its resources through negotiation instead of through war [which is an assumption that we could incorporate into $v(C)$ but that is easier to handle as a side-constraint on preferences], then 1 prefers to have 3 cede it 10 units than to opt for the outcome $((150, 150, 0), \{1, 2\}, \{3\})$.

Notice that if the distribution of resources is (150, 100, 50) and if we incorporate the preceding reasoning, then the characteristic function for this game is not the same as before. First, we must let $v(1) = 150$, because nation 1 is sufficiently powerful to thwart any attempt by 2 and

3 to absorb its resources. Second, calculating $v(C)$ by supposing that winning coalitions can get anything and losing coalitions can ensure nothing ignores the fact that if 1 receives one additional unit of resources, then the outcome (300, 0, 0) prevails ultimately. Thus, neither 2 nor 3 has any incentive to coalesce with 1, and 1 has no incentive to coalesce with 2 or 3, since it knows that neither will agree to any increase in its resources. What of the possibility, though, that 2 tries to absorb 3? If we make the plausible assumption that 1 can also make war on 3 if 2 attacks 3, then 2's attack only guarantees 1's ability to secure an overall majority of resources and to secure eventual dominance as well.

This discussion, which supposes that national leaders look ahead and anticipate the possibility that one nation might become dominant, requires a characteristic function that differs from the one that we have considered thus far. Since any single player in our example can end the game by ceding enough resources so that another player controls one-half of the resources, the security value of individual nations equals their current resources, less those that they might have to relinquish to let some other nation secure one-half of the resource total. Thus,

$$v(i) = r_i - [150 - \max_{j \neq i} (r_j)].$$

And since no coalition can get more than what it already has, plus those resources that might be ceded to its largest member,

$$v(i, j) = r_i + r_j + [150 - \max_{k \in \{i,j\}} (r_k)].$$

If the game begins with the resource distribution (140, 100, 60), then

$$v(1) = 90, \ v(2) = 90, \ v(3) = 50$$
$$v(1, 2) = 250, \ v(2, 3) = v(1, 3) = 210.$$

Thus, solution theory cannot predict that a country is eliminated (that its resources are set equal to zero), since the security value of any individual country exceeds zero. And the competitive solution for this game is the following set of proposals:

$$((125, 125, 50); \{1, 2\}, \{3\})$$
$$((125, 90, 85); \{1, 3\}, \{2\})$$
$$((90, 125, 85); \{2, 3\}, \{1\})$$

This argument seems to correspond to what some scholars mean by a balance of power. Indeed, with the distribution (150, 100, 50), no wars occur and no coalitions may form. Instead, the threat of coalitions and the knowledge that, if one state gains more than 150 units, then all other states will be absorbed together maintain the system's stability. Each nation's ability to predict the outcome from the game that prevails if one of the nations is eliminated is sufficient to maintain the balance.

We can extend this discussion to any number of players, but our objective is not to present theorems about specific and highly stylized models. Instead, we want to show how a sequential dynamic perspective can yield predictions that differ significantly from those that derive from a static perspective, and how that difference often provides the key ingredient to understanding political events. Our argument asserts simply that our models should incorporate all those features of reality that play an important role in determining people's strategies. Thus, although we should not try to model too much of reality, we should also not model too little.

10.3 A finitely repeated game: the chain store paradox

The preceding sections assume that all players have complete information about a game's structure, including the other players' preferences. But in the context of elections and sophisticated voting in committees, we have already seen how games with incomplete information differ from their counterparts with complete information. This difference gains special prominence for repeated games, because early choices may reveal something about preferences that an opponent can use in subsequent choices, and because players may have an incentive to choose strategies that reveal one thing about their preferences rather than another. The "he-thinks-that-I-think ..." regress becomes even more complex, because now we must also consider regresses of the form "if I choose ... he will believe ... and choose ..., in which case I should believe ... and choose...."

The finitely repeated prisoners' dilemma provides an especially good illustration of analytic possibilities. In section 10.1 we assumed that the prisoners' dilemma is repeated infinitely many times. This is a critical assumption, because if the game is finite, then both persons should defect on the last trial, and thus the unique subgame-perfect equilibrium requires defection on every trial. But one problem with this conclusion is that cooperation emerges in many finitely repeated experimental prisoners' dilemmas. Although patterns of choice vary considerably, it is empirically false to suppose that infinite repetition of the prisoners' dilemma is necessary to induce cooperative action. We also should recognize that in many sequential real-life situations, people value and cultivate certain reputations, such as a reputation for sanctioning opponents by implementing threats or a reputation for abiding by

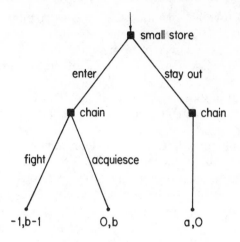

10.3 Weak chain store.

vote-trading agreements, even if, in the short term, such agreements are not personally optimal. By demonstrating a willingness to carry out a threat that is mutually disagreeable in the initial stages of a finitely repeated game, a person might hope to induce choices that depart from the apparent subgame perfect equilibrium. Such strategies are well known to legislative leaders, and they may even provide the glue that holds cooperative agreements together. The question, then, is what explanation can we give to the experimental evidence and what rationale is there for the emergence and cultivation of reputation in politics? Can we devise a basis for deducing the cultivation of reputation as an equilibrium strategy for some noncooperative game that we might use later to explain the formation and maintenance of political coalitions?

We warn the reader that the study of such questions is only beginning to receive scholarly attention, and all that we can do here is to review what appears to be the most promising direction of inquiry. In this spirit, Kreps, Milgrom, Wilson, and Roberts, after applying a modification of Selten's perfect-equilibrium concept to a finitely repeated game that shares many features of the prisoners' dilemma, conclude that a "little bit of uncertainty" yields choices that might form the basis of cooperative action. The game they examine is referred to as "the chain store paradox," which is labeled a paradox because it concerns the cultivation of a reputation that cannot survive if everyone knows the game's true extensive form. The extensive form in Figure 10.3 and the corresponding normal form in Figure 10.4 describe a situation in which a

Small store

	Enter	Stay out
Chain: Acquiesce	0, b	a, 0
Fight if entered	−1, b−1	a, 0

10.4 Weak chain store normal form.

small regional store must decide whether to enter a market monopolized by a chain with outlets in N cities. If the small store decides not to compete, then its payoff is 0 and the chain continues to earn an amount $a > 1$. But if the small store enters the market, then the chain either can fight by cutting prices or it can acquiesce. If the chain fights, then its profits decline to -1, and the small store's profits become $b - 1$, $0 < b < 1$. If the chain acquiesces, then its profits are zero and that of the small store are b.

This game's normal form has two equilibrium outcomes, the upper left- and the lower right-hand cells. But the second of these equilibria is not subgame perfect. Looking at the game's extensive form and working backward up the tree, we find that the players will only reach the equilibrium-strategy pair (acquiesce, enter). If 1 enters, then 2 should acquiesce. Thus, by entering, 1 guarantees himself a payoff of $b > 0$, but by staying out, 1 receives nothing. Hence 1 enters and 2 acquiesces.

Suppose, however, that there are n successive plays of this game, with one play in each of the chain store's markets by a different potential competitor. The chain store prefers a reputation for toughness, which keeps potential competition out of its markets. But everyone can reason in a way that parallels our analysis of the finitely repeated prisoners' dilemma: The chain will acquiesce in the nth market, and because it has nothing to gain from playing tough in the $(n - 1)$th market, it will acquiesce there as well. This argument leads to the inevitable conclusion that the chain store should acquiesce in the first market and in each market thereafter. Because this is the unique subgame perfect equilibrium to this sequential game, there appears to be no logical place for a strategy of toughness.

Consider potential entrants, however, who remain uncertain about the chain store's true preferences and who cannot eliminate with certainty the possibility that they are playing the game in Figure 10.5, which describes a tough chain. That is, each potential entrant must

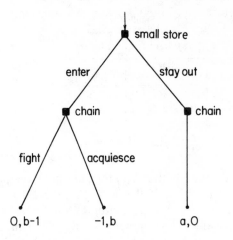

10.5 Tough chain store.

contemplate the possibility that the chain store derives some intrinsic pleasure out of playing tough, in which case it fights if challenged.

This problem shares certain similarities to Ledyard's analysis of voting (see Chapter 5). There, each potential voter knows his cost of voting, but he does not know the costs of others; he knows only the distribution from which these costs are drawn. Each voter must then select a strategy that is consistent with this information and consistent with how he believes others will respond to the information that they have and their beliefs about him. We take a similar perspective in our analysis of binary agendas in Chapter 6 with respect to incomplete information. Here, although information is one-sided, we can think of the chain store as being either of two types: weak or tough. The chain store knows its type and its utility function (it is of the first type), but it also knows that others remain uncertain. The present analysis, though, is more complex, in that entrants can use the chain store's past actions to revise whatever initial priors they might share about the chain's type.

The numerous parallels of the chain store paradox to situations in politics render the method of analysis important. For example, imagine a nation or a union leader whose resolve others continually challenge. Negotiations with terrorists for the release of hostages in an airplane hijacking, coupled with a concern that one concession will encourage other terrorists, fits this scenario almost perfectly. Also, promises to abide by a vote trade or to join some coalition to secure a particular outcome may require the belief that certain persons will abide by their agreements or that they will be willing, at some later date, to sanction those who fail to act accordingly. Looking closely at the solution that Kreps and Wilson propose, then, we begin by noting its three parts: (1) a rule for potential entrants to update their guess as to the likelihood that the chain store will fight entry; (2) a strategy for the chain store that details its responses at every stage in the sequence, taking account of all possible histories; and, (3) a strategy for each potential entrant that details what this player ought to do, given what has happened in the past.

We first consider how potential entrants might condition their beliefs about the chain store on the basis of their past observations. Letting p_i be the probability that a potential entrant at stage i believes that the chain store is tough and will fight if challenged, and letting p_o be p_i's initial value (that is, $p_1 = p_o$), then Kreps and Wilson propose the following three-component updating rule:

1. If there is no challenge at stage i, then $p_{i+1} = p_i$. That is, if there is no entry, then a potential entrant can learn no new information about the chain store, so the previous prior about the chain's type is maintained.

2. If $p_i = 0$, or if there is entry at stage i and the chain acquiesces, then $p_{i+1} = 0$. That is, if the potential entrant already believes that chain store is weak and prefers to acquiesce when challenged, or if potential entrants observe that it acquiesces, then potential entrants all assume at the next and all successive stages that the chain store will not fight.

3. if $p_i > 0$, if there is entry at stage i, and if the chain challenges this entry, then p_{i+1} equals b^{n-i-1} or p_i, whichever is greater (where b refers to the entrant's payoff if it enters and the chain acquiesces).

Component 1 seems straightforward. Component 2 is a specific hypothesis about beliefs, and certainly we can imagine other possibilities. It states that once the chain has acquiesced, its reputation is ruined and it can take no future actions to change the belief that it will always acquiesce when challenged. If we substitute alternative assumptions for 2, then equilibrium strategies different from the one that Kreps and Wilson detail doubtless will prevail. Component 3 is less straightforward. It derives from a Bayesian perspective on how people update priors, given new information.

Before we describe this derivation, though, we state the strategies for the chain store and for the potential entrant that we subsequently show are equilibrium strategies. We emphasize that we cannot find these strategies in any deductive or simple way. Indeed, they may seem to be anything but obvious.

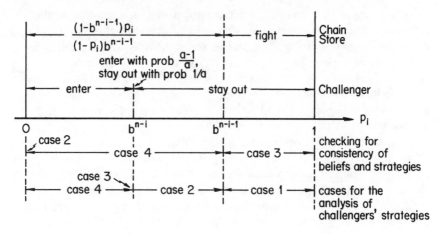

10.6 Chain store paradox strategies.

Chain store strategy: If entry occurs at the last stage, n, then acquiesce. For $i < n$ and $p_i \geqslant b^{n-i-1}$, then fight. If $p_i < b^{n-i-1}$, then fight with probability

$$\frac{(1 - b^{i-i-1})p_i}{(1 - p_i)b^{n-i-1}} \tag{10.4}$$

and acquiesce with 1 minus this probability.

Strategy of potential entrant: if $p_i > b^{n-i}$, then do not enter. If $p_i < b^{n-i}$, then enter. If $p_i = b^{n-i}$, then stay out with probability $1/a$ and enter with probability $(a - 1)/a$.

Figure 10.6 portrays these strategies, to assist in identifying the cases that we discuss later.

The proof that these strategies and the three-part updating rule for priors are an equilibrium has two parts. First, we must show that the presumed rules governing how people adjust their beliefs about probabilities, components 1, 2, and 3, are consistent with the players' strategies. For example, we should not require the chain store to fight entry if there is no hope that fighting will benefit its reputation. And since all strategies are common knowledge, the implementation of a strategy should not cause a potential entrant to change his beliefs in a way that differs from what components 1, 2, or 3 require. Second, starting anywhere in the game, we must show that if the updating rule and the strategies are common knowledge, then no player has an incentive to defect unilaterally from his strategy.

First, to prove that the rules for updating beliefs are consistent with the players' strategies, we have four cases: The first case corresponds to no entry. The remaining three cases correspond to the values of p_i that we indicate in Figure 10.6.

Case 1: No entry takes place in stage i: Thus, as its strategy specifies, the chain store neither fights nor acquiesces, in which case we can learn nothing new about whether the chain is tough or weak. This result is consistent with component 1, namely, that $p_{i+1} = p_i$.

Case 2: Entry occurs in stage i, and $p_i = 0$: Component 2 asserts that once the chain loses its reputation for toughness, it is gone forever; that is, $p_{i+1} = 0$, and it remains zero in all subsequent periods. Clearly, then, the chain should acquiesce and not incur any unnecessary costs, because it cannot gain anything by fighting entry. Expression (10.4) is consistent with this assumption, because it reduces to zero for $p_i = 0$; that is, (10.4) requires the chain to acquiesce.

Case 3: Entry occurs in stage i, and $p_i \geqslant b^{n-i-1}$: The chain store's strategy specifies that it should fight with certainty. But since we must suppose that a potential entrant knows the chain's strategy, it must know for this value of p_i that the chain will fight if challenged, even if it is weak. Hence, nothing new can be learned about whether the chain is indeed tough or weak by observing it not acquiescing in this stage; component 3 requires this by setting $p_{i+1} = p_i$.

Case 4: Entry occurs at stage i, and $0 < p_i < b^{n-i-1}$: In this instance the chain store's strategy is to fight with the probability that (10.4) specifies. If it acquiesces, then it must be weak, and in accordance with component 2, $p_{i+1} = 0$. If it fights, then component 3 states that the challenger in the next stage should update its estimate that the chain is tough, p_{i+1}, to b^{n-i-1}. We derive this new belief about the likelihood that the chain is strong from *Bayes's rule* as follows:

$$p_{i+1} = \Pr[2 \text{ is tough} / 2 \text{ fights}]$$

$$= \frac{\Pr[2 \text{ is tough and fights}]}{\Pr[2 \text{ fights}]}$$

$$= \frac{\Pr[\text{fights} / \text{tough}] \, \Pr[\text{tough}]}{\Pr[\text{fights} / \text{tough}] \, \Pr[\text{tough}] + \Pr[\text{fights} / \text{weak}] \, \Pr[\text{weak}]}$$

In this instance, we know from the chain's strategy that:

Pr[fights/tough] = 1 from Figure 10.5;

Pr[tough] = p_i by definition;

$$Pr[\text{fights/weak}] = \frac{(1 - b^{n-i-1})p_i}{(1 - p_i)b^{n-i-1}}$$

And finally, Pr[weak] = 1 − Pr[tough] = 1 − p_i. Substituting these identities into Bayes's rule yields $p_{i+1} = b^{n-i-1}$, which is what component 3 implies for this case. (Parenthetically, we note with respect to case 3, that in this instance Pr[fights / weak] = 1, in which case the previous derivation reduces to $p_{i+1} = p_i$, as component 3 requires.) Thus, beliefs (or, more precisely, the rule for updating beliefs) and strategies remain consistent.

The second step in the analysis consists of verifying that these strategies are an equilibrium, given the updating rule for beliefs. Looking first at potential challengers, what we must show is that, even if a challenger knows the chain's strategy, it has no incentive to choose differently. That is, we must show that the challenger's strategy is a best response to the chain store's strategy. Before we begin, though, we should point out that if the challenger enters, and if q is the probability that the chain acquiesces, then the challenger's expected utility is

$$bq + (1 - q)(b - 1) = b - 1 + q, \tag{10.5}$$

in which case, entry is profitable only if $b - 1 + q$ exceeds zero, or equivalently, only if $q > 1 - b$. Keeping this fact in mind, we must again consider four cases, which the value of p_i, relative to b^{n-i-1} and b^{n-i} determines. These four cases correspond to the second set of intervals that Figure 10.6 depicts. (In the analysis that follows, we should keep in mind that since b is a constant and since $1 > b > 0$, then $b^{n-i} < b^{n-i-1}$).

Case 1: $p_i \geqslant b^{n-i-1}$. The challenger's strategy is not to enter, which is a best response to the chain store's strategy, because in this case the chain's strategy is to fight. The challenger can gain nothing by entering.

Case 2: $b^{n-i} < p_i < b^{n-i-1}$. The challenger's strategy is not to enter, and the chain store's strategy is to fight with the probability that expression (10.4) gives. The potential entrant does not know with certainty whether the chain store is tough or weak, but he believes that it is weak with probability $1 - p_i$. Thus, from the perspective of the challenger, the probability that the chain store will fight equals

$$\text{Pr[tough]} + \text{Pr[weak]} \times \text{Pr[fights/weak]}$$

$$= p_i + (1 - p_i) \left[\frac{(1 - b^{n-i-1})p_i}{(1 - p_i)b^{n-i-1}} \right]$$

$$= p_i/b^{n-i-1}. \tag{10.6}$$

.To show that entry never can be profitable in this case, notice that in the assumed range of p_i, expression (10.6) reaches a minimum if we assume that p_i equals the lower bound of its permissible values, b^{n-i}. That is, the probability of fighting is smallest and of acquiescing greatest if $p_i = b^{n-i}$, which yields $b^{n-i}/b^{n-i-1} = b$ as the minimum probability of fighting and $1 - b$ as the maximum probability of acquiescing. These probabilities are the best that a potential entrant can hope for in this case. Thus, in the range of p_i, the probability of acquiescence is less than $1 - b$, so from expression (10.5), we find that the challenger's maximum expected value of entering is negative. Hence, the challenger should stay out, as the challenger's strategy requires.

Case 3: $p_i = b^{n-i}$. The challenger's strategy is a lottery between entering and staying out. The challenger believes that the chain will acquiesce if it is weak, with probability $1 - b^{n-i}$, and if we substitute this probability into expression (10.6), we learn that the challenger believes that the chain will fight with probability b and acquiesce with probability $1 - b$. Substituting $1 - b$ for q in expression (10.5) shows that the challenger is indifferent between entering and staying out. Hence, any lottery between entering and staying out suffices as a best response.

Case 4: $p_i < b^{n-i}$. For any value of p_i in this interval, the value of expression (10.6), the probability that the challenger believes that the chain will fight, is less than b, so the challenger's belief about the probability that the chain will acquiesce is greater than $1 - b$. From expression (10.5), we conclude that the challenger should enter.

This discussion of the four cases confirms that the challenger's strategy is a best response to the chain store's strategy. What remains is for us to show that the chain's strategy is a best response to the challengers' strategies. Notice that in some circumstances the challenger's best response to the chain's strategy is to stay out, which is what the chain hopes to accomplish for as many periods as possible. That is, by fighting when challenged in early periods even though it is weak, the chain can maintain its uncertain reputation for possibly being tough. But such a strategy is not costless for the chain. Suppose that in the ith stage, a challenger enters. The chain store has two options. Either it can acquiesce or it can fight. If it fights, then its payoff is -1 in this round, but if it acquiesces, then its payoff is zero in this and all succeeding rounds, because all challengers learn that the chain store is weak. The question is whether the chain store's gains from foreclosed entry compensates for the loss incurred by fighting now. Answering this question is equivalent to ascertaining whether the chain's strategy is a best response to each challenger's strategy.

Consider a specific numerical example, which helps to show why the chain's strategy is a best response. With this example we must demonstrate that whenever the chain incurs a loss of -1 by fighting entry, this loss is offset in successive stages, given the way in which acting strongly affects future potential entrants' beliefs and decisions. Let the challenger's initial estimate of the probability that the chain store is strong, p_o, be $\frac{1}{10}$, let $b = \frac{1}{2}$, and let the total number of markets (playing periods), n, be 7. We can now proceed by stages, keeping in mind that in each market the chain observes the potential entrant's decision before it must choose:

Stage i = 1: In this stage $p_1 = \frac{1}{10}$ and $b^{7-i-1} = 1/2^5 = \frac{1}{32}$. Hence, $p_1 > b^{n-i-1}$, so referring to Figure 10.6, we find that the chain store fights any entry. Because the challenger knows that the chain will fight regardless of whether the chain is strong or weak, it does not enter. Hence, the reward to the chain is a.

Stage i = 2: Because no entry occurred at stage 1, $p_2 = p_1 = \frac{1}{10}$. Again, because $p_2 > b^{n-i} = b^5 = 1/2^5$, entry does not occur, and the reward to the chain is a in this stage also. But if entry should occur, then the chain store would fight since $p_2 > b^{n-i-1} = 1/2^4 = \frac{1}{16}$.

Stage i = 3: Because there was no entry at stage 2, a potential entrant's belief about the probability that the chain is tough does not change. That is, $p_3 = p_2 = \frac{1}{10}$. As in stage 2, $p_2 > b^{n-i} = b^4 = 1/2^4$, so entry does not occur. This reveals the rationality of the chain store's strategy in stage 1. If entry had occurred there, and if it had acquiesced, its payoffs would be zero in every stage of the game. But if it fights in stage 1, then it receives a payoff of -1 there but a payoff of a in stages 2 and 3, which more than compensates for the loss in the first period. To be certain that the chain should fight in this stage if entry occurs, notice that $p_3 < b^{n-i-1} = 1/2^3$. Using expression (10.4) to calculate the probability that the chain fights if it is challenged, we get

$$[(1 - \tfrac{1}{8})/10]/[(1 - \tfrac{1}{10})/8] = \tfrac{7}{9}$$

and it acquiesces with probability $\frac{2}{9}$. But this does not give the challenger incentive to enter, because entering yields the expected payoff

$$(b - 1)p_3 + (1 - p_3)[(b - 1)\tfrac{7}{9} + b2/9] = -\tfrac{3}{10}.$$

Stage i = 4: Because no entry occurred in the previous stage, $p_4 = \frac{1}{10}$. But $p_4 < b^{n-i} = \frac{1}{8}$, so the challenger enters. The chain store fights with probability,

$$[(1 - \tfrac{1}{4})/10]/[(1 - \tfrac{1}{10})/4] = \tfrac{1}{3}$$

and it acquiesces with probability $\frac{2}{3}$. The expected reward of entry for the challenger is $\frac{1}{4}$.

Stage i = 5: If the chain store acquiesced in the previous stage, then $p_i = 0$, entry occurs, and the chain store acquiesces in this and in all subsequent stages. But if the chain store fought entry in the previous stage, $p_5 = b^{n-i} = 1/2^2 = \frac{1}{4}$. With this belief, the challenger is indifferent between entering and staying out. Thus, it randomizes using the lottery $(1/a, (a - 1)/a)$. If the challenger stays out, then the chain receives a. If the challenger enters, then the chain store randomizes between fighting and acquiescing with probabilities $\frac{1}{3}$ and $\frac{2}{3}$. Conditional on the challenger entering, then, the chain's expected payoff is $-\frac{1}{3}$. Thus, the chain's net expected reward becomes $a(1/a) - \frac{1}{3} = \frac{2}{3}$, which makes up in part for the loss incurred in the previous stage.

Stage i = 6: Assuming that the challenger entered in the previous stage, $p_2 = b^{7-6} = \frac{1}{2}$, so the challenger stays out in this round. Thus, the chain receives a, which again more than compensates for the loss from fighting in stage 4. If the challenger stayed out in stage 5, so that the chain received a to compensate for the loss in stage 4, $p_6 = \frac{1}{4}$. But now p_6 is less than $b^{7-i} = \frac{1}{2}$, and the challenger enters. In this event, the chain acquiesces with certainty.

Stage i = 7: The challenger enters, and the chain acquiesces.

The reader should notice a subtle but important feature of this analysis. We not only check whether the chain's strategy is a best response against each entrant's decision as dictated by the entrant's strategy, but we also check the implications of what happens if the chain finds itself on a part of the game tree that it cannot reach if potential challengers abide by their strategy. In stages 1 and 2, for example, we also look at the chain's response if it is challenged and we find that the chain's strategy not only keeps a challenger from entering in the current

stage, but it also keeps future challengers from learning its true type
(that it *is* weak) if someone mistakenly enters. This is not much different
from showing that the equilibrium that we have found is also a perfect
equilibrium (see Section 3.9). Kreps and Wilson's analysis actually uses
a refinement of this equilibrium concept, which they call a sequential
equilibrium.

We cannot present a satisfactory treatment of this refinement here; it
is difficult enough to apply in the current version of the chain store
paradox. Furthermore, even subtle changes in this game alter the
analysis significantly. For example, the assumption that the potential
challenger is different at each stage is crucial. If only two players are
involved, we must consider a more complex strategy for the challenger,
one that permits him to test the waters, so that he could try to learn as
quickly as possible whether the chain is weak or tough. Such a situation
might correspond to a legislature that must decide in each budget period
whether an agency is overestimating budgetary requirements or whether
the agency's requests legitimately reflect its mandated objectives. The
agency presumably knows what appropriation it requires, but the
legislature must decide either to grant the agency's request and later
decide whether to monitor the agency (with some cost to the legislature)
or to fund at a lower level and take the chance that some constituents
will be dissatisfied. This sequential game, although similar to the chain
store example, should pose different analytical problems.

In spite of the simplifying assumption that challengers are distinct
players at each stage, the formulations of equilibrium strategies and
consistent beliefs here represent an extreme example of creative thinking, and the ideas illustrated are at the forefront of game theory.
Minimally, they may tell us something new about how people learn the
preferences of others. Game theory traditionally takes preferences as
common knowledge, but the preceding analysis reveals the development of a science of learning and of strategic-preferences revelation. We
should be aware of the existence of such developments, even if we must
leave applications at this point to specialists.

10.4   The Shapley value and the power index

The boundary between descriptive and normative concepts often remains obscure in game theory. Owing to its mathematical complexity
and to people's cognitive limitations, some scholars prefer to interpret
game theory as a normative science, as a tool that people can use to find
out how best to achieve their goals. Similarly, because much of game
theory tries to interpret formally the meaning of "rational action," and
because we cannot deduce this meaning directly from the preference

axioms in Chapter 1, proposed solution concepts necessarily convey an ad hoc flavor. Consequently, some scholars believe that we can interpret only in a limited, normative way, even those concepts that models of negotiating and bargaining explicitly motivate. Of course, distinctions between descriptive and prescriptive concepts are necessarily unclear, because any prescription about action requires a descriptive component, namely, of how others will act or otherwise respond to one's decisions.

The research motivations for the concepts that we review in this and in the next section seem more ambiguous than do other parts of game theory. Generally, the spirit of their definitions is more in keeping with answering the question "What is a fair outcome?" compared with "What payoff will each person receive?" But recent research tends to blur distinctions further by offering descriptive counterparts to prescriptive definitions. This blurring seems especially characteristic of theories that model the bargaining process between persons taken two at a time. Although these theories contain several ambiguities, their asserted applications to politics are numerous, including models of negotiations in international relations, evaluations of biases in constitutional rules, and, in particular, calculations of the degree to which weighted voting schemes comply with one-man one-vote criteria.

The Shapley value is a decidedly normative scheme that tries to define a "fair" utility n-tuple, but without asserting that such an outcome will ever actually prevail in a game. But we begin a discussion of the Shapley value with two caveats. First, although we can derive it from a set of axioms that seek to give it some normative credibility, those axioms do not seem intuitively appealing, as do Arrow's for example. One axiom is that if v and v' are the characteristic functions of any two games, and if q and q' are a player's Shapley values in v and v', respectively, then the player's value in the composite game, $v + v'$, is $q + q'$. But the characteristic function is a slippery concept, and the substantive meaning of "composite game" is unclear. The second caveat is that although we can extend the definition of the Shapley value to games without transferable utility, this extension is difficult to interpret in terms of any descriptive model of negotiations, and thus appears to be little more than a mathematical trick. And without transferable utility, it loses many of its nice mathematical properties, including uniqueness. Hence, we restrict our discussion to games with transferable utility.

To motivate the Shapley value, consider this three-person transferable utility game:

$$v(1) = 0; \ v(2) = 2; \ v(3) = 3$$
$$v(1, 2) = 3; \ v(1, 3) = 6; \ v(2, 3) = 7$$
$$v(1, 2, 3) = 11.$$

Suppose that coalitions form sequentially – first, person i, then person j, and so on, up to the coalition of the whole – so that in a three-person game any one of the following six sequences is possible: [1, 2, 3], [1, 3, 2], [2, 1, 3], [2, 3, 1], [3, 1, 2], and [3, 2, 1]. (For [1, 2, 3], read "person 2 joins {1} to form {1, 2}, and then 3 joins {1, 2} to form {1, 2, 3}"). Consider the sequence [1, 2, 3], and this argument: By joining the empty coalition, person 1 contributes $v(1) = 0$ to the value of the resulting coalition, {1}. With the addition of person 2, the value of the resulting coalition, {1, 2}, increases to 3; that is $v(1, 2) - v(1) = 3$. Hence, 2's net contribution is 3. Finally, if we add person 3 to form the coalition of the whole, then we can attribute an increment of 8 to 3's action. That is, $v(1, 2, 3) - v(1, 2) = 8$. Summarizing, we have:

Marginal contribution of

Sequence	Person 1	Person 2	Person 3
[1, 2, 3]	0	3	8

Expanding this table to include all possible sequences, we derive these numbers:

Marginal contribution of

Sequence	Person 1	Person 2	Person 3
[1, 2, 3]	0	3	8
[1, 3, 2]	0	5	6
[2, 1, 3]	1	2	8
[2, 3, 1]	4	2	5
[3, 1, 2]	3	5	3
[3, 2, 1]	4	4	3

Thus, a person's marginal contribution to a coalition depends on what sequence occurs.

We are developing a scheme for identifying a fair outcome when we know nothing about the players except what the characteristic function tells us. Therefore, we suppose that all sequences are equiprobable and that people receive rewards in accordance with their expected or average marginal contribution. In the preceding example, person 1's average marginal contribution is 2 [that is, $(0 + 0 + 1 + 4 + 3 + 4)/6 = 2$]. Similarly, persons 2 and 3 have expected marginal contributions of 3.5 and 5.5, respectively. So the Shapley value for this game is the 3-tuple (2, 3.5, 5.5).

More generally,

Shapley value: The Shapley value, q_i, for person i in an n-person, cooperative game in characteristic-function form with transferable utility is

$$q_i = \sum_{i \in C, C \subset N} (c - 1)! \, (n - c)! \, [v(C) - v(C - \{i\})]/n!$$

Where c is the size of C and "!" denotes factorial, that is, $(c - 1)! = 1 \times 2 \times 3 \times \ldots \times (c - 2) \times (c - 1)$. [Notice that $\sum_{i=1}^{n} q_i = v(N)$.]

To explore the Shapley value's properties further, consider this three-person, transferable utility game with a core:

$v(i) = 0$ for $i = 1, 2, 3$
$v(1, 2) = 5000;$ $v(1, 3) = 7000;$ $v(2, 3) = 0$
$v(1, 2, 3) = 7000.$

In this instance, the core consists of any 3-tuple in which person 1 gets at least 5,000, and 1 and 3 together get 7,000. But we compute the Shapley value thus:

| | Marginal contribution of | | |
Sequence	Person 1	Person 2	Person 3
[1, 2, 3]	0	5,000	2,000
[1, 3, 2]	0	0	7,000
[2, 1, 3]	5,000	0	2,000
[2, 3, 1]	7,000	0	0
[3, 1, 2]	7,000	0	0
[3, 2, 1]	7,000	0	0
Total:	26,000	5,000	11,000

Hence, $\mathbf{q} = (q_1, q_2, q_3) = (4333, 833, 1833)$.

Notice that $\mathbf{q}$ does not correspond to any payoff 3-tuple in the core (or to any other solution concept described earlier), so we find no reason to regard $\mathbf{q}$ as a descriptive notion. But the usual interpretation of $\mathbf{q}$ is that it represents what a person should expect to receive on the average if his reward corresponds to his marginal contribution. Therefore, $\mathbf{q}$ corresponds to a fair division of the game's value. Why we should regard an outcome outside of the core (which presumably takes account of all relevant strategic advantages and disadvantages) as fairer than one in the core remains unclear. Still, the Shapley value yields an interesting and informative analysis for at least one class of games, simple games. Consider a symmetric, strong simple n-person game with transferable utility. The important feature of q for such games reflects that, because the game is symmetric, people's labels are irrelevant and everyone is indistinguishable. Consequently, all Shapley values are equal; that is, $q_1 = q_2 = \cdots = q_n$. Since all q_i's must sum to $v(N)$, and assuming a $(0, 1)$ normalization [that is, $v(N) = 1$], we get $q_i = 1/n$, for all i. We can also derive this result from a second property of symmetric simple games. With n odd, the only person who adds any value to a coalition is the person who converts that coalition from winning to losing. And since all sequences are equiprobable, all persons have the same chance, $1/n$, of being pivotal.

Quite directly, then, the Shapley value draws our attention in simple games to the notion of players who pivot (those who can convert losing coalitions into winning coalitions) and to the possibility of measuring a person's "power" or value by the probability that he is the pivotal member of a winning coalition. In this context we refer to q as the power index.

Power index: For simple games with a $(0, 1)$ normalization,

$$q_i = \sum_{i \in C \in W^*} (c - 1)!(n - c)/n!$$

where W^* is the set of all minimum-winning coalitions.

The power index has found widespread use in the evaluation of the fairness of various voting procedures and it is used to ascertain whom a particular procedure might advantage or disadvantage. To illustrate, consider a town council in which each member represents some historically defined district. Suppose that each district's population has changed over the years, and that now the populations of the districts are thus:

district	population
1	4,900
2	3,700
3	1,400
total	10,000

Suppose that the town council is opposed to redefining neighborhoods and districts and that it is unwilling to add more members, but that in the interest of fairness, it is willing to institute *weighted voting*. If the council uses majority voting, and if each member receives one vote per 100 people in the district, then the representative from district 1 has 49 votes, and so on. A motion requires a majority of 51 votes to carry. Listing all six possible sequences that form the coalition of the whole, we circle below the pivotal person in the sequence, the one who contributes the 51st vote and converts a losing coalition into a winning one:

$$[1, ②, 3] \quad [1, ③, 2]$$
$$[2, ①, 3] \quad [2, ③, 1]$$
$$[3, ①, 2] \quad [3, ②, 2]$$

Thus, every member is pivotal as many times as any other, so that $q_1 = q_2 = q_3 = \frac{1}{3}$. This is certainly a strange result in the light of the unequal weighting of votes. Since the intent of weighted voting is to compensate for an unequal population distribution, the corresponding simple game seems unfair.

For an even more extreme example, consider this scenario: Jones, president and chairman of Consolidated Smoke, anticipates a takeover bid by Allied Chimney. Jones owns 24% of Consolidated's shares, and five heirs of the firm's founder hold the rest, with these weights: 27%, 27%, 18%, 2%, and 2%. Jones guesses that if Allied succeeds in securing majority control, then he will lose his job. But he also fears the other stockholders will accept any reasonable offer from Allied. Jones's only hope, then, is to bribe one or more of the heirs to his cause. Whom should he bribe and whom should be ignore?

In this instance the voting weights yield the following power indices:

Weight:	27	27	24	18	2	2
q_i:	$\frac{1}{3}$	$\frac{1}{3}$	$\frac{1}{3}$	0	0	0

Thus, stockholder 4, with 18% of the vote, cannot be pivotal, so he shares a zero power index with stockholders 5 and 6, who each control but 2% of the vote. Jones, then, can safely ignore heirs 4, 5, and 6, because each is never pivotal in determining control of Consolidated. Instead, he should focus his attention on an heir with 27% of the vote. The reader may believe that this example is fanciful. But the six-member legislature of Nassau County, N.Y., to accommodate inequities of population, instituted a weighted voting scheme with nearly identical voting weights, thereby rendering three members irrelevant as critical coalition partners.

For a more complicated example, consider the 1965 United Nations reform of its Security Council. As originally established in 1945, the

Security Council consisted of five permanent members and six temporary members. To pass a motion, all five permanent members and any two temporary members had to agree. Hence, for a temporary member, say i, to be a pivot, he must be preceded in the sequence leading to $\{N\}$ by all five permanent members and by one temporary representative. Given the temporary member who precedes i, there are 6! ways to permute the six members who must precede i. There are 4! ways to permute the temporary members succeeding i. Additionally, any one of five temporary members can precede i. Hence, a temporary member is pivotal in exactly $5(6!)(4!) = 86,400$ sequences. Since there are 11! = 39,916,800 possible sequences, then q_{temp} must equal $86400/39916800 = .0022$. But a permanent member is pivotal only if four other permanent members and two or more temporary members follow him. In this instance, some messy calculations yield $q_{\text{perm}} = .1974$. The total power of the temporary members, then, is $6(.0022) = 13/1000 = .0132$, and that of the permanent members is $5(.1974) = .987$.

Under the guise of redressing this imbalance, the United Nations in 1965 increased the Security Council's size from 11 to 15 by adding four new temporary members. The minimum majority also increased from 7 to 9, but, as before, the permanent members retained their veto. The power indices under this new arrangement are, $q_{\text{temp}} = .0018$ and $q_{\text{perm}} = .196$. Thus, with this reform the total power of the nonpermanent members rose "impressively" from .013 to $10(.0018) = .018$. Such calculations support the suspicions either that the reformers were befuddled or that they operated under different motives.

Another favorite target for application of the Shapley value is the electoral college. If we assume that presidential elections correspond to a 51-person, majority-rule-weighted game (50 states plus the District of Columbia), the electoral college exhibits a slight but insignificant bias in favor of larger states. But we might properly ask what it means for a state to be a pivotal member of a coalition, because states do not join the winning side, one at a time. Nevertheless, we can give this application some justification. Suppose that the election concerns a single issue, and that everyone's preferences on that issue are single peaked. Then we know from Theorem 4.3' that the social preference order within each state under majority rule is transitive, with an ideal point at that state's median-voter's preference. From Theorem 4.3, we know that the election equilibrium corresponds to the preference of the median state (the median "citizen's" preference), weighted by the electoral college vote. That is, if we order the states on the issue by the ideal preferences of their median voters, then, counting either from the left or the right, the median preference of the state corresponding to the 269th electoral vote is the election equilibrium. If all orderings of medians on the issue are equiprobable, and if we calculate the probability that any given state's median preference is the equilibrium, then that number is the state's power index.

Scholars have proposed the use of a variety of other indices to evaluate the fairness of voting systems. For example, the Banzhaf-Coleman index looks at the *swings* of each player. Briefly, if C is a winning coalition with $i \in C$, then i swings if $C - \{i\}$ is losing.

> *Banzhaf-Coleman index:* If S_i denotes the number of coalitions for which i swings, and if n is the number of players, then $B_i = S_i / \Sigma_{i=1}^n S_j$ is the normalized Banzhaf-Coleman index.

For example, if only the coalitions $\{1, 2\}$, $\{1, 3\}$, and $\{1, 2, 3\}$ are winning, then $\mathbf{B} = (\frac{3}{5}, \frac{1}{5}, \frac{1}{5})$, because player 1 has three swings but 2 and 3 have only one swing each. Assuming a $(0, 1)$ normalization, their power indices are $\mathbf{q} = (\frac{2}{3}, \frac{1}{6}, \frac{1}{6})$. Thus, the power index and the Banzhaf-Coleman index are not equivalent.

In another example of the application of these indices, Miller and Straffin both evaluate a scheme proposed for Canada to allow amendments to its constitution. This scheme required that any amendment be approved by the provinces of Quebec and Ontario, by any two of the four Maritime provinces, and by British Columbia and any two central provinces, or if British Columbia objects, by all three central provinces. Their power and Banzhaf-Coleman indices are as follows:

Province	Power index	Banzhaf-Coleman	% of population (1970)
Ontario	.3155	.2178	34.9
Quebec	.3155	.2178	28.9
British Columbia	.1250	.1634	9.4
Maritime (total)	.1192	.2376	7.9
Central (total)	.1251	.1635	17.0

Thus, regardless of which index we use, British Columbia appears to have an advantage over all other provinces.

Both the Shapley and Banzhaf-Coleman indices have been extended to compound games, for which the researcher seeks to compute, for example, the probability that a randomly selected voter in Alaska is pivotal in a presidential election, compared with the probability that a voter from California is pivotal. At this point, however, the relationship of these analyses to game theory grows ever more speculative. The indices reflect no particular game or the modeling of specific sets of individual decisions, and it is difficult, if not impossible, to interpret the differences between the indices in a particular context, or to appreciate their implications, if any, for policy. Curiously, though, the indices are not unrelated mathematically to various approaches to modeling two-person bargaining, to which we now turn.

10.5   The Nash bargaining model

An alternative but nevertheless related approach to defining each player's value in a game derives from Nash's attempt to formulate a fair arbitration scheme that models negotiations between two persons. This approach is similar to that proposed by Arrow in that it formulates a set of seemingly reasonable axioms for such arbitration schemes. These axioms then imply a specific procedure for resolving disputes between persons. Specifically, suppose that if two negotiators fail to reach any accord, then the status quo, x_o, prevails, and let $u_i(x_o) = v_i$ for $i = 1, 2$. Nash then poses the bargaining problem as the search for a rule (mathematical function) that operates over the set of feasible utility outcomes, U, and (v_1, v_2) to yield a specific outcome in U, (u_1^*, u_2^*), which represents a negotiated and fair outcome. What we want to find, then, is a function, $\mathbf{g}$, such that,

$$\mathbf{g}(U, v_1, v_2) = (u_1^*, u_2^*) \in U.$$

The five specific axioms that Nash imposes are

N1.   Individual rationality: $u_i^* \geqslant v_i$, for $i = 1, 2$.

N2.   Pareto-optimality: There is no point, $(u_1, u_2) \in U$, such that $u_i \geqslant u_i^*$, for $i = 1, 2$.

These two assumptions are nothing more than the requirement that the rule, $\mathbf{g}$, choose a point in the core.

N3.   Independence from linear transformations: Let the feasible outcome set, U', be obtained from the set U by linearly transforming each person's utility function thus:

$$u_i' = a_i u_i + b_i, \, a_i > 0, \, i = 1, 2.$$

If $\mathbf{g}(U, v_1, v_2) = (u_1^*, u_2^*)$, then

$$\mathbf{g}(U', v_1', v_2') = (a_1 u_1^* + b_1, a_2 u_2^* + b_2) \in U'.$$

Thus, if we transform each person's utility function in a positive linear way, then we transform the outcome of the arbitration process in the same way.

N4. Independence from irrelevant alternatives: Let (u_1^*, u_2^*) and (v_1, v_2) both be in T, which is a subset of U. Then,

$$(u_1^*, u_2^*) = \mathbf{g}(U, v_1, v_2) \text{ implies } (u_1^*, u_2^*) = \mathbf{g}(T, v_1, v_2).$$

We should not confuse this axiom with Arrow's far weaker independence axiom. N4 requires that, if the solution to a game in which T is the set of feasible outcomes is (u_1^*, u_2^*), and if we add outcomes to T to form U without changing the security levels of either person, then the solution with U should be either (u_1^*, u_2^*) or one of the outcomes that has been added. N4 seeks to eliminate this tongue-in-cheek possibility: The patron of a restaurant is told that two soup dishes are available today, vegetable and chicken, whereupon he orders chicken soup. A short time later, the waiter returns to announce that noodle soup is available also, at which point the patron changes his order from chicken soup to vegetable soup.

Actually, this assumption is not innocuous. We have already seen that if we apply it to groups, then voting systems with three or more persons can violate it. In the two-person case, suppose that the parties must divide a dollar, and that if they reach no agreement, then each person gets nothing. As we shall see later, the arbitrated solution to this game is (.50, .50). But suppose that we are informed that person 2 cannot receive more than .50. N4 requires that (.50, .50) remains the solution. Is this result fair? After all, person 1's opportunities for gain appear to exceed 2's opportunities. Should not the solution take account of 2's diminished opportunities and award more to 1? The counterargument defending N4 is that outcomes in which person 2 gets more than .50 are irrelevant to 2, because he cannot attain them even if they are theoretically available.

N5. Symmetry: If $(u, u') \in U$ implies $(u', u) \in U$ and if $v_1 = v_2$, then $u_1^* = u_2^*$.

Thus, if the game is symmetric, in the sense that both person's security levels are identical and U is symmetric [if, for example, the payoff 2-tuple (5, 3) is available, then (3, 5) is available as well], then the arbitrated outcome should award equal payoffs to both persons.

Nash's result is that *the unique function that satisfies axioms N1–N5 is the function that picks (u_1^*, u_2^*) out of U, so as to maximize the product*

$$(u_1 - v_1)(u_2 - v_2).$$

We call any outcome that maximizes this product the *Nash arbitration point*.

For example, suppose that two bargainers must decide how to divide

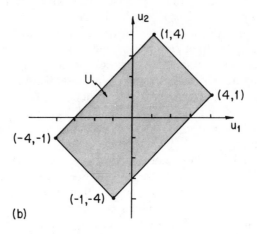

b_1. b_2

	b_1	b_2
a_1	1, 4	-1, -4
(a) a_2	-4, -1	4, 1

(b)

10.7(a) A game with nonsymmetric threats. (b) A symmetric game
with nonsymmetric threats.

$100 and that their respective utility functions for money are

$$u_1(x_1) = x_1 \text{ and } u_2(x_2) = (x_2)^{1/2},$$

where $x_1 + x_2 = 100$. Finally, suppose that the status quo is $v_1 = v_2 = 0$.
We obtain the Nash-arbitration solution to this problem by maximizing
u_1 times u_2, subject to the constraint that the two people cannot divide
more than $100. Letting $x_2 = 100 - x_1$, the problem, then, is to
maximize

$$x_1(100 - x_1)^{1/2}.$$

Taking the derivative of this with respect to x_1, and setting the result
equal to zero, we get

$$\frac{100 - 3x_1/2}{(100 - x_1)^{1/2}} = 0,$$

which implies that $x_1 = 200/3 = \$66.67$ and that $x_2 = \$33.33$. This
solution, then, reflects that person 2's utility rises with money at a less
rapid rate than does 1's utility. In a sense, then, 1 is more sensitive to

variations in his wealth than is 2, and the Nash arbitration point accommodates this sensitivity.

Naturally, as with any normative scheme, there are objections to Nash's axioms. Consider the normal-form game in Figure 10.7a and the corresponding graph of U in Figure 10.7b. The maxmin strategies are the mixed strategies $(\frac{1}{5}, \frac{4}{5})$ for 1 and $(\frac{1}{2}, \frac{1}{2})$ for 2, with corresponding security levels $v_1 = v_2 = 0$. Figure 10.7b also shows that U satisfies the symmetry requirement of N5. Hence, the solution to this game must be symmetric. In this instance, it follows that $(u_1^*, u_2^*) = (2.5, 2.5)$.

But is this game truly symmetric in the sense that no one has a bargaining advantage? Suppose that, during the negotiations 2 threatens to play b_1, to induce 1 to choose a_1. If the threat backfires, then 2 gets -1, and person 1 receives his lowest possible payoff, -4. But person 1 has no equivalent threat. He can threaten a_2, to induce 2 to play b_2, but if that threat backfires, then the player making the threat, 1, receives his lowest possible payoff. Thus, in terms of threat capabilities, players 1 and 2 are in dissimilar situations. If each person uses his threat strategy, then the outcome is $(-4, -1)$. And if we use this outcome as our status quo point instead of $(0, 0)$ to calculate the Nash arbitration point, we get $(1, 4)$, which is an outcome that perhaps better reflects 2's threat advantage.

To modify the Nash scheme so that it accommodates differential threat potentials, suppose that each person chooses a threat strategy, t_1 or t_2, respectively, that tries to redefine the status quo point in ways that are most favorable to each. If we now interpret the status quo as the outcome that prevails if both persons use their threat strategies, and if we assume that the Nash arbitration scheme selects the final outcome, then we can represent the situation as a noncooperative game, in which by choosing t_1, 1 seeks to maximize u_1^* (t_1, t_2), and by choosing t_2, 2 seeks to maximize $u_2^*(t_1, t_2)$, where u_1^* and u_2^* maximize

$$[u_1 - v_1(t_1, t_2)][u_2 - v_2(t_1, t_2)].$$

The players can gain nothing by cooperating in their selection of threat strategies, since the Nash scheme necessarily chooses a Pareto-optimal point in U. Hence, any increase in 1's utility yields a decrease in 2's utility.

It is possible to show that *every finite two-person game has interchangeable and equivalent equilibrium-threat strategies*. To illustrate this proposition, suppose that utility is transferable between the players in the ratio 1:1 on the Pareto frontier. That is, let $u_1 + u_2 = k$. Then, we must maximize

$$b_1 \qquad\qquad b_2$$

	b_1	b_2
a_1	$1 - 4 = -3$	$- 1 + 4 = 3$
a_2	$-4 + 1 = -3$	$4 - 1 = 3$

10.8 Computing threat strategies.

$$[u_1 - v_1(t_1, t_2)][u_2 - v_2(t_1, t_2)].$$

Substituting k for $u_1 + u_2$ and differentiating with respect to u_1 yields

$$k - 2u_1 + v_1(t_1, t_2) - v_2(t_1, t_2).$$

Setting this expression equal to zero to solve for a maximum gives us

$$u_1^* = \frac{[k + v_1(t_1, t_2) - v_2(t_1, t_2)]}{2}.$$

Similarly, differentiating with respect to u_2 instead yields

$$u_2^* = \frac{[k + v_2(t_1, t_2) - v_1(t_1, t_2)]}{2}.$$

Hence, person 1 seeks to maximize $v_1(t_1, t_2) - v_2(t_1, t_2)$, and 2 seeks to minimize this quantity. The solution, of course, is the minmax solution to the corresponding zero-sum game.

For the game in Figure 10.7a, we form the appropriate game by subtracting 2's payoff from 1's in each cell, and we assume that the resultant zero-sum game, which Figure 10.8 depicts, is the one that confronts the players. Referring to Figure 10.8, notice that, in accordance with our earlier conclusion that player 1 does not have an effective threat strategy, 1 is indifferent between any strategy as a threat, and 2's optimal threat strategy is the pure strategy, b_1.

What is especially appealing about this analysis is that instead of defining security levels exogenously, these levels are now endogenous. The concept of a status quo is problematical otherwise, because it supposes that we can define payoff n-tuples if the players refuse to play the game. But if refusing to play the game is an available option, then we should represent it as a strategic choice in the situation's normal form. Thus, the preceding development is general because it leaves open the possibility that each player has a strategy denoted "don't play" and that if either chooses that strategy, then the payoff 2-tuple (v_1, v_2) prevails.

10.6 Relating the Shapley value to Nash's scheme

Although the discussion of the Shapley value, q, focuses on n-person games, we should hope for some correspondence between q and the Nash arbitration model for at least the special case of two-person games. If no correspondence exists, then we have two essentially normative ideas that remain at best unrelated, and at worst, inconsistent. Fortunately, the Shapley value and Nash's scheme are closely related. To see the relationship, consider the case of transferable utility, so that $u_1 + u_2 = k$ describes the Pareto-frontier of U. If we then let $u_2 = k - u_1$ and maximize the product of utilities,

$$[u_1 - v_1][k - u_1 - v_2],$$

then, as we showed earlier, the Nash arbitration point solves to,

$$u_1^* = \frac{k + v_1 - v_2}{2}; \; u_2^* = \frac{k + v_2 - v_1}{2}.$$

Now consider the Shapley value. In characteristic-function form the implicit game is

$$v(1) = v_1; \; v(2) = v_2; \; v(1, 2) = k.$$

Setting up the table of marginal contributions, which we use for the calculation of q, we have these entries:

	Marginal contribution of	
Sequence	Person 1	Person 2
[1, 2]	v_1	$k - v_1$
[2, 1]	$k - v_2$	v_2

Thus, since q_1 is the average of 1's marginal contributions, and q_2 is the average of 2's marginal contributions, we get $q = ((k + v_1 - v_2)/2, (k + v_2 - v_1)/2)$, which corresponds identically to the Nash arbitration values.

We can find other equivalences between Nash and Shapley values for games without transferable utility and for n-person games. Indeed, value theory remains a principal research interest of game theorists and a vast literature has emerged. This literature contains several extensions of Nash's original idea, applications to market games, and alternative axiom systems and arbitration schemes. And in at least one respect, these schemes tell us to formulate public policy quite differently than the prescriptions that planners ostensibly follow. The 1960s saw the emergence and widespread teaching and application of cost–benefit analysis in the area of public policy planning and implementation. From the assumption that the objective of such policy is to maximize society's net monetary wealth, the appropriate rule for policy selection appeared to be "select only those policies for which the *summed* benefits across all persons exceed *summed* costs, and if we must make particular choices, select the policy that maximizes the difference between these benefits and costs."

Even if we ignore the monstrous task of measuring people's benefits and costs in terms of money, we can justify this prescriptive normatively only if we can redistribute net benefits to compensate those who suffer a net loss if we select that policy. But cost–benefit analysis, at least in its applications, often ignores an analysis of redistribution opportunities. The Shapley and Nash values provide a different prescription: Maximize the *product* of all individual net benefits. The disadvantage of this prescription is that every member of society has a veto over policy. Furthermore, if anyone's benefits are zero, then the product of utilities over all of society is zero, no matter how great the benefits to others might be. But the advantage is that uncompensated losers enjoy a veto as well.

We can cite one interesting and informative application of these normative value schemes. Briefly, Dyer and Miles sought to devise a fair mechanism, whereby the Cal-Tech Jet Propulsion Lab (JPL) could choose trajectories for the two 1977–78 Jupiter-Saturn flyby space probes that produced those spectacular pictures of Saturn's rings and Jupiter's moons. Laden with scientific experiments, JPL had to select a pair of trajectories, knowing that different trajectories benefited different experiments, and that the 15 science teams in charge of these experiments held different preferences over alternative pairs. Cost-benefit analysis seemed inappropriate because it was meaningless to place dollar values on the relative successes of different experiments, and also because various teams might not have regarded such an attempt as appropriate. Furthermore, Dyer and Miles wanted to design a procedure that not only would make a choice, but also that all interested parties would regard as fair, no matter what trajectory it might choose. After narrowing the list to 35 feasible pairs, Dyer and Miles, with JPL's cooperation, interviewed the 15 teams to secure a rough approximation of their cardinal utility functions. They then ranked the pairs according to various normative criteria, including variants of Nash's arbitration scheme. Three alternatives, which we denote by A, B, and C, consistently ranked first through third by the various criteria, with A generally ranking first. Dyer and Miles reported these results to JPL's final selection committee, which chose B over A.

It appears that JPL used Dyer and Miles's analysis to narrow its list of options, but it seemed to reject their advice in favor of the second ranked alternative after that alternative had been modified slightly. Curiously, this second-ranked alternative satisfied an additional property: *Over the set of all pairs, it was the unique Condorcet winner*. Thus, we cannot reject the hypothesis that normative criteria were of secondary (or of no) importance, and that JPL chose alternative B because a majority coalition could not form to overturn it. (We can only speculate that A would have been chosen if it too had been modified, but then we must answer why A was not so modified.) This case study suggests that our confidence in schemes such as the Nash arbitration solution would considerably increase if we could relate them to some game-theoretic model of the bargaining process. With such a model we could judge better the appropriateness of its application, and perhaps even its reasonableness as a normative concept.

Actually, several bargaining models exist that yield the Nash arbitration points as their prediction. But the difficulty with such models is that each must accommodate the complexity of a situation's extensive form in what seems to be an ad hoc way, and extensions to the n-person case appear to defy analysis. To appreciate the difficulty here, suppose that we let each player first propose a demand that lies on the Pareto frontier of possibilities. Next, suppose that we confront each person with three strategies: Accept the demand of the other or otherwise make some concession to the opponent; hold out for one's own demand; and break off negotiations and implement the mutually distasteful status quo or threat outcome. At this point we must decide whether the players choose their strategies simultaneously or whether bargaining proceeds sequentially. If they choose simultaneously, then we must specify the outcome that prevails if they choose inconsistent strategies. If they choose sequentially, then we must formulate a rule that specifies who moves first. In either case, we must also decide whether bargaining can proceed indefinitely, whether time is a real cost for players, and what happens if they reach no agreement. It is evident, moreover, that incomplete information is important in negotiations. Negotiations permit people to assess the preferences of others, to ascertain their willingness to compromise if the threat of no agreement is real, and to measure each other's beliefs about these parameters. This view is strongly reminiscent of the strategic character of the chain store paradox. Thus, bargaining models are not only difficult to formulate, but also difficult to solve.

10.7   Reciprocity and repeated games

Instead of reviewing an assembly of alternative bargaining models, we can give some insight into how analysis might proceed with an example that is especially germane to politics. Our example comes from some recent research by Calvert and from the observation that legislators engage in considerable cooperation without explicit enforcement mechanisms. Legislators trade on bills and approve benefits for the constituencies of others, and members of one legislative committee do not question the substantive expertise and actions of another committee, often with only an implicit promise of reciprocal acquiescence. In short, "norms of universalism (among members) and reciprocity (between committees)" appear to pervade Congress. The research question that such norms engender is: Can we rationalize universalism and reciprocity as both a cooperative outcome when viewed as the product of some negotiation process and as a noncooperative equilibrium when viewed from the perspective of individual legislators?

Consider two persons who must play the same game an infinite number of periods. In each period there is some probability, p_i, that person i is in a position to ask a favor of the other that is worth b_i to him. Hence, in any period, either no one, one player, the other, or both may have a reason to ask a favor. A person who has a favor asked of him can either ignore the request, or expend the effort q_j, $0 \le q_j \le 1$, at a cost of $q_j c_j$, in trying to grant the request. Thus, if $q_j = 0$, then the person refuses the request, whereas if $q_j = 1$, then he grants the favor. Nature then decides whether i receives the benefit, b_i, and we suppose that the probability that i realizes b_i is q_j.

Our analysis of this game proceeds thus: First, we ascertain the conditions under which the strategy, "always accede to a request" when chosen by both players, yields an efficient outcome. Efficiency is a minimal requirement for an outcome to satisfy if it is to correspond to any cooperative solution of bargaining hypotheses such as the Nash arbitration scheme, and if an outcome fails to satisfy this condition, then we cannot rationalize it in terms of any such hypothesis. Second, we establish the conditions under which "always accede" is an equilibrium strategy for both players. Finally, using the Nash arbitration point as an example, we derive the conditions that an outcome must satisfy if it is to be such a point. If these three sets of conditions are consistent, then we can conclude that at least one cooperative outcome, the Nash point, is sustained as a noncooperative equilibrium, which rationalizes that outcome as an *implicit* cooperative agreement.

We begin, as in our discussion of the repeated prisoners' dilemma, by describing the players' payoffs, and we weight future payoffs by the discount parameter w. Thus, expression (10.1) states a person's current evaluation of a stream of payoffs that extend into an indefinite future, so if each player, i, always supplies the level of effort q_i whenever a request is made, then 1's discounted expected payoff is

$$\sum_{t=1}^{\infty} w^{t-1}(q_2 p_1 b_1 - q_1 p_2 c_1) = \frac{q_2 p_1 b_1 - q_1 p_2 c_1}{(1-w)}, \qquad (10.7a)$$

and for player 2 this expected payoff is

$$\frac{q_1 p_2 b_2 - q_2 p_1 c_2}{(1-w)}. \qquad (10.7b)$$

A strategy is a rule for granting favors, given what has happened in the past, which is the same interpretation of a strategy as in the infinitely repeated prisoners' dilemma. Indeed, this game is identical to the game that Section 10.1 analyzes if we set $b_1 = b_2 = b$, $c_1 = c_2 = c$, $p_1 = p_2 = 1$; constrain q to be either 0 or 1; and referring to Figure 10.1, use the identities $w = 0$, $y = b$, $z = -c$, and $x = b - c$. The differences between this new game and the repeated prisoners' dilemma is that here the players' costs and benefits need not be symmetric, players can expend varying amounts of effort acceding to a request, and both players need not seek a favor in each period nor need they seek them with the same probability.

In spite of these differences, the purpose of granting a favor is the same as cooperating in the repeated prisoners' dilemma, the hope that such action leads the other person to grant favors in the future. Our first step in the analysis now is to be certain that the strategies that we consider provide each player with greater benefits than if he never reciprocated a favor. Indeed, we must first establish the conditions under which full reciprocation is not only beneficial, but also Pareto-efficient. We begin by considering the conditions under which always granting favors is mutually advantageous ($q_1 = q_2 = 1$). Notice that expressions (10.7a) and (10.7b) are both positive whenever

$$\frac{p_1 b_1 - p_2 c_1}{1 - w} > 0$$

$$\frac{p_2 b_2 - p_1 c_2}{1 - w} > 0$$

or equivalently, if

$$\frac{c_1}{b_1} < \frac{p_1}{p_2} < \frac{b_2}{c_2}. \qquad (10.8)$$

If we can satisfy expression (10.8), then both players prefer a system in which each always accedes to a request over what they would receive if neither granted any favor. Satisfying this expression is a minimum requirement for the game to be interesting. To show now that the reciprocal granting of favors is *efficient*, notice that the particular sequence of moves chosen by chance is a random variable. One possible sequence is "1 and 2 ask a favor in period 1, 1 but not 2 asks a favor in period 2,...," and the probability that we would observe this sequence depends on p_1 and p_2. Because we know these two probabilities, we also know the density function that they define over the set of all possible sequences. Now suppose that we reduce player 1's effort in period k from 1 to $1 - r_k(s)$. We let r be a function of s to indicate that we must condition 1's decisions on the particular sequence. We cannot assume that 1 reduces his effort in a period in which he is not asked for a favor. Similarly, let us reduce 2's efforts by the amount $t_k(s)$. Notice that whenever 1 reduces his effort, he gains, because this reduction saves costs, but he loses whenever 2 reduces his effort. Hence, the overall expected gain to 1 is

$$c_1 E\left[\sum_{k=1}^{\infty} r_k(s)w^{k-1}\right] - b_1 E\left[\sum_{k=1}^{\infty} t_k(s)w^{k-1}\right],$$

where "E" denotes expected value with respect to the random variable, s. We rewrite this expression as $c_1 E_1 - b_1 E_2$, and we suppose that it is positive and player 1 gains from the change in choices. That is, let $c_1/b_1 < E_2/E_1$. But we must also satisfy expression (10.8), since we want full reciprocity to be beneficial. Thus, $E_2/E_1 < b_2/c_2$, or equivalently, $c_2 E_2 - b_2 E_1 < 0$. The left-hand side of this inequality, though, is the expected change in 2's utility if we reduce both persons' efforts. Thus, if 1 gains by a change from always granting favors, then 2 must lose. Always granting favors on the part of both players is Pareto-optimal.

Since we want ultimately to rationalize the Nash arbitration point as a noncooperative equilibrium outcome, we impose expression (10.8) throughout the remainder of the analysis, and we proceed to the second step, showing that always granting favors can be supported as a noncooperative equilibrium.

To this point we have been concerned with the outcomes of the game, but we have not yet specified the players' strategy sets. That is, we have not specified how a player might condition his actions on the other person's actions. And as with the infinitely repeated prisoners' dilemma, there are an infinity of strategies that we might inspect. Instead of proceeding by focusing on a few obvious alternatives, here we will take advantage of a result that Abreu established, that *if there is a way to enforce a sequential strategy, then one way is to make punishments as severe as possible.* Thus, suppose that both players adopt the strategy of *permanent retaliation*: "Grant every favor until the other person fails to grant a favor, at which point never grant a favor again."

Notice that if both persons choose this strategy, then full reciprocity prevails. The question, then, is whether such strategies are an equilibrium pair. To show that they are such a pair, we must also show that permanent retaliation is a best response to permanent retaliation. Suppose, then, that player 1 must decide in the current period whether to grant 2 a favor. The expected benefit of granting the favor, assuming that 2 is abiding by the permanent retaliation strategy, is the cost of this favor, $-c_1$, plus $w[(p_1b_1 - p_2c_1)/(1 - w)]$, which is the discounted stream of benefits that begin in the next period (any switch in strategy affects player 2's strategy beginning with the next period). Because one refusal ends cooperation forever, the expected utility of refusing to grant a favor pays 0 in this and all subsequent periods. Hence, the permanent-retaliation strategy is a best response for 1 if and only if

$$-c_1 + w\left[\frac{p_1b_1 - p_2c_1}{1 - w}\right] > 0.$$

Algebraic manipulation changes this inequality to

$$c_1/b_1 \le \frac{wp_1}{1 - w(1 - p_2)}. \tag{10.9a}$$

A parallel analysis yields this condition for 2:

$$c_2/b_2 \le \frac{wp_2}{1 - w(1 - p_1)}. \tag{10.9b}$$

Notice that as either the discount rate, w, or the probability of needing of favor, p_1, approaches zero, both parts of expression (10.9) become more difficult to satisfy. Thus reciprocity of favors prevails if people care about the future and anticipate the need for favors. Also, this reciprocity does not require that benefits from a favor, b_1, exceed the costs of granting a favor, c_1. If $p_1 = p_2 = 1$, then expression (10.9a) requires that $c_1/b_1 \le w/(1 - w)$, which we satisfy if the discount rate is low.

These conclusions are like those that Section 10.1 derives, and it is also possible to show that a strategy of tit for tat supports full reciprocity as well. But now we turn to the final step in the analysis, showing that full reciprocity is a Nash arbitration point.

Consider the product of the discounted expected payoffs of both players if player i expands the effort q_i in every period. From expressions (10.7a) and (10.7b) we find that this product equals

$$\left[\frac{q_2 p_1 b_1 - q_1 p_2 c_1}{1 - w}\right]\left[\frac{q_1 p_2 b_2 - q_2 p_1 c_2}{1 - w}\right].$$

To find the Nash arbitration point, we maximize this product with respect to both q_1 and q_2, subject to the constraint that both q's are positive and not greater than one. What we want to know is when and if this maximum occurs when both players always grant favors. Taking the derivative of the preceding expression first with respect to q_1, and ignoring $(1 - w)^2$, because it is a constant, yields

$$p_2(b_1 b_2 + c_1 c_2) - 2p_1 b_1 c_2.$$

Notice that this expression does not depend on q_1. Thus, we maximize the product of utilities for the permitted limits on the q's for q_1 equal either to 0 (if the derivative is negative) or 1 (if the derivative is positive). Because we want a solution at $q_1 = 1$, always granting favors is a Nash arbitration point only if this expression is always positive. Thus, setting the expression greater than 0, and rearranging terms gives us

$$\frac{p_1}{p_2} < \frac{b_1 b_2 - c_1 c_2}{2 b_1 c_2}. \tag{10.10a}$$

If we differentiate the product of utilities with respect to q_2 instead, and proceed as we just have done, we get

$$\frac{p_2}{p_1} < \frac{b_1 b_2 + c_1 c_2}{2 b_2 c_1}$$

or equivalently,

$$\frac{p_1}{p_2} > \frac{2 b_2 c_1}{b_1 b_2 + c_1 c_2}. \tag{10.10b}$$

Combining expressions (10.10a) and (10.10b) yields

$$\frac{2 b_2 c_1}{b_1 b_2 + c_1 c_2} < \frac{p_1}{p_2} < \frac{b_1 b_2 + c_1 c_2}{2 b_1 c_2}. \tag{10.11}$$

Thus, if we can satisfy expressions (10.9a), (10.9b), and (10.11), then always granting favors is the unique Nash arbitration point and it is supported as well as a noncooperative equilibrium. That is, if we can satisfy all of these expressions, then the Nash bargain does not require explicit enforcement to be maintained. To evaluate these conditions, suppose that $c_i/b_i = k$, $i = 1$ and 2. Then expression (10.11) requires that

$$\frac{2k}{1 + k^2} < \frac{p_1}{p_2} < \frac{1 + k^2}{2k},$$

or,

$$p_1 > \frac{2p_2 k}{1 + k^2} \quad \text{and} \quad p_2 > \frac{2p_1 k}{1 + k^2}.$$

Rewriting expressions (10.9a) and (10.9b) now in terms of k gives us

$$p_1 \geqslant \frac{(1 - w)k}{w} + kp_2 \quad \text{and} \quad p_2 \geqslant \frac{(1 - w)k}{w} + kp_1.$$

Notice that as k increases, as the cost of granting a favor increases relative to the benefit of having a favor granted, all four constraints, become more severe. Similarly, as p_1 increases, p_2 must increase as well, and vice versa.

Instead of exploring these constraints further, we wish merely to emphasize that we can satisfy all four constraints simultaneously. Thus, the Nash arbitration point can be the equilibrium of a noncooperative supergame. Accordingly, we can imagine a scenario in which the legislators agree at some initial stage of a legislative session to norms of reciprocity and universalism. They need not engage third parties to enforce that agreement, because it is an equilibrium to the ensuing noncooperative game.

10.8   Summary

Little in politics occurs as an isolated event. A vote trade in a legislature; a treaty among nations; an agreement by interest groups to lobby collectively for some benefit; and a willingness by all persons in society to tolerate religious diversity – each occurs in an environment of a continuing game. Generally it is advisable to simplify reality by isolating events, so that we do not become bogged down in the impossible task of seeking to model everything at once. But we must be cautious not to exclude important considerations. Classical solution theory provides an abstraction of one extreme sort, an abstraction that ignores the issue of enforcement. Sequential game analysis provides one means of rationalizing cooperation without explicitly enforceable contracts.

The model in Section 10.7, in particular, although it uses several special assumptions, nevertheless offers a powerful perspective. Norms of different sorts pervade society and give social and political processes a coherence that permits other forms of activity to occur. Democratic systems function, in large part, because the losers in elections know that they can compete again. If it were otherwise, then democracy would evolve into dictatorship. The simple norm of not excluding losers and their parties from the political process, then, assures the empirical relevance of the subjects that this book discusses throughout.

Such norms generally exist only implicitly, and they work best if there is no need for explicit enforcement. Therefore, the topoics that this chapter covers remain fundamental for understanding democratic systems. Admittedly, this chapter focuses on the simpler problems: finitely and infinitely repeated prisoners' dilemmas and the chain store paradox. Those readers who are encountering the chain store paradox and games of incomplete information for the first time might regard the discussion as anything but simple: however, the analysis proceeds straightforwardly, because each of these problems seems well structured. The models define precisely the periods and the choices available to all players in each period. But actual bargaining between two or more persons is unstructured, and portraying its extensive form seems to be an impossible task. Today, therefore, the solution theory that Chapters 8 and 9 review provides the mainstay of analysis. But researchers will continue to probe for a more nearly integrated theory, one in which we can ultimately rationalize all hypotheses about cooperative action as some kind of noncooperative equilibrium.

Despite the profound importance of the ideas that this chapter discusses, we have introduced few new terms and concepts. The exceptions are the repeated prisoners' dilemma, present value, the chain store paradox, the Shapley value, the power index, the Banzhaf-Coleman index, the Nash arbitration point, and threat strategies. As research proceeds, however, the reader should expect many more new terms and concepts to be added to any such list. Kreps and Wilson's modification of the concept of a perfect equilibrium, a sequential equilibrium, will probably be refined further, and new ways to treat cooperation and bargaining as an extensive-form game will be developed. Also, the reader who is familiar with some parts of game theory may wonder why we have not introduced other ideas in various chapters such as the kernel, the nucleolus, and the ε-core. These ideas, however, are tangential to the main developments of the theory we seek to explore; they are inventions that satisfy criteria other than substantive meaning. And as political scientists become better acquainted with the political theory that this volume reviews, as the set of substantive applications of this theory expands, other concepts will be discarded and new ones invented.

References and a guide to the literature

Chapter 1

The applicability of the "rational choice" paradigm to politics has seen considerable debate. For general comments, see William H. Riker and Peter C. Ordeshook, *An Introduction to Positive Political Theory*, Englewood Cliffs, NJ: Prentice-Hall, 1973; William H. Riker, *Liberalism Against Populism*, San Francisco: W. Freeman, 1983; and John C. Harsanyi, "Rational Choice Models of Political Behavior vs. Functionalist and Conformist Theories," *World Politics*, 21, 1969. For an introductory treatment suitable for lower-division undergraduate courses, see Peter H. Aranson, *American Government*, Boston: Little, Brown, 1981; and Robert Abrams, *Foundations of Political Analysis*, New York: Columbia Univ. Press, 1979. With respect to the details of this paradigm, the literature on utility theory and preferences is considerable. The properties of ordinal and cardinal utility, alternative axiom systems, and the treatment of subjective probabilities are complete scholarly fields, with branches in psychology, operations research, statistics, and economics. The expected utility theorem, Theorem 1.1, should be attributed, however, to the seminal volume on game theory, John von Neumann and Oskar Morgenstern, *The Theory of Games and Economic Behavior*, Princeton: Princeton Univ. Press, 1945. An excellent introductory discussion of probability and decision making is provided in Howard Raiffa, *Decision Analysis*, Reading, MA: Addison-Wesley, 1968. The student seeking a more advanced treatment should consult Morris DeGroot, *Optimal Statistical Decisions*, New York: McGraw-Hill, 1970. For a discussion of the "as if" principle and its role in economics, see Milton Friedman, *Essays in Positive Economics*, Chicago: Univ. of Chicago Press, 1953. The quotation on the nature of purposive choice is taken from Arthur Rosenbluth, Norbert Wiener, and V. Bigelow, "Behavior, Purpose and Teleology," *Philosophy of Science*, 10, 1943: 18. Other sources for an exposition of the axioms of preference include the game theory texts cited in Chapter 3 as well as the social choice texts cited in Chapter 2. The examples cited with respect to the methods of postulated preference are David R. Mayhew, *Congress: The Electoral Connection*, New Haven, CT: Yale Univ. Press, 1974; Morris P. Fiorina, *Congress: Keystone of the Washington Establishment*, New Haven: Yale Univ. Press, 1977; William H. Riker, *Federalism: Origin, Operation, Maintenance*, Boston: Little, Brown, 1964; and William Niskanen, *Bureaucracy and Representative Government*, Chicago: Aldine, 1971. With respect to our discussion of indifference curves and utility maximization, almost any introduction to microeconomic theory surveys these topics, including the mathematical preliminaries required for an understanding of optimization with Lagrange multipliers. But spatial preferences, with ideal points that are internal to the feasible set of outcomes, are somewhat unique to political science. Perhaps the first to use the city-block metric in political models was Douglas Rae and Michael Taylor, "Decision Rules and Policy Outcomes," *British Journal of Political Science*, 1, 1971: 71–90. As this chapter indicates, though, preferences based on Euclidean distance play a special role in formal political models. For additional discussion of the properties of such preferences as well as a view on how they might be estimated using survey data, see James M. Enelow and Melvin J. Hinich, *Spatial Analysis of Elections*, New York: Cambridge Univ. Press, 1984. The discussion of the calculus of voting follows William H. Riker and P. C. Ordeshook, "The Calculus of Voting," *American Political Science Review*, 62, 1968: 25–42. A significant literature has developed dealing with the apparent paradox that people vote even though the chance that their vote is decisive is nearly zero in most elections. For a current bibliography, see the Palfrey and Rosenthal essay cited in Chapter 5. Also, the reader should consult John Ferejohn and Morris Fiorina, "The Paradox of Not Voting: A Decision Theoretic Analysis," *American Political Science Review*, 69, 1975: 525–36; and "Closeness Counts Only in Horseshoes and Dancing," *American Political Science Review*, 69, 1975: 920–5. There the authors propose a model based on a decision criterion called "minimax-regret," which does not presume that people form subjective estimates of the likelihood of creating or breaking ties and which can predict, with an appropriate definition of states of nature, significantly higher turnout levels than does the expected utility model. This criterion, however, can yield silly predictions. For example, if every citizen believes that having a fatal accident on the way to the polls is a conceivable possibility (the issue of how remote is irrelevant), then no one should vote. And if the states of nature are defined appropriately, we also can predict that if people do vote, then they should all vote for themselves.

Chapter 2

The central result of this chapter is Arrow's general impossibility theorem, which Kenneth J. Arrow presents and discusses in *Social Choice and Individual Values*, New Haven, CT: Yale University Press, 1963 (1st ed. 1951). Arrow's volume is seminal and unusually lucid for such a work. A considerable body of research builds on this result to prove other impossibility theorems, to strengthen Arrow's original result, and to interpret these results for economic and political theory as well as public policy. An excellent survey for the beginning student is presented in Charles R. Plott, "Axiomatic Social Choice Theory: An Interpretation and Overview," *American Journal of Political Science*, 20, 1976, 511–96. Texts include Peter C. Fishburn, *The Theory of Social Choice*, Princeton: Princeton Univ. Press, 1973; Amartya K. Sen, *Collective Choice and Social Welfare*, San Francisco: Holden Day, 1970; and Thomas Schwartz, *The Logic of Collective Choice*, New York: Columbia Univ. Press, 1986. The bibliography in Schwartz's book in particular is inclusive and up to date. The computations that Figure 2.1 reports come from several sources, but see Richard G. Niemi and Herbert Weisberg, "A Mathematical Solution for the Probability of the Paradox of Voting," *Behavioral Science*, 13, 1968: 317–23; and Mark B. Garman and Morton I. Kamien, "The Paradox of Voting: Probability Calculations," *Behavioral Science*, 13, 1968: 306–17; and William V. Gehrlein and Peter C. Fishburn, "The Probability of the Paradox of Voting," *Journal of Economic Theory*, 13, 1976: 14–25. The proof of Arrow's theorem presented here is taken from Thomas Schwartz, "Impossibility Theorems: A Nonformal Introduction to Proof Strategies," working paper, Univ. of Texas at Austin, 1985. Most of the examples of paradoxes are from Peter C. Fishburn, "Paradoxes of Voting," *American Political Science Review*, 68, 1974: 537–46. The substantive illustration of the DePew amendment as well as of the Powell amendment are from William H. Riker, *Liberalism Against Populism*, cited earlier. Although Riker introduces these examples in several of his earlier published essays, this volume also summarizes his current thinking on the implications of the ideas presented in this chapter and is a useful interpretation for political scientists. Theorem 2.2 is proved formally in Richard D. McKelvey, "Intransitivities in Multidimensional Voting Models and Some Implications for Agenda Control," *Journal of Economic Theory*, 16, 1976: 472–82 and is generalized in his 1979 *Econometrica* paper, "General Conditions for Global Intransitivities in Formal Voting Models" (pp. 1085–1112). McKelvey's result places no limit on the size of the steps that can be taken in forming an appropriate agenda, but Norman Schofield, in "Instability of Simple Dynamic Games," *Review of Economic Studies* 45, 1976: 575–94, shows that if there are three or more dimensions, then the social preference relation under majority rule is cyclic even locally. For a discussion of the implications of these results to the study of politics, see the essays in P. C. Ordeshook and Kenneth Shepsle, *Political Equilibrium: A Delicate Balance: Essays in Honor of William H. Riker*, Boston: Kluwer-Nijhoff, 1982. The two seminal papers on Theorem 2.3 are Allan Gibbard, "Manipulation of Voting Schemes: A General Result," *Econometrica*, 41, 1973: 587–601; and Mark A. Satterthwaite, "Strategy-Proofness and Arrow's Conditions: Existence and Correspondence Theorems for Voting Procedures and Social Welfare Functions," *Journal of Economic Theory*, 1975: 187–217. A lucid discussion of Theorem 2.3 is presented in Allan M. Feldman, *Welfare Economics and Social Choice Theory*, Boston: Kluwer-Nijhoff, 1980. The theorem we offer is an adaptation proved by Thomas Schwartz of a more general result about nonresolute choice functions that he presents in "No Minimally Reasonable Collective-Choice Process Can Be Strategy-Proof," *Mathematical Social Sciences*, 3, 1982: 57–72. This essay, however, is not recommended reading for anyone but specialists in the field. For the beginning student, the best starting point for understanding strategic voting and the ideas that we will be discussing in this context in subsequent chapters is Robin Farquharson, *Theory of Voting*, New Haven, CT: Yale University Press, 1969. Finally, although a more complete discussion of vote trading is postponed until later chapters, the proof that sincere voting and separable preferences yield a

Condorcet winner, if such a winner exists, comes from Thomas Schwartz, "Collective Choice, Separation of Issues and Vote Trading," *American Political Science Review*, 71, 1977: 999–1010. Vote trading and its relationship to the Condorcet paradox became a lively research area in the 1970s, and Schwartz's essay contains the relevant references, summarizes this research, and answers most of the pertinent questions. We note in passing that although we do not make a strong distinction, some scholars prefer to reserve the term "vote trading" for the activities of pairs of persons and to refer to trades among more than two persons as "logrolling." Finally, one possibility that this chapter does not discuss is that socialization renders people's preferences sufficiently similar on many issues that the preferences on those issues are not intransitive and social decisions are not manipulable. The pervasiveness of such issues is, of course, an empirical question, but for a theoretical investigation with respect to cyclic social preferences, see Richard Niemi and Herbert Weisberg, "A Mathematical Solution for the Probability of the Paradox of Voting," *Behavioral Science*, 13, 1968: 317–23; Richard Niemi, "Majority Decision Making with Partial Unidimensionality," *American Political Science Review*, 63, 1969: 489–97; Peter C. Fishburn, "Voter Concordance, Simple Majorities, and Group Decision Methods," *Behavioral Science*, 18, 1973: 364–73; and, with respect to nonmanipulability, see Joseph Greenberg, "Consistent Majority Rules over Compact Sets of Alternatives," *Econometrica*, 47, 1979: 627–36.

Chapter 3

Two texts that present general introductions to noncooperative game theory are Guillermo Owen, *Game Theory*, New York: Academic Press, 1983; and Ewald Burger, *Introduction to Game Theory*, Englewood Cliffs, NJ: Prentice-Hall, 1975. Mastery of these volumes requires a background in calculus and probability and set theory, and they are challenging for the beginning student. Another introduction to the formal development of noncooperative game theory is Herve Moulin, *Game Theory for the Social Sciences*, New York: New York Univ. Press, 1982. Moulin's book is especially interesting because of the imaginative examples and exercises it offers and the links it develops between noncooperative and cooperative theory. Perhaps the classic text in game theory, however, is R. Duncan Luce and Howard Raiffa, *Games and Decisions*, New York: Wiley, 1957. Although quite readable and comprehensive, it is unfortunate that this text has not been updated to incorporate the considerable developments since its writing. Martin Shubik's recent book *Game Theory in the Social Sciences*, Cambridge, MA: MIT Press, 1982, offers perhaps the most comprehensive and updated bibliography. Discursive texts that are especially valuable for teaching game-theoretic reasoning include Thomas Schelling, *The Strategy of Conflict*, Cambridge, MA: Harvard Univ. Press, 1971; Henry Hamburger, *Games as Models of Social Phenomena*, San Francisco: Freeman, 1979 (this volume makes an excellent undergraduate text); Gary Brewer and Martin Shubik, *The War Game*, Cambridge, MA: Harvard Univ. Press, 1979; and the series of volumes by Stephen Brams, *Game Theory and Politics*, New York: The Free Press, 1975; *The Presidential Election Game*, New Haven, CT: Yale Univ. Press, 1978; *Biblical Games*, Cambridge, MA: MIT Press, 1980; and *Superpower Games*, New Haven, CT: Yale Univ. Press, 1985. Two collections of essays that are of particular interest to political scientists are Steven J. Brams, Andrew Schotter, and G. Schwodiauer, eds., *Applied Game Theory*, Wurzburg: Physica Verlag, 1979; and P. C. Ordeshook, *Game Theory and Political Science*, New York: New York Univ. Press, 1978. Although several authors have offered different proofs of Theorem 3.2, the proof that we offer is adapted from J. B. Rosen, "Existence and Uniqueness of Equilibrium Points for Concave *n*-Person Games," *Econometrica*, 43, 1975: 520–34. The equilibrium concept that is the focus of noncooperative game theory, the Nash equilibrium, comes from John F. Nash, "Equilibrium Points in *n*-Person Games," *Proceedings of the National Academy of Science*, 36, 1950: 48–49. A useful discussion of the advantages and potential problems with Nash equilibria, as well as proposed resolutions, is presented in David G. Pearce, "Rationalizable Strategic Behavior and the Problem of Perfection," and B. Douglass Bernheim, "Rationalizable Strategic Behavior," both in *Econometrica*, 52, 1984: 1007–50. Our discussion of perfect equilibria is taken from Reinhardt Selten, "Reexamination of the Perfectness Concept for Equilibrium Points in Extensive Games," *International Journal of Game Theory*, 4, 1975: 25–55.

Chapter 4

The analysis of campaign resource allocations in Section 4.3 as well as an analysis of several other resource allocation models can be found in Steven Brams's volume, *The Presidential Election Game*, cited earlier. The discussion of equivalent candidate objectives is taken from Peter H. Aranson, Melvin J. Hinich, and P. C. Ordeshook, "Election Goals and Strategies: Equivalent and Nonequivalent Campaign Objectives," *American Political Science Review*, 58, 1974: 135–52. A number of other models posit alternative objectives that incorporate policy preferences for candidates. For an analysis of electoral competition with this assumption, see Donald Wittman, "Candidate Motivation: A Synthesis of Alternatives," *American Political Science Review*, 77, 1983: 142–57; and Randall L. Calvert, "Robustness of the Multidimensional Voting Model: Candidate Motivations, Uncertainty and Convergence," *American Journal of Political Science*, 29, 1985: 69–95. The seminal analysis of voting in one dimension with single-peaked preferences, the subject of Section 4.5, is Duncan Black's *Theory of Committees and Elections*, Cambridge: Cambridge Univ. Press, 1958, and his earlier volume with R. A. Newing, *Committee Decisions with Complementary Valuation*, Glasgow: Hodge, 1951. Although not concerned explicitly with two-candidate elections, Black first proposed the concept of a single-peaked preference as a means of avoiding the Condorcet paradox, and the Black and Newing essay anticipates much of what follows in Chapter 6. The seminal volume on the application of these ideas to elections with a sensitivity to political issues is Anthony Downs, *An Economic Theory of Democracy*, New York: Harper & Row, 1957. The formal literature on multidimensional election competition is considerable and it begins with two essays: Otto A. Davis and Melvin J. Hinich, "A Mathematical Model of Policy Formation in a Democratic Society," in J. L. Bernd, ed., *Math. Applications in Political Science*, 2, Dallas: Southern Methodist Univ. Press, 1967; and Charles R. Plott, "A Notion of Equilibrium under Majority Rule," *American Economic Review*, 57, 1967: 787–806. Like Black, Plott is not specifically concerned with two-candidate elections, but his essay presents an important sufficient condition for a Condorcet winner in a multidimensional space assuming generalized utility functions (which condition Figure 4.17 summarizes). Actually, we can interpret Plott's condition, which assumes as a precondition that no more than one person most prefers the presumed Condorcet point, as establishing *generically* that Condorcet winners are rare. By generically we mean that the precondition has a small if not zero probability of being a real constraint on the applicability of the result. The Davis and Hinich essay is specifically concerned with elections, and it anticipates many of the subsequent developments in this area. Borrowing from the concept of a loss function in statistics, it is perhaps the first to introduce Euclidean preferences, which have since become a standard assumption of most election models. One advantage of this assumption is that it lends itself nicely to the development of statistical methodologies for the estimation of preferences from survey research data. The best general survey of the formal literature is by James M. Enelow and Melvin J. Hinich, *Spatial Analysis of Elections*, cited earlier. Numerous alternative models are discussed there, including a variety of stochastic models, models that consider the relationships between election issues and specific parameters of public policy, and statistical procedures for estimating appropriate parameters. Although the research of Black, Davis and Hinich, and Plott provided sufficient conditions for the existence of Condorcet winners in a spatial structure, derivations of necessary and sufficient conditions are essentially refinements and generalizations of their work. Our text focuses on the theorem proved in Otto A. Davis, Morris DeGroot, and Melvin J. Hinich, "Social Preference Orderings and Majority Rule," *Econometrica*, 40, 1972: 147–57, even though it is restricted to Euclidean preferences, because such preferences succumb to geometric intuition; but readers should also consult Steven Slutsky, "Equilibrium under α-Majority Voting," *Econometrica*, 47, 1979: 1113–25, and Judith Sloss, "Stable Outcomes in Majority Voting Games," *Public Choice*, 14, 1973: 19–48. Sloss's essay is also an excellent survey of the relationships among various concepts and, in addition, treats indifference contours that are not thin, but that admit intransitive indifference. The discussion in Section 4.7 of mixed strategy solutions to resource allocation games is taken from Owen's text, *Game Theory*, and readers who might be interested in an application of this class of games should see H. P. Young, "A Tactical Lobbying Game," in Ordeshook, *Game Theory and Political Science*, cited earlier. The concept of the uncovered set has several sources, but its first extensive application can be found in Nicholas Miller, "A New Solution Set for Tournaments and Majority Voting," *American Journal of Political Science*, 24,

1980, 68–96, and a nearly equivalent idea is developed for two-candidate elections in Richard D. McKelvey and P. C. Ordeshook, "Symmetric Spatial Games without Majority Rule Equilibria," *American Political Science Review*, 70, 1976; 1172–84. The results relating to the bounds on the uncovered set using Euclidean preferences are developed by Richard D. McKelvey, "Covering, Dominance and Institution Free Properties of Social Choice," *American Journal of Political Science*, forthcoming, May, 1986. We emphasize again that the bound that this essay places on the uncovered set is most important. Countless students of political theory mistakenly believe that if a Condorcet winner does not exist in an election model, then we cannot say anything about candidates' strategies and even that the concept of popular control of public policy is meaningless. Much of this book argues against that proposition directly or indirectly, and McKelvey's results here are a case in point. A general and very readable survey of the literature relating to the uncovered set is presented in Nicholas Miller, "Agenda Formation in Voting Bodies: A Preliminary Survey," paper presented at the 1985 Annual Meeting of the Public Choice Society, New Orleans. The centralizing tendency that the uncovered set justifies in two-candidate elections provides a theoretical basis for Gordon Tullock's assertion that real social processes do not exhibit the instabilities that precise mathematical models seem to convey ("Why So Much Stability," *Public Choice*, 37, 1981: 189–202). Indeed, we can credit Tullock's book *Towards a Mathematics of Politics*, Ann Arbor: Univ. of Michigan Press, 1968, as being the first to raise the substantive issue of why democratic systems generally do not appear to exhibit the instabilities that Arrow's and McKelvey's theorems (Theorems 2.1 and 2.2) seem to imply. The experiments reported in this section are taken from Richard D. McKelvey and P. C. Ordeshook, "Two-Candidate Elections without Majority-Rule Equilibria," *Simulation and Games*, 13, 3, 1982: 311–35. The model of incomplete information in elections described in Section 4.8 is taken from Richard D. McKelvey and P. C. Ordeshook, "Elections with Limited Information: A Fulfilled Expectations Model Using Contemporaneous Poll and Endorsement Data as Information Sources," *Journal of Economic Theory*, 36, 1985: 55–85, and "Some Experimental Results Relating to a Multidimensional Model of Elections with Incomplete Information," *Public Choice*, 44, 1984: 61–102. The treatment of incomplete information offered in these essays follows the development of rational expectations models in economics, but the literature there is specialized and largely inaccessible. Several papers examine election models with probabilistic voting and these are reviewed in Enelow and Hinich's text, but the reader might also consult Melvin J. Hinich, John Ledyard, and P. C. Ordeshook, "Nonvoting and the Existence of Equilibria under Majority Rule," *Journal of Economic Theory*, 4, 1972: 144–53; as well as Peter Coughlin and S. Nitzan, "Electoral Outcomes with Probabilistic Voting and Nash Social Welfare Maxima," *Journal of Public Economics*, 15, 1981: 113–21. Those students interested in probabilistic voting models should read Peter Coughlin's most recent essays and working papers. Our discussion of multi-candidate competition is taken largely from Gary Cox, "Electoral Equilibrium in Multicandidate Elections: Plurality vs. Approval Voting," working paper,

Univ. of Texas at Austin, 1983. For a discussion of the effects of nonvoting on multicandidate equilibria, see Melvin J. Hinich and P. C. Ordeshook, "Plurality Maximization vs. Vote Maximization: A Spatial Analysis with Variable Participation," *American Political Science Review*, 64, 1970: 722–91. For a more general yet formal discussion of the number of parties, see William H. Riker, "The Number of Political Parties: A Re-examination of Duverger's Law," *Comparative Politics*, 9, 1976: 93–106. The spatial modeling literature examines a far greater variety of institutions and electoral considerations than we review in this chapter. On the issue of how candidates might adjust their positions to control the entry of third-party candidates see Thomas Palfrey, "Spatial Equilibrium with Entry," *Review of Economic Studies*, 51, 1984: 139–56. For analyses of ambiguity see Kenneth Shepsle, "The Strategy of Ambiguity, Uncertainty and Competition," *American Political Science Review*, 66, 1972: 551–68; and Richard McKelvey and Jeff Richelson, "Cycles of Risk," *Public Choice*, 18, 1974: 41–66. Although this chapter is devoted almost exclusively to single-member district elections and plurality rule, see, for example, Gary Cox, "Electoral Equilibria under Approval Voting," *American Journal of Political Science*, 29, 1985: 112–18. Gerald Kramer, in "A Dynamical Model of Political Equilibrium," *Journal of Economic Theory*, 16, 1977: 310–34, offers a model of elections in which, if candidates sequentially adjust their strategies to maximize their votes, then the candidates will converge near the "center" of the electorate's preference density even if there is no Condorcet winner. Thomas Romer and Howard Rosenthal, in "Bureaucrats vs. Voters: On the Political Economy of Resource Allocation by Direct Democracy," *Quarterly Journal of Economics*, 1979: 563–87, examine referendum voting when public officials, rather that campaigning as candidates, control the motions that appear on the ballot. A game-theoretic treatment of interest-group activities in campaigns is provided by David Austen-Smith, "Interest Group Campaign Contributions and Spatial Voting," working paper, Univ. of Rochester, 1984. The reader should also consult another essay by David Austen-Smith, "The Spatial Theory of Electoral Competition: Instability, Institutions, and Information," *Environment and Planning*, 1, 1983: 439–59, which includes an excellent bibliography. The problems occasioned by candidates having to secure the nomination of party activists before they can compete in a general election is explored in John H. Aldrich, "A Downsian Spatial Model with Party Activism," *American Political Science Review*, 77, 1983: 974–90. For an application of spatial theory that takes theoretical advantage of the difference between the median and the mean in nonsymmetric probability densities to model the growth of government, see Allan H. Meltzer and Scott F. Richard, "A Rational Theory of the Size of Government," *Journal of Political Economy*, 89, 1981: 914–27. For the development of methodologies for estimating the parameters of a spatial elections model see Enelow and Hinich's *Spatial Analysis of Elections*, cited earlier; and John H. Aldrich and Richard D. McKelvey, "A Method of Scaling with Applications to the 1968 and 1972 Presidential Elections," *American Political Science Review*, 71, 1977: 111–30. Finally, for a general interpretive essay on the perspectives of spatial models, see Benjamin Page, *Choices and Echoes in Presidential Elections*, 1976, Chicago: Univ. of Chicago Press.

Chapter 5

With the possible exception of two-candidate elections, the prisoners' dilemma is the most widely studied example of a substantive noncooperative game, and it is impossible to review even a part of the literature that pertains to it. Because the prisoners' dilemma illustrates the free-rider problem associated with public goods in economics and political theory, much of the public goods literature can be viewed as an elaboration of the perspective that this dilemma provides. The seminal work on the relationship between the free-rider problem and group action is Mancur Olson, *The Logic of Collective Action*, Cambridge, MA: Harvard Univ. Press, 1968; but the reader should also consult Olson's latest book, *The Rise and Decline of Nations*, New Haven, CT: Yale Univ. Press, 1982. For a general overview of some of the issues this chapter addresses, see Russell Hardin, "Collective Action as an Agreeable *n*-Person Prisoners' Dilemma," *Behavioral Science*, 16, 1971: 472–81, as well as Hardin's book, *Collective Action*, Chicago: Univ. of Chicago Press, 1975. For a critique of Olson's assertion that larger groups are disadvantaged with respect to smaller groups in their ability to organize for collective action, see Norman Frohlich and Joe A. Oppenheimer, "I Get By with a Little Help from My Friends," *World Politics*, 23, 1970: 104–20. In that essay, Frohlich and Oppenheimer argue, moreover, that the relationship between group size and the likelihood with which groups fail to provide collective or public goods does not follow formally unless we incorporate transaction costs into the analysis. An especially useful text for undergraduate teaching in this area is their volume, *Modern Political Economy*, Englewood-Cliffs, NJ: Prentice-Hall, 1978. The literature on public goods and externalities is also substantial, and a good introduction is provided in two volumes by James M. Buchanan, *The Demand and Supply of Public Goods*, Skokie, IL: Rand McNally, 1968, and *The Bases of Collective Action*, New York: General Learning Press, 1971. A valuable survey and text for political scientists that also reviews some of the literature in election and committee models is Dennis Mueller, *Public Choice*, Cambridge: Cambridge Univ. Press, 1979. The discussion of government failure owing to interest group activity is taken from Peter H. Aranson and P. C. Ordeshook, "A Prolegomenon to a Theory of the Failure of Representative Democracy," in Richard Auster and Barbara Sears, eds., *American Re-Evolution*, Tucson: Univ. of Arizona, 1977, and "Public Interest, Private Interest, and the Democratic Policy," in Roger Benjamin and Stephen L. Elkin, eds., *The Democratic State*, Lawrence: Univ. of Kansas Press, 1985. This volume also contains an excellent bibliography that should be consulted for those wishing to study the field of political economy. For a discussion of vote trading as a prisoners' dilemma, see William H. Riker and Stephen Brams, "The Paradox of Vote Trading," *American Political Science*

Review, 67, 1973: 1235–47. However, for a more appropriate game-theoretic analysis in which it is shown that if legislators anticipate potential vote trades, then sophisticated voting yields the Condorcet winner, see James M. Enelow, "Non-cooperative Counter-threats to Vote Trading," *American Journal of Political Science*, 23, 1979: 121–38. The example of arms races relies on R. Harrison Wagner, "The Theory of Games and the Problem of International Cooperation," *American Political Science Review*, 77, 1983: 330–46. For a more extensive treatment of the concept of political entrepreneurship, see Norman Froelich, Joe A. Oppenheimer, and Oran R. Young, *Political Leadership and Collective Goods*, Princeton: Princeton Univ. Press, 1971. Although the discussion of voting in a rational choice context begins with Anthony Downs, *An Economic Theory of Democracy*, cited earlier, and William H. Riker and P. C. Ordeshook, "A Theory of the Calculus of Voting," cited earlier, the specific theoretical models presented here rely on John Ledyard, "The Pure Theory of Two-Candidate Elections," *Public Choice*, 44, 1984: 1–60; and Thomas Palfrey and Howard Rosenthal, "Voter Participation and Strategic Uncertainty," *American Political Science Review*, 79, 1985: 62–78. The equilibrium notion used in these essays is called a Bayesian Equilibrium, and more technically advanced readers should consult Roger B. Myerson's "Bayesian Equilibrium and Incentive Compatibility: An Introduction," discussion paper 48, Center for Mathematical Studies, Northwestern Univ. 1983. Myerson presents a less formal and quite readable discussion in his 1984 discussion paper, "An Introduction to Game Theory." With respect to demand revelation mechanisms, the most advanced research also uses Bayesian equilibrium notions, but the literature here is generally accessible only to specialists. Two readable sources in the context of complete information are the special issue of *Public Choice* edited by Gordon Tullock and Nicholas Tideman (vol. 29, 1977); and Allan M. Feldman, *Welfare Economics and Social Choice Theory*, cited earlier. Our formal development follows Feldman's presentation, but for a critique of this procedure as a practical device in politics, see William H. Riker, "Is 'A New and Superior Process' Really Superior?" *Journal of Political Economy*, 87, 1979: 875–90.

Chapter 6

Much of what follows in this chapter's first three sections was anticipated by Duncan Black and R. A. Newing in *Committee Decisions with Complementary Valuation*, cited earlier. Curiously, it took theorists some 20 years to recognize fully the intuition presented in this small monograph and to formalize its insight. Perhaps no book more profoundly stimulated the perspective that institutions and rules are creatures of human choice and ought to be investigated from the perspective of the outcomes they yield than James Buchanan and Gordon Tullock's *The Calculus of Consent*, Ann Arbor: Univ. of Michigan Press, 1962. In terms of the formal results presented in this chapter, Sections 6.1 and 6.2 rely on Gerald H. Kramer, "Sophisticated Voting over Multidimensional Choice

Spaces," *Journal of Mathematical Sociology*, 2, 1972: 165–80. The substantive application of this perspective, based on the observation that the institutional structure of Congress can induce equilibria and have other direct effects on outcomes, can be found in Kenneth Shepsle, "Institutional Arrangements and Equilibrium in Multidimensional Voting Models," *American Journal of Political Science*, 23, 1979: 27–59; and Shepsle and Barry Weingast, "Structure-Induced Equilibrium and Legislative Choice," *Public Choice*, 37, 1981: 503–19. Because they begin but do not end research, these essays rely on separability and the assumption that legislators are sincere voters. The seminal essay on sophisticated voting is, of course, Robin Farquharson, *Theory of Voting*, cited earlier. An exploration of the backward reduction method as a simple definition of sophisticated voting is given in Richard D. McKelvey and Richard G. Niemi, "A Multistage Representation of Sophisticated Voting for Binary Procedures," *Journal of Economic Theory*, 18, 1978: 1–22. McKelvey and Niemi conjectured that their representation of sophisticated voting was equivalent, in terms of final outcomes (although not necessarily in terms of actual strategies), to Farquharson's definition, and this has been confirmed by Rodney Gretlein, "Dominance Elimination Procedures on Finite Alternative Games," *International Journal of Game Theory*, 12, 1983: 107–13. The problems that anticipation occasions for stability with issue-by-issue voting is discussed extensively in Black and Newing's monograph, but our discussion was stimulated by Arthur Denzau and R. Mackay, "Structure Induced Equilibrium and Perfect Foresight Expectations," *American Journal of Political Science*," 25, 1981: 762–79. Section 6.4's discussion of amendment agendas relies on several sources: The theorem about sophisticated agenda equivalents is from Kenneth Shepsle and Barry Weingast, "Uncovered Sets and Sophisticated Voting Outcomes with Implications for Agenda Institutions," *American Journal of Political Science*, 28, 1984: 49–74. Figure 6.17 comes from this essay. The original analysis of agendas and the uncovered set is Nicholas Miller's "A New Solution Set for Tournaments and Majority Voting," cited earlier. See also Thomas Schwartz and P. C. Ordeshook, "Agendas and the Control of Political Outcomes," working paper, Univ. of Texas at Austin, 1986, for a formal analysis of congressional agenda procedures. The specific game modeling endogenous amendment agendas is taken from Richard D. McKelvey, "Covering, Dominance, and Institution Free Properties of Social Choice," which we cited in the previous chapter, but the reader should also see Jeffrey Banks, "Sophisticated Voting Outcomes and Agenda Control," *Social Choice and Welfare*, 1985. The spatial experiments on parliamentary procedure are reported in Richard D. McKelvey and P. C. Ordeshook, "The influence of Committee Procedures on Outcomes: Some Experimental Evidence," *Journal of Politics*, 46, 1984: 182–205. Other experimental essays on the effects of procedures on outcomes include Mark R. Isaacs and Charles R. Plott, "Cooperative Game Models on the Influence of the Closed Rule in Three Person Majority Rule Committees," in P. C. Ordeshook, *Game Theory and Political Science*, cited earlier; Cheryl L. Eavey and Garry J. Miller, "Bureaucratic Agenda Control: Imposition or Bargaining," *American*

Political Science Review, 78, 1984: 719–33; and Keith Krehbiel, "Sophistication, Myopia, and the Theory of Legislatures: An Experimental Study," Calif. Inst. of Tech., working paper, 1984. For some useful experiments that document the power of agendas when voters vote sincerely as well as an interesting example of an attempt to manipulate outcomes by the judicious selection of an agenda, see Charles Plott and Michael Levine, "A Model of Agenda Influence on Committee Decisions," *American Economic Review*, 68, 1978: 146–60; and Levine and Plott, "Agenda Influence and Its Implications," *Virginia Law Review*, 63, 1977: 561–604. The discussion of killer and saving amendment strategies in Congress is taken from James Enelow and David H. Koehler, "The Amendment in Legislative Strategy: Sophisticated Voting in the U.S. Congress," *The Journal of Politics*, 42, 1980: 396–413. For additional examples (as well as a reanalysis of the Powell amendment that we discuss in Chapter 2), see James M. Enelow, "Saving Amendments, Killer Amendments, and an Expected Utility Theory of Sophisticated Voting," *Journal of Politics*, 43, 1981: 1062–89. Also see Bo Bjurulf and Richard Niemi, "Strategic Voting in Scandinavian Parliaments," *Scandinavian Political Studies*, 1, 1978: 5–22.

Chapter 7

Any discussion of the characteristic function for cooperative games, indeed of game theory itself, begins, of course, with James von Neumann and Oskar Morgenstern, *The Theory of Games and Economic Behavior*, cited in Chapter 1. This volume remains one of the most readable sources on $v(C)$ and on the von Neumann-Morgenstern solution hypothesis (which we call the V-set in Chapter 9). The game in Figure 7.1 is analyzed in Richard D. McKelvey and P. C. Ordeshook, "An Experimental Test of Solution Theories for Cooperative Games in Normal Form," in P. C. Ordeshook and Kenneth Shepsle, *Political Equilibrium*, cited earlier. The characteristic function for the example "Valley of the Dump" as well as the example of firms polluting a lake are from Lloyd Shapley and Martin Shubik, "On the Core of an Economic System with Externalities," *American Economic Review*, 1969: 678–84. For specific applications to Mancur Olson's analysis of collective action, see Norman Schofield, "A Game Theoretic Analysis of Olson's Game of Collective Action," *Journal of Conflict Resolution*, 1975: 441–61. Market games are, of course, another area in which there is a considerable literature, and the text and bibliography of Martin Shubik's textbook, cited earlier, should be consulted. Indeed, this area has given rise to a class of games called "nonatomic games" in which the number of players (consumers and firms) is an uncountable infinity, but such games do not appear to have any apparent application to politics. Further details on the beta core are presented in Robert Aumann, "A Survey of Cooperative Games without Sidepayments," in Martin Shubik, ed., *Essays in Mathematical Economics in Honor of Oskar Morgenstern*, Princeton: Princeton Univ. Press, 1967. The initial critique of the characteristic function is taken from J. C. C. McKinsey, *Introduction to the Theory of Games*, New York: McGraw-Hill, 1952. For additional discussion of difficulties with $v(C)$ see Robert Rosenthal,

"Cooperative Games in Effectiveness Form," *Journal of Economic Theory*, 51, 1972: 88–101, which is also the source of the flower bed example.

Chapter 8

The seminal essay on the Coase theorem is, of course, Ronald Coase, "The Problem of Social Cost," *Journal of Law and Economics*, 3, 1960: 1–44. The subsequent literature on this idea fills volumes and the reader can consult almost any contemporary book on political economy or the law and economics for further details and implications. Perhaps the most readable essay on balanced games, cores, and markets, and the source of our theorem on balanced games is Herbert E. Scarf, "The Core of an *N*-Person Game," *Econometrica*, 35, 1972: 50–69. Our discussion of the Genossenschaften is drawn entirely from Alvin Klevorick and Gerald Kramer, "Social Choice on Pollution Management: The Genossenschaften," *Journal of Public Economics*, 2, 1973: 101–46. The analysis of majority rule and the city-block metric, and the proof that a Condorcet always exists with this metric in two dimensions comes from Douglas Rae and Michael Taylor's essay, cited in Chapter 1. The experiments reported in Figure 8.15a and b are taken from Janet Beryl, Richard D. McKelvey, P. C. Ordeshook, and Mark Winer, "An Experimental Test of the Core in a Simple N-Person, Cooperative, Nonsidepayment Game," *Journal of Conflict Resolution*, 20, 1976: 453–79. The results reported in Figure 8.16 are taken from Morris Fiorina and Charles Plott, "Committee Decisions under Majority Rule," *American Political Science Review*, 72, 1978: 575–98. The experimental logrolling outcomes are reported in two essays by Richard D. McKelvey and P. C. Ordeshook: "Vote Trading: an Experimental Study," *Public Choice*, 35, 1980: 151–84, and "Experiments on the Core: Some Disconcerting Results," *Journal of Conflict Resolution*, 25, 1981: 709–24. The experimental outcomes reported in section 8.8 with respect to the game portrayed in Figure 7.1 are from McKelvey and Ordeshook, "An Experimental Test of Solution Theories for Cooperative Games in Normal Form," in Ordeshook and Shepsle, cited earlier. With respect to problems occasioned by the characteristic-function representation of markets, see Robert Aumann, "Disadvantageous Monopolies," *Journal of Economic Theory*, 6, 1975: 1–11. Andrew Postlewaite and Robert W. Rosenthal, "Disadvantageous Syndicates," *Journal of Economic Theory*, 9, 1974: 324–26. A game-theoretic approach to studying cartels abstractly is provided in an interesting essay by Abraham Charnes and S. C. Littlechild, "On the Formation of Unions in *n*-Person Games," *Journal of Economic Theory*, 6, 1973: 1–11. The proof of Theorem 8.5 is presented originally in Joseph Greenberg, "Consistent Majority Rules over Compact Sets of Alternatives," *Econometrica*, 47, 1979: 627–736. For further discussion, and for the source of our examples in Section 8.9, see James D. Laing, S. Nakabayashi, and Benjamin Slotznick, "Winners, Blockers and the Status Quo: Simple Collective Decision Games and the Core," *Public Choice*, 1983: 263–79. For more discussion of the noncooperative interpretation of the core see Herve Moulin, *Game Theory for the Social Sciences*, cited earlier.

Chapter 9

Any of the game theory texts cited in earlier chapters are a source of additional reading on the V-set and bargaining set theory. William F. Lucas's counterexamples to the existence of V-sets are given in "A Game with No Solution," *Bulletin of the AMS*, 74, 1968: 237–9; and M. Rabie, "Games with No Solutions and Empty Cores," tech. report 474, School of Operations Research, Cornell Univ., 1980. Although the literature on the bargaining set is extensive, it was first introduced by Robert J. Aumann and Michael Maschler in "The Bargaining Set for Cooperative Games," in Dresher et al., eds., *Advances in Game Theory*, Princeton: Princeton Univ. Press, 1964. The existence theorem for $\mathcal{M}_1$ with transferable utility in B. Peleg, "Existence Theorem for the Bargaining Set $\mathcal{M}_1(i)$," in Martin Shubik, ed., *Essays in Mathematical Economics*, Princeton: Princeton Univ. Press, 1967. The size principle comes from the earlier cited work of William H. Riker, *The Theory of Political Coalitions*, and this is also the source of our discussion of the 1824 U.S. presidential election. Although this hypothesis is not deduced from any cooperative solution concept, its relationship to the concept of a coalitionally rational proposal (which is itself a hypothesis that should be deduced ultimately from some more general idea) is identified in Richard D. McKelvey and Richard Smith, "Internal Stability and the Size Principle," working paper, Carnegie-Mellon University, 1976; and Norman Schofield, "The Bargaining Set in Voting Games," *Behavioral Science*, 25, 1980: 120–9. For applications of the size principle (and other hypotheses) to parliamentary systems see Abraham DeSwann, *Coalition Theory and Cabinet Formations*, Jossey-Bass, San Francisco, 1973; and Howard Rosenthal, "Size of Coalitions and Electoral Outcomes in the French Fourth Republic," in Sven Groennings, et al., eds., *The Study of Political Coalitions*, New York: Holt, Rinehart, and Winston, 1970. For a comprehensive bibliography and review, see Norman Schofield, "Bargaining Set Theory and Stability in Coalition Governments," *Mathematical Social Sciences*, 3, 1982: 9–31. The strong bargaining set is introduced by Robert Wilson, "Stable Coalition Proposals in Majority-Rule Voting," *Journal of Political Economy*, 3, 1971: 254–71. Wilson also proves in this essay the useful result that elements of a main-simple V-set must be Pareto-optimal. Richard D. McKelvey then shows in "Covering, Dominance, and the Institution Free Properties of Social Choice," which we cited in Chapter 4, that the V-set must also be contained in the uncovered set. The five-person counterexample portrayed in Figure 7.5 and the definition of the competitive solution come from Richard D. McKelvey, P. C. Ordeshook, and Mark Winer, "The Competitive Solution for *n*-Person Games without Transferable Utility with an Application to Committee Games," *American Political Science Review*, 72, 1978: 599–615. The experiments on parliamentary systems are more fully reported in P. C. Ordeshook and Mark Winer, "Coalitions and Spatial Policy Outcomes in Parliamentary Systems: Some Experimental Results," *American Journal of Political Science*, 24, 1980: 730–52, and those on vote trading appear in Richard D. McKelvey and P. C. Ordeshook, "Vote Trading: An Experimental Study," *Public Choice*, 35, 1978:

151–84. The experimental anomalies are taken from Richard D. McKelvey and P. C. Ordeshook, "Some Experimental Results That Fail to Support the Competitive Solution," *Public Choice*, 40, 1983: 281–92; and Gary J. Miller and Joe A. Oppenheimer, "Universalism in Experimental Committees," *American Political Science Review*, 76, 1982: 561–74. An apparent counterexample to *ℋ*'s existence in spatial games is offered in James D. Laing and Scott M. Olmstead, "An Experimental and Game-Theoretic Study of Committees," in P. C. Ordeshook, ed., *Game Theory and Political Science*, cited earlier. The defensible set is described in Rodney Gretlein, "The Defensible Set," Ph.D. diss., Carnegie-Mellon Univ., 1980, and the bargaining equilibrium is discussed in Elaine Bennett, "The Bargaining Equilibrium for Finite Alternative Voting Games," paper presented at the 1984 annual meeting of the Public Choice Society.

Chapter 10

Experiments with the prisoners' dilemma comprise a vast literature. Perhaps the best starting point is Anatol Rapoport and A. M. Chammah, *Prisoners' Dilemma*, Ann Arbor: Univ. of Michigan Press, 1965. A thorough analysis of the infinitely repeated prisoners' dilemma, and the source of our discussion, is Michael Taylor, *Anarchy and Cooperation*, New York: Wiley, 1976. The sociobiological analogue comes from Robert Axelrod, *The Evolution of Cooperation*, New York: Basic Books, 1984. Our discussion of the balance of power is taken from Emerson M. S. Niou and P. C. Ordeshook, "A Theory of the Balance-of-Power in International Systems," *Journal of Conflict Resolution*, December, 1986. Sequential equilibria are defined in David M. Kreps and Robert Wilson, "Sequential Equilibria," *Econometrica*, 50, 1982: 863–94. This essay, however, is anything but easy reading for the beginning student. The chain store paradox and the finitely repeated prisoners' dilemma are analyzed with this concept by David Kreps, Paul Milgrom, John Roberts, and Robert Wilson, in various combinations, in "Rational Cooperation in the Finitely Repeated Prisoners' Dilemma," "Reputation and Imperfect Information," and "Predation, Reputation and Entry Deterrence," all in *The Journal of Economic Theory*, 27, 1982: 245–312. The Shapley value was originally developed by Lloyd Shapley, "A Value for *n*-Person Games," in H. Kuhn and A. W. Tucker, eds., *Contributions to the Theory of Games*, vol. 2, Princeton: Princeton Univ. Press, and the power index was developed by Lloyd Shapley and Martin Shubik, "A Method for Evaluating the Distribution of Power in a Committee System," *American Political Science Review*, 48, 1954: 787–92. The Banzhaf-Coleman index has two sources: John F. Banzhaf, "Weighted Voting Doesn't Work," *Rutgers Law Review*, 19, 1965: 317–43; and James S. Coleman, "Control of Collectivities and the Power of a Collectivity to Act," in Bernard Lieberman, ed., *Social Choice*, New York: Gordon and Breach, 1971. The discussion of the proposed revision to the Canadian constitution comes from D. R. Miller, "A Shapley Value Analysis of the Proposed Canadian Constitutional Amendment

500   References and a guide to the literature

Scheme," *Canadian Journal of Political Science*, 6, 1973: 140–3; and Philip D. Straffin, Jr., "Homogeneity, Independence, and Power Indices," *Public Choice*, 7, 1978: 113–23. For additional discussion of the electoral college, spatial models, and the power index, see Melvin J. Hinich and P. C. Ordeshook, "The Electoral College: A Spatial Analysis," *Political Methodology*, 1, 1974: 1–31. An insightful approach to the normative analysis of rules and procedures that does not rely on power indices and the like is presented in Michael L. Balinski and Peyton Young, *Fair Representation*, New Haven, CT: Yale University Press, 1982. For a comprehensive review of the bargaining literature and the general modeling problem, see John C. Harsanyi, *Rational Behavior and Bargaining Equilibrium in Games and Social Situations*, Cambridge: Cambridge Univ. Press, 1977. This volume contains citations to Nash's original essays, to Harsanyi's own seminal contributions, which are the basis for Bayesian equilibrium notions in game theory, as well as to the recent literature. One interesting approach that we do not review here, however, but that comes from the same analytic perspective as the chain store paradox is Ariel Rubinstein, "Equilibrium in Supergames with the Overtaking Criterion," *Journal of Economic Theory*, 30, 1983: 74–97. See also Roger B. Myerson, "Analysis of Two Bargaining Problems with Incomplete Information," working paper 582, Northwestern Univ., 1983. But again, these essays are not easy reading. The report on the JPL space probe is from James S. Dyer and Ralph E. Miles, Jr., "An Actual Application of Collective Choice Theory to the Selection of Trajectories for the Mariner Jupiter/Saturn 1977 Project," *Operations Research*, 24, 1976: 220–44. For a more general discussion of the implication of value theory for the formulation of public policy, see Horace W. Brock, "A New Theory of Social Justice Based on the Mathematical Theory of Games," in Ordeshook, *Game Theory and Political Science*, cited earlier. Our discussion of reciprocity is from Randall L. Calvert, "Reciprocity among Self-Interested Actors: Asymmetry, Uncertainty, and Bargaining," working paper, Washington Univ., 1985. The reference to Dilip Abreu is "Repeated Games with Discounting: A General Theory and an Application to Oligopoly," working paper, Princeton Univ., 1982. Those who would like to see a more thorough treatment of sequential games, especially as applied to economics, should consult James W. Friedman, *Oligopoly and the Theory of Games*, Amsterdam: North-Holland, 1977.

Index

501